18th International Scientific Conference on Current Problems of the Corporate Sector (ICRTEG 2021)

SHS Web of Conferences Volume 115 (2021)

Bratislava, Slovakia
20 May 2021

Editors:

Dana Hrušovská
Maria Kmety Barteková
Maria Trúchliková
Monika Raková

ISBN: 978-1-7138-3473-1

Printed from e-media with permission by:

Curran Associates, Inc.
57 Morehouse Lane
Red Hook, NY 12571

Some format issues inherent in the e-media version may also appear in this print version.

For additional information, please contact EDP Sciences – Web of Conferences
at the address below.

EDP Sciences – Web of Conferences
17, Avenue du Hoggar
Parc d'Activité de Courtabœuf
BP 112
F-91944 Les Ulis Cedex A
France

Phone: +33 (0) 1 69 18 75 75
Fax: +33 (0) 1 69 28 84 91

contact-edps@webofconferences.org

Additional copies of this publication are available from:

Curran Associates, Inc.
57 Morehouse Lane
Red Hook, NY 12571 USA
Phone: 845-758-0400
Fax: 845-758-2633
Email: curran@proceedings.com
Web: www.proceedings.com

TABLE OF CONTENTS

UNIVERSITY OF ECONOMICS IN BRATISLAVA

Faculty of Business Management

Department of Business Economy

CURRENT PROBLEMS OF THE CORPORATE SECTOR 2021

18th International Scientific Conference

20th May 2020
Bratislava, Slovak Republic

ORGANIZED BY

FOUNDATION MANAGER
UNIVERSITY OF ECONOMICS IN BRATISLAVA, Faculty of Business Management, Department of Business Economy

EDITED BY

Ing. Dana Hrušovská, PhD.
PhDr. Mária Kmety Barteková, PhD.
Ing. Monika Raková, PhD.
Ing. Mária Trúchliková, PhD.

PROGRAMME COMMITTEE

ORGANIZING COMMITTEE

CURRENT PROBLEMS OF THE CORPORATE SECTOR is an international scientific conference, thematically focused on the various current themes of the corporate sector and the topics linked to the business environment. Its main aim is the presentation of current business issues that concern managers and entrepreneurs operating in CEE markets, presentation of scientific research outcomes by the conference participants and supporting the knowledge transfer by finding the links between scientific research and business practice and the cooperation with domestic and foreign educational and research institutions.

The condition of publishing the paper in the proceeding is its originality, its contribution, presentation of empirical knowledge both at home and abroad. The 18th volume of the conference focuses on following fundamental issues:

1. Corporate Problems in the Digital World
2. Current Financial Problems of the Corporate Sector
3. Current Managerial Trends in the Corporate Sector

edp sciences

Web of Conferences

Statement of Peer review

In submitting conference proceedings to *Web of Conferences*, the editors of the proceedings certify to the Publisher that

1. They adhere to its **Policy on Publishing Integrity** in order to safeguard good scientific practice in publishing.
2. All articles have been subjected to peer review administered by the proceedings editors.
3. Reviews have been conducted by expert referees, who have been requested to provide unbiased and constructive comments aimed, whenever possible, at improving the work.
4. Proceedings editors have taken all reasonable steps to ensure the quality of the materials they publish and their decision to accept or reject a paper for publication has been based only on the merits of the work and the relevance to the journal.

Title, date and place of the conference

Current Problems of the Corporate Sector 2021
20 May 2021, Bratislava, Slovakia (online)

Proceedings editor(s):

Dana Hrusovska
Maria Kmety Bartekova
Maria Truchlikova
Monika Rakova

Date and editor's signature

29/6/2021

Digital Transformation in the Context of the European Union and the Use of Digital Technologies as a Tool for Business Sustainability

Diana Bednarčíková[1,*], *Radka* Repiská[2]

[1]University of Economics in Bratislava, Faculty of Business Management, Department of Information Management, 852 35 Bratislava, Slovakia
[2]University of Economics in Bratislava, Faculty of Business Management, Department of Business Economy, 852 35 Bratislava, Slovakia

Abstract. The potential of digitalization, digital transformation and digital technologies is important for environmental, economic and social sustainability in businesses and can be applied in various industries. In this scientific article, main author´s aim is to examine the green digital transformation in a theoretical generalization, then in the context of the European Union and to examine the use of digital technologies in the field of sustainable development in selected companies in the world. We used explanatory general theoretical methods. In the literature review, we defined the main concepts related to digital transformation, mapped the digital transformation in the context of the European Union and highlighted the link between sustainability and digital technologies. Examples of the implementation of sustainable technologies in various companies and compliance with sustainability are found in the results of the work and discussion. At the end of the scientific article, we propose definitions: Green digital transformation is the implementation of digitalization and interaction of green technologies into business processes, activities, products and models, when the company becomes environmentally sustainable. Green technology represents the integration and subsequent use of digital technologies to innovate business processes and are used in the field of sustainable environmental development of the organization.

Introduction

The global pandemic of the coronavirus COVID-19 has confirmed the importance of digitization in the world and many businesses have been forced to digitally transform in order to create their business and enterprise activities. Digital transformation in companies represents a change of business processes towards efficiency, innovation, competitiveness, sustainability and overall economic growth of the company. The application of digitization,

* Corresponding author: diana.bednarcikova@euba.sk

digitalization, the use of ICT (Information and Communication Technologies) and the interaction of digital technologies in business: processes, activities, models lead to the digital transformation of the business and may ultimately become completely digital. Any digital technologies are: Artificial Intelligence, Cloud Computing, Internet of Things, Robotics Process Automation, etc. referred to as Industry 4.0 (Fourth Industrial Revolution) technologies, they are widely used in all areas of society and connect the physical, digital and biological worlds. The development of digital technologies is key to environmental sustainability, as well as serving as a catalyst for companies, for example, where digital technologies: reduce pollution and emissions, improve energy efficiency, degrade the environment, lead to better resource allocation and efficiency.

The paper consists of four main parts. The next section describes the literature review consisting of three main parts (overview of definitions in the field of green and digital transformation, the European context and sustainability of digital technologies). The second section defines research goals and methodology, the third section clarifies the results of the review on specific companies and the fourth part closes paper with the main implications of the review and sets the direction for future research.

The aim of the scientific article is to examine the green digital transformation in a theoretical generalization, then in the context of the European Union and to examine the use of digital technologies in the field of sustainable development in selected companies in the world.

1 Literature review

1.1 Definition of digitalization, digital transformation, digital technologies and sustainability

Digitalization is identified as the most significant technological trend for society and business [1]. Digitalization precedes **digitization**, which is a technical process for converting analog information streams to digital bits 1 with 0 with discrete and discontinuous values [2, 3]. Bican and Brem [4] state that digitization refers to a technical process in which analog information is transformed by coding and programming by information technology lenses into a digital format, and this process represents the integration of digital technologies into everyday life. **Digitalization** is defined by Linkov et al. [5] as increased interconnection and networking of digital technologies to improve communication, services and trade between people, organizations and things. According to Bican and Brem [4], digitalization is a socio-technological process of applying digitization techniques - digitization in broader social and institutional contexts that create the infrastructure of digital technologies. Digitalization is the impact of digitization on society and refers to the application of digital technologies that bring about changes in societies between businesses and in the business markets caused by digitization [3]. Digitalization consists in introducing digital technological innovations into existing (organizational, industrial, social) systems in a way that changes their functioning [6].

Digital transformation is a profound change and accelerates business activities, process, competencies, models in order to take full advantage of changes and opportunities in digital technology and its impact on society in a strategic and priority way [4]. Digital transformation refers to unprecedented disruptions to society, industry, organizations stimulated by advances in digital technologies such as: Artificial Intelligence, Big Data Analytics, Cloud Computing and The Internet of Things [7]. The digital transformation of business models is transforming consumer preferences and consumption as industries introduce digital technologies to increase their competitiveness in order to change customer relationships, internal processes and value prepositions [3].

The first prerequisite for digital transformation is **digital technologies** that transform the way environmental sustainability issues are measured and managed [7]. Digital technologies are the technologies of the fourth industrial revolution, also called Technologies Industry 4.0. Digital technology under Bican and Brem [4] is the basis for innovation in platforms and digital environments, and these technologies are highly interconnected, enabling and increasing information processing capacity. Feroz, Zo & Chiravuri [7] further state that digital technologies offer organizations unique opportunities to develop new environmental business models or to adopt digital technologies to incorporate environmentally sustainable practices into existing business models. Digital technologies are forcing companies to reimagine their customer value offerings as new social and mobile applications, analytics, the Internet of Things, artificial intelligence, biometrics, blockchain, cloud and edge computing, and many other advances allow them to deliver value in ways that simply were not possible [8].

1.2 Green Digital Transformation in the European Context

Digitalization has significant social and environmental implications, and various organizations in the world (e.g.: OECD, UN, EU) have considered the need for institutions and practices in the field of sustainable management. The European Union also contributes to the promotion and surveillance of **sustainability** and **digital transformation** in the world and in business through its legal frameworks, policies, programs and instruments.

On *4 December 2019*, the European Environment Agency (EEA) published its report ***"The European Environment - State and Outlook 2020"*** (SOER 2020) and its results make us think. Current environmental, climate and sustainability challenges have reached unprecedented scale and urgency and effectiveness, and immediate and coordinated action and systemic solutions. The Union's environmental policy has brought significant benefits in recent decades without affecting the EU's long-term competitiveness, with the acquis focusing even more on long-term visions, targets and strategic framework policies effectively reducing environmental pressures over the last few years.

However, the loss of biodiversity, the deterioration of ecosystem services, climate change and its impacts, as well as the unsustainable use of resources, environmental pollution and related risks to human health and living conditions, nature, ecosystems and the economy, require further decisive action in the EU and worldwide. The European Commission responded to these challenges and adopted the **European Green Deal** on *11 December 2019* - a new growth strategy for dual, **green** and **digital transformation**, which aims to transform the Union into a fair and prosperous society with a competitive, climate-neutral and resource-efficient economy. A Regulation of the European Parliament and of the Council (EU) enshrines the goal of the EU to become the **first climate-neutral continent by 2050** to protect, preserve and enhance the EU's natural capital and protect citizens' health and well-being from environmental risks and impacts.

According to the SOER 2020 report ***"The European Environment - State and Outlook 2020"***, the coming years are a unique opportunity for the Union to demonstrate its leadership in sustainability and to address the urgent challenges in this area that require systemic solutions [9]. The changes in the global climate and ecosystems that can be observed since the 1950s are unparalleled in recent decades to millennia. The world's population has tripled since 1950, while the number of people living in cities has quadrupled. With the current growth model, environmental pressures are expected to increase further, with direct and indirect harmful effects on human health and well-being.

This is especially true for the sectors with the greatest impact on the environment - food, mobility, energy, as well as infrastructure and buildings.

On *10 March 2020*, the European Commission laid the foundations for an industrial strategy that should support the **dual transformation** to a **green** and **digital economy**, increase the global competitiveness of EU industry and strengthen Europe's **open strategic autonomy**. Updating the **2020 Industrial Strategy: towards a stronger Single Market for Europe's recovery** included a list of measures to support the green and dig**ital transformation of EU industry**, many of which have already been adopted or launched. However, the pandemic drastically affected the speed and extent of this transformation. Companies that strive for sustainability and digitization are likely to be among the leaders of tomorrow.

The **European Green Deal** forms the basis for the Recovery Instrument proposed by the European Commission on *27 May 2020* "**Next Generation EU**", which supports investment in key green sectors needed to build resilience and create growth and jobs in a fair and inclusive society. A mechanism to support recovery and resilience, which, together with the Union budget for **2021-2027**, will drive the Union's economic recovery from the coronavirus crisis, while making this recovery lasting and ensuring prosperity for future generations in the transition to a **sustainable way of life** and advancing **digital transformation**.

On *27 May 2020*, the European Commission published the report „**The EU budget powering the recovery plan for Europe**" where is proposing to harness the full potential of the EU budget. Next Generation EU of € 750 billion as well as targeted reinforcements to the long-term EU budget for 2021-2027 will bring the total financial firepower of the EU budget to € 1.85 trillion. The funds raised from the **Next Generation EU** instrument will be invested in three pillars which are shown in Figure 1:

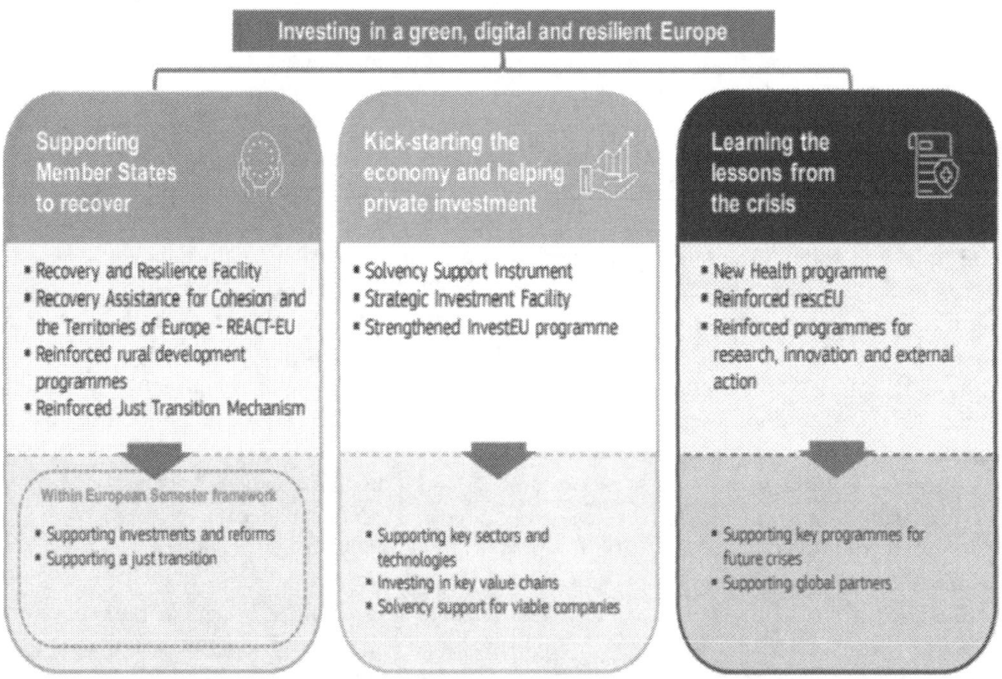

Fig. 1. Next Generation EU: investing in a greener and more digital EU (3 pillars).

The new package will be reflected in particular in the field of **green and digital transformation**, with the aim of modernizing the economies of the Member States as well as strengthening the businesses affected by the COVID-19 pandemic. Relaunching the economy does not mean going back to the status quo before the crisis, but bouncing forward. We must repair the short-term damage from the crisis in a way that also invests in our long-term future.

All of the money raised through **Next Generation EU** will be channelled through EU programmes in the revamped long-term EU budget to the activities shown in the Table 1.

Table 1. The policy fundamentals of the recovery.

A. The European Green Deal as the EU's recovery strategy	• Support for projects in the field of energy from renewable sources (especially wind and solar energy), start-up of a green hydrogen economy • Support for the renovation of buildings and infrastructure • Greener transport and logistics • Strengthening the Fair Transformation Fund
B. Strengthening the Single Market and adapting it to the digital age	• Investing in bigger and better connectivity (5G network) • Stronger industrial and technological presence in strategic sectors (Artificial Intelligence, Cyber Security, Supercomputers, Cloud and others) • Building a real data economy • Increased cyber resilience
C. A fair and inclusive recovery for all	• Support for workers and enterprises through the European Unemployment Insurance Scheme • A Skills Agenda for Europe and a Digital Learning Action Plan • Fair minimum wages and binding remuneration transparency measures • Fight against tax evasion

The majority of proposed partnerships aim to **accelerating the transitions towards a climate-neutral, circular society and economy, and achieving the Sustainable Development Goals** (SDG). Consequently, many will team up with the private sector to decarbonise the energy, mobility and steel sectors, or with Member States to provide solutions for the sustainable management of resources (e.g. water, soil) and preserve biodiversity. Others aim at **building greater resilience** of Europe's health care systems and industry, including by ensuring the EU's **technological leadership** and autonomy in critical areas, such as Smart Networks and Services, **Artificial Intelligence**, Data and Robotics and Photonics. Built4People, Driving Urban Transitions, Clean Energy Transition, and the EIT KICS are also relevant for the New European Bauhaus.

In ministerial declaration „**A Green and Digital Transformation of the EU**" states that digital technologies are an important part of the EU's green growth strategy (EU Green Deal) [10]. Businesses, governments and NGOs need to work together to implement new technologies, innovative solutions and progressive public policies that reflect our shared vision of a more prosperous and sustainable future. Therefore, we can say that we are proud of the 26 large international companies that formed the **European Green Digital Coalition** "EGDC" and signed the *Declaration in Support of the EU's Green and Digital Transformation*, which was launched on *Digital Day 2021* on *19 March* as a virtual event. The Member States of the European Union, together with Norway and Iceland, have signed a declaration on accelerating the use of **green digital technologies** for the benefit of the **environment**, thereby committing themselves to key **digital initiatives**.

As part of the creation of a EGDC in fewer of their companies, the directors of 26 international companies have committed themselves to take action in areas such as:

- Invest in the development and deployment of greener digital technologies and services that are more energy and material efficient.

- Develop methods and tools to measure the net impact of green digital technologies on the environment and climate by joining forces with NGOs and relevant professional organizations.
- International companies, together with representatives of other sectors, work together to develop recommendations and guidelines for the green digital transformation of these sectors that are beneficial to the environment, society and the economy.

The European Green Digital Coalition will not only help the technology sector to become more sustainable, circular and zero polluters, but will also support the sustainability goals of other priority sectors such as energy, transport, agriculture and construction, while contributing to an innovative, inclusive and resilient society [10-13].

1.3 Sustainability of digital technologies

Sustainability consists of three dimensions such as: environmental, economic and social and form the so-called sustainable triangle. The dimensions are interrelated and are defined according to Wut et al. as follows: environmental sustainability (environmental) is defined as "the continuous preservation of the ecosystem and its functions; economic sustainability means the lasting ability of the economic system to meet all human needs and social sustainability promotes the well-being of all people; such as access to food, medicine, education and recreation [14]. Overall, sustainability aims to protect our natural environment and human health, while promoting the balance of ecosystems and innovation, and economic behavior must be improved in the interests of human and environmental sustainability; for this purpose, technological innovation is a support tool [15]. Wut et al. also defines digital sustainability, which in the narrower sense is the sustainable use of digital resources (data and computer files, source codes and technologies) - the ability of a digital system to be constantly maintained, resp. Renew and, more broadly, represent organizational activities that seek to achieve the goals of sustainable development through the creative introduction of technologies that: create, use, transmit or obtain electronic data [14].

Digitalization increases the likelihood of social and environmental sustainability challenges and threats, including the carbon footprint associated with increased demand for electricity generation, cybersecurity weaknesses and social disparities due to the growing gap in access to ICT, commonly referred to as digitization. The "digital divide" [5]. Without the transformation of existing businesses, the economic and environmental challenges of the future cannot be addressed in a sustainable way, and therefore resource optimization for more sustainable businesses is driven by digital technologies that form the basis of digital innovation in platforms and digital environments [4]. Digital technologies (Artificial Intelligence, Big Data, Internet of Things, Social Media Analytics, Cloud Computing, and Mobile Technologies) are increasingly used to improve environmental sustainability, transform various aspects of environmental sustainability and enable transformations in various areas of environmental sustainability such as: pollution control, waste management, sustainable production and urban sustainability [7]. Digital technologies support the development of value proposals that combine environmental, social and economic value and can make a significant contribution to the goals of sustainable development [16].

In the COVID-19 era, **Industry 4.0 technologies** are also expected to accelerate the transition to sustainable business processes and to achieve sustainable development, especially as a priority of economic recovery, which will have to address biodiversity loss, climate change and promote circular economy principles [17]. Industry 4.0 and digital technology, respectively - Artificial Intelligence (AI) is associated with positive and negative impacts on humans and the planet, and much attention is currently being paid to the analysis and evaluation of these impacts [18]. Digitalization supports the development of Artificial Intelligence (AI) and, thanks to it, computer science enables the creation of intelligent machines (algorithms and software) that function and react like humans [5]. Demartini, Evans and Tonelli state that companies currently rely on AI, IoT and big data analysis to implement sustainable business practices that include reducing carbon emissions and minimizing other environmental impacts [19].

2 Research objectives and methodology

The aim of the scientific article is to examine the green digital transformation in a theoretical generalization, then in the context of the European Union and to examine the use of digital technologies in the field of sustainable development in selected companies in the world.

The starting point for the elaboration of a scientific article was: study, analysis, synthesis and comparison of foreign literature in the form of: books, professional articles, studies, surveys, laws, policies, etc. Deduction and induction methods were also used.

The research subjects were: the European Union and business entities such as: Amazon, Carbon Tracker, Google, IBM, Microsoft, Ørsted, Patagonia, Walmart and Xcel Energy. Due to the scope of the scientific article, it was not possible to examine all the theoretical definitions and areas of the use of digital technologies in the business processes of selected companies.

3 Results and discussion

PWC company estimates that using artificial intelligence for environmental applications could contribute up to $ 5.2 trillion to the global economy by 2030 [20]. The use of Artificial Intelligence (AI) for environmental applications in four sectors such as agriculture, energy, transport and water is expected to have the potential to increase global GDP to 4.4% by 2030 (the estimate for Europe is 5.4%). At the same time, the application of AI leverage could reduce global greenhouse gas (GHG) emissions by 4% in 2030, equivalent to 2.4 Gt $CO2e$ - equivalent to the annual emissions of Australia, Canada and Japan by 2030. While improving productivity, AI could create 38.2 million net new jobs in the global economy that would offer more skilled occupations as part of this transition.

Examples of the implementation of sustainable technologies in various companies and compliance with sustainability are as follows:

- **Amazon** - uses Artificial Intelligence to meet its long-term goal of powering its global infrastructure with 100% renewable energy. This includes its AWS cloud platform.
- **Carbon Tracker** - a think-tank focused on combating climate change, uses AI to monitor emissions from coal-fired power plants using satellite imagery. Using satellite data, they help drive investments toward lower footprint businesses.
- **Google** - uses the AI model to reduce the energy load of its data centers - the Google Cloud Platform resource-hungry and reduce the energy cost of cooling by 40%.
- **IBM** - uses AI for better weather forecasting, making their forecasts 30% more accurate. This helps renewable energy companies better manage their power plants, maximize renewable energy production and reduce carbon emissions.
- **Microsoft** - is committed to being carbon-negative by 2030, using Artificial Intelligence to run huge public data centers (Cloud Deals) called Microsoft Azure. Microsoft's cloud computing has already enhanced energy efficiency and reduced

material waste. Increased availability of serverless and open-source software minimizes cooling, ventilation, and air conditioning in fewer data centers. The addition of power management to Microsoft products has enabled intelligent power consumption on end devices such as monitors and hard drives.

- **Ørsted** - a provider of wind technology and bioenergy from Denmark and in 2020 this company ranked as the *most sustainable company in the world*.

  It was to work with Microsoft, where together they would successfully face environmental challenges. Ørsted's biggest effort is to build a "completely green world" with a 100% carbon-free energy supplement by 2025. The company is limiting oil and coal-based activities in favor of clean energy systems. Ørsted owns more than a thousand offshore wind turbines equipped with sensors that continuously generate valuable data. Microsoft's advanced predictive analytics and technology is part of Ørsted's digital strategy for sustainable data and time-saving data transformation.

- **Patagonia** - a sustainable clothing company uses organic materials, sells worn and reworked clothing and organic accessories. The company also provides global funding through online banking and runs the online blog The Cleanest Line, where articles are devoted to environmental crises and solutions. Patagonia, which is consciously aware, has implemented a number of innovations in the company's management, such as: recycled building materials with laminated windows that prevent overheating, solar panels with photosensors and motion detectors, LED lighting, new heating, ventilation and air conditioning systems controlled by intelligent network. In addition, the company has a central Artificial Intelligence workstation that automatically manages all operating systems from indoor to outdoor irrigation. Bathrooms and toilets are equipped with water control sensors. The landscape and plants around the buildings are also selected and designed to reduce water consumption. Workers are advised to use electric cars and then receive financial compensation for commuting by bicycle or public transport.

- **Walmart** - one of the largest retail companies is deploying digital transformations multiple times, working to eliminate waste and energy consumption and ensure supply chain management. First and foremost, the numerous built-in IoT sensors and shelf scanning robots are proving to be sustainable in terms of energy savings and customer experience. In e-retail, it allows its customers to shop at home, reducing transport utilization and $CO2$ emissions. In 2018, the company patented the idea of a robobee - an unmanned drone for pollinating crops equipped with cameras and sensors. This tool also makes it possible to detect agricultural problems and gain sufficient control over the Walmart food supply chain, which in turn minimizes food waste.

- **Xcel Energy** - a coal-fired and nitrous oxide-emitting energy company, uses Artificial Intelligence to better predict energy consumption patterns and adapt its operating systems, significantly increasing efficiency (~ 20%).

According to Bican and Brem (2020), business models need to be analyzed in terms of their future readiness and impact on sustainability, as both economic and environmental sustainability are interrelated challenges for businesses that cannot be addressed in isolation.

4 Conclusion

The current problem in companies is outdated technologies that are not efficient and then the lack of information about the impact of technologies on the environment, so companies are not environmentally friendly and unknowingly damage the environment.

Digitalization has a long-term high potential in companies and can be applied in almost every industry to various business processes [21]. Digital transformation is a process driven by digital technologies and plays an important role in meeting the goals of environmental, economic and social sustainability. If companies interact with digital technologies and use digitalization in the areas of environmental sustainability, then the process of green digital transformation occurs in companies and digital technologies are referred to as green, resp. sustainable.

We did not find a definition of green digital transformation and green technology from the available sources, and therefore we propose the following definitions:

- ***Green digital transformation*** *is the implementation of digitalization and interaction of green technologies into business processes, activities, products and models, when the company becomes environmentally sustainable.*
- ***Green technology*** *represents the integration and subsequent use of digital technologies to innovate business processes and are used in the field of sustainable environmental development of the organization.*

Based on the analyzed documents from the European Commission, we can confirm that it is determined to realize its main ambition, which is to convert the states and businesses of the European Union through a green and digital transformation. This is also confirmed by the European Green Digital Coalition (EGDC), which is made up of the 24 states of the European Union together with Norway and Iceland, where the states signed a declaration on accelerating the use of green technologies for the benefit of the environment and thus committed to key digital initiatives. EGDC will not only help the technology sector to become more sustainable, circular and zero polluters, but will also support the sustainability goals of other priority sectors such as energy, transport, agriculture and construction, while contributing to an innovative, inclusive and resilient society [22-27].

The application and use of digital technologies can have a positive impact on companies in: saving money, reducing greenhouse gas emissions, identifying products offering excellent energy efficiency, reducing the cost of consumables and public services, increasing profits, taking business to the next level, facing competition with new ones technologies, the offer of new services, the simplification of the flow of operations, innovation, the increase in organizational performance, the transition from the old business approach to the digital, the development and creation of new skills and the indirect maintenance of labor markets. On the other hand, they also have negative impacts and impacts on organizations in the areas of environment and sustainability. Restricting digital technologies and innovation would not be an environmental solution, as digital technologies have higher benefits than consequences and protect the environment and contribute to sustainability in the areas concerned. Elimination of harmful effects could be avoided by assessing the energy efficiency of digital technologies according to traditional parameters such as: functionality, security, scalability and accessibility, which would also include green practices and targets as performance monitoring criteria. However, the potential of digital technologies to contribute to addressing environmental issues far exceeds the unintended negative effects it can cause. Innovations in data generation and analysis combined with Artificial Intelligence (AI) and advanced robotics, as well as cloud computing, have opened the door to many positive impacts on sustainability.

Acknowledgement

The paper was elaborated within Project of young teachers, researchers and doctoral students in full-time study No. I-21-105-00 Digitalization of enterprises and application of digital technologies in business processes in proportion 100 %.

References

1. J. Reis, M. Amorim, N. Melão, Y. Cohen, M. Rodrigues, Digitalization: A Literature Review and Research Agenda. In: Proceedings on 25th International Joint Conference on Industrial Engineering and Operations Management – IJCIEOM, IJCIEOM 2019. Lecture Notes on Multidisciplinary Industrial Engineering. Springer, Cham, 443-456 (2020)
2. J. S. Brennen, D. Kreiss, Digitalization, The International Encyclopedia of Communication Theory and Philosophy, John Wiley & Sons, Inc., New Jersey, USA. (2016)
3. T. Ritter, C. L. Pedersen, Digitization capability and the digitalization of business models in business-to-business firms: Past, present, and future, Industrial Marketing Management, 86, 180-190 (2020)
4. P. M. Bican, A. Brem, Digital Business Model, Digital Transformation, Digital Entrepreneurship: Is There A Sustainable "Digital"?. *Sustainability*, **12**, 5239 (2020)
5. I. Linkov, B. D. Trump, K. Poinsatte-Jones, M.-V. Florin. Governance Strategies for a Sustainable Digital World. *Sustainability*, **10**, 440 (2018)
6. S. Fielke, B. Taylor, E. Jakku, Digitalisation of agricultural knowledge and advice networks: A state-of-the-art review. *Agricultural Systems*, **180**, 102763 (2020)
7. K. Feroz, H. Zo, A. Chiravuri, Digital Transformation and Environmental Sustainability: A Review and Research Agenda. Sustainability, 13, 3, 1530 (2021)
8. J. W. Ross, C. M. Beath, M. Mocker, "Creating Digital Offerings Customers Will Buy,". *MIT Sloan Management Review*, **61**, 64-69, (2019)
9. European Environment Agency (EEA) (2019). The European Environment — State and Outlook 2020. Luxembourg: Publications Office of the European Union, 2019. ISBN 978-92-9480-115-9. Available at: <doi: 10.2800/48006>
10. COM(2021), DECLARATION of the European Green Digital Coalition Members "In support of the Green and Digital Transformation of the EU". Available at: < https://bit.ly/2Q6XwEd > [Accessed 19-03-2021]
11. European Council, Council of the European Union (2020). Digitalisation for the Benefit of the Environment. Available at: <https://data.consilium.europa.eu/doc/document/ST-13957-2020-INIT/en/pdf> [Accessed 11-12-2020]
12. European Council, Council of the European Union (2020). Infographic - Multiannual financial framework 2021-2027 and Next Generation EU. Available at: <https://www.consilium.europa.eu/en/infographics/mff2021-2027-ngeu-final/> [Accessed 21-04-2021]
13. European Council, Council of the European Union (2021). Digital Europe Programme and repealing Decision (EU) 2015/2240. Available at: <https://eur-lex.europa.eu/legal-content/EN/TXT/PDF/?uri=CONSIL:ST_6789_2020_REV_1&from=EN> [Accessed 18-03-2021]
14. T. M. Wut, D. Lee, W. M. Ip, S. W. Lee, Digital Sustainability in the Organization: Scale Development and Validation. *Sustainability*, **13**, 3530 (2021)
15. C.-T. Tseng, S. S. C. Shang, Exploring the Sustainability of the Intermediary Role in Blockchain. *Sustainability*, **13**, 1936 (2021)

16. P. Gregori, P. Holzmann, Digital sustainable entrepreneurship: A business model perspective on embedding digital technologies for social and environmental value creation. Journal of Cleaner Production, 272, 122817 (2020)
17. D. Lepore, A. Micozzi, F. Spigarelli, Industry 4.0 Accelerating Sustainable Manufacturing in the COVID-19 Era: Assessing the Readiness and Responsiveness of Italian Regions. *Sustainability*, **13**, 2670 (2021)
18. H. S. Sætra, AI in Context and the Sustainable Development Goals: Factoring in the Unsustainability of the Sociotechnical System. *Sustainability*, **13**, 1738 (2021)
19. D. Melissa, S. Evans, F. Tonelli, Digitalization Technologies for Industrial Sustainability. *Procedia Manufacturing*, **33**, 264-271 (2019)
20. PWC, How can enable a sustainable future. Available at: <https://www.pwc.co.uk/services/sustainability-climate-change/insights/how-ai-future-can-enable-sustainable-future.html>[Accessed 05-05-2021] (2021)
21. O. Mulhern, Artificial Intelligence: Can it help achieve Environmental Sustainability? <https://earth.org/data_visualization/ai-can-it-help-achieve-environmental-sustainable> (2021)
22. COM(2019) 640 final, The European Green Deal. Available at: <https://ec.europa.eu/info/sites/default/files/european-green-deal-communication_en.pdf> [Accessed 11-12-2019]
23. COM(2020) 102 final, A New Industrial Strategy for Europe. Available at: <https://eur-lex.europa.eu/legal-content/EN/TXT/PDF/?uri=CELEX:52020DC0102&from=EN> [Accessed 10-03-2020]
24. COM(2020) 442 final, The EU budget powering the recovery plan for Europe. Available at: <https://eur-lex.europa.eu/resource.html?uri=cellar:4524c01c-a0e6-11ea-9d2d-01aa75ed71a1.0003.02/DOC_1&format=PDF> [Accessed 27-05-2020]
25. COM(2020) 456 final, Europe's moment: Repair and Prepare for the Next Generation. Available at: <https://eur-lex.europa.eu/legal-content/EN/TXT/PDF/?uri=CELEX:52020DC 0456&qid=1620598808861&from=EN> [Accessed 27-05-2020]
26. COM(2020) 652 final, Proposal for a Decision of the European Parliament and of the Council on a General Union Environment Action Programme to 2030. Available at: <https://www.slov-lex.sk/legislativne-procesy/SK/LPEU/2020/637> [Accessed 14-10-2020]
27. Official Journal of the European Union (2020). Council conclusions on shaping Europe's digital future (2020/C 202 I/01). Available at: <https://eur-lex.europa.eu/legal-content/EN/TXT/PDF/?uri=OJ:C:2020:202I:FULL&from=ET> [Accessed 16-06-2020]

Procurement Transformation in Digital World

Josef Číž [1], *Michaela* Krechovská[1], and *Alena* Palacká [1*]

[1]University of West Bohemia, Faculty of Eonomics, Department of Finance and Accounting, Univerzitni 22, 306 14 Pilsen, Czech Republic

Abstract. Technological development and progress in digitization permeate all areas of life, of course not excluding companies and their activities and individual business processes. In this paper, we focus on the area of corporate procurement as one of the key areas for fulfilling corporate goals and its transformation in the age of digitalization. The paper aims to identify the changing function of procurement and possible opportunities of digitalization for the development of procurement and its role in organization. As the results shown, procurement is a value-creating process that is changed by the digitalization. It is moving from a traditional focus on cost reduction to value creation through dynamic cooperation and building relationships with business partners. Procurement is beginning to play an increasingly strategic role in organizations. It becomes a partner to support business efficiency and the creation of new business models.

1 Introduction

Procurement includes functions like as purchasing and supply management, along with managing inbound and outbound logistics [1]. Capgemini [2] define procurement as a complete process from sourcing to accounts payable (known also as Source-to-Pay). It is one of the key business areas that affects corporate profitability and the fulfilment of corporate goals. Yang et al. [3] confirm the evidence of a link between procurement strategy and corporate strategy in European and U.S. firms.

In recent years, one of the main topics addressed in the field of corporate procurement is the issue of cost reduction. Procurement is considered as the main cost driver for the organization [4] and cost reductions in the procurement process can have a high effect on the company's profit [5]. However, the traditional procurement and its function is evolving. Recently, we can come across concepts such as Digital procurement or Procurement 4.0. It is important to respond to the coming changes and think about new opportunities and possibilities for companies and procurement evolution.

The aim of the paper is to identify the changing function of procurement and possible opportunities of digitalization for the development of procurement and its role in organization.

* Corresponding author: astankov@fek.zcu.cz

2 Methods

First, we will focus on the analysis of procurement development, then we will discuss its transformation through digitalization. Attention will be focused also on the necessary technical solution. The analysis of the results of published studies and research will be widely used to identify the development of procurement and the opportunities that digitization brings in this area. Subsequent synthesis of knowledge will lead to the identification of these trends. Last but not least, a case study of the implementation of a new IT system in a selected multinational company will be presented. The case study will complement the paper as an example of a solution for procurement that enables the connectivity of international teams.

3 Results and Discussion

3.1 Evolution of procurement

The development of procurement since the early 1980s is shown in the Figure 1. When we look at procurement evolution, procurement has become more sophisticated and his function's role was changed from transactional management to strategic sourcing. An interesting turning point occurs mainly in the second stage of e-procurement with the onset of electronic inquiries, e-auctions, e-catalogs, etc. Today process monitoring and reporting are improving and digital procurement enabling to improve efficiencies with advanced analytics, data-driven optimization and automation tools. The focus of procurement will be on collaboration with external partners to create new, innovative business models [6]. Detailed characteristics of developmental stages can be seen in the following figure.

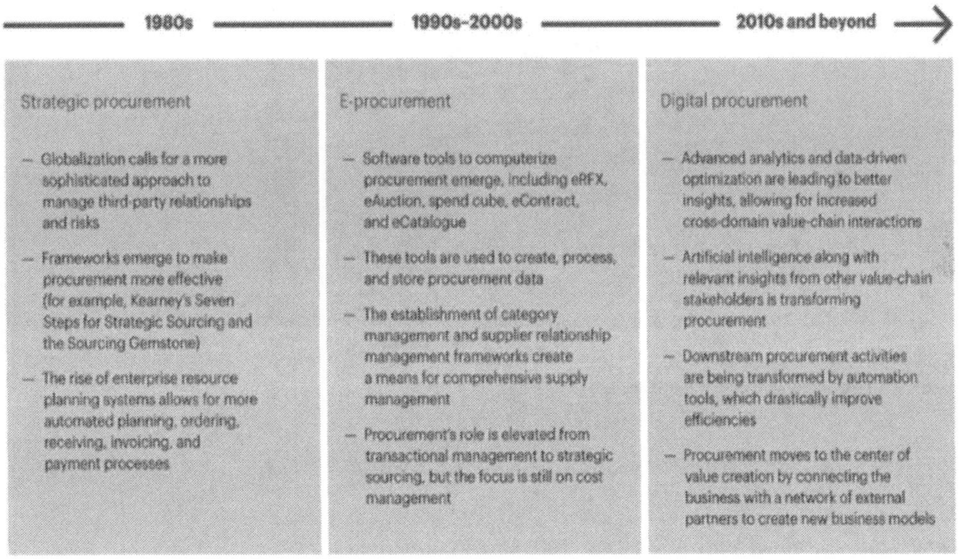

Fig. 1. Procurement evolution [6]

The development of procurement and its role goes hand in hand with the development of IT systems used for procurement. According to the Glass and Kleemann [7], The application of IT systems has developed since the 1970s, where the demand for production materials used MRP system known as "Material Requirements Planning". This system enabled to connect production and procurement. In 1975, about 700 companies had implemented MRP, in 1981 it was already around 8,000 companies [8, 9]. Processing took place on mainframe computers, usually once a week, data were entered using punched labels and people received the results on printed reports. Another improvement was MRP II. In fact, the functionality of the existing system has been improved.

The next stage of IT use in procurement was ERP phase, known as "Enterprise Resource Planning", that offered mainly other options and modules and was so universal that their functions covered the needs of most companies. The main advantage was in stronger cross-company integration of IT systems, providing a common basis for all major business functions across a company, spanning from sales over finance, again production (and others) to procurement [7].

E-Procurement is a term used for electronic purchasing solutions. It is software that allows you to address and select suppliers and possibly the purchase itself. Overhead materials, commodities and services are most often purchased using the e-Procurement solution. It takes place in a business relationship between two companies (B2B). The system is widely used for E-Auctions, E-Inquiries, Request for quotation. E-Procurement has been broadly adopted in organizations [10].

Reflecting the effects of Industry 4.0, new technologies and data management, we get to other development stage and Procurement 4.0 is appeared. Procurement 4.0 is a relatively new term and was introduced recently [11]. Simply said, Procurement 4.0 means application of Industry 4.0 tools in procurement.

The development of IT systems for procurement is summarized in Figure 2.

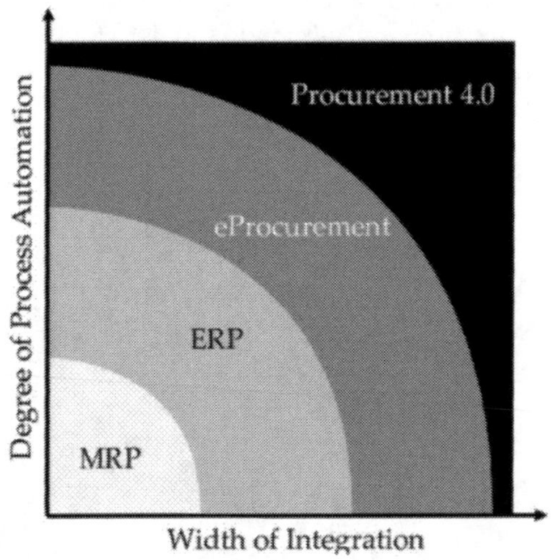

Fig. 2: The evolution of procurement IT systems [7]

3.2 Procurement transformation in digital environment

3.2.1 Opportunities and changing role of procurement

We can find several definitions of digitalization, but as Parida et al. [12] points out commonly agreed definition of digitalization is currently missing. The authors view digitalizaton as *"the use of digital technologies to innovate a business model and provide new revenue streams and value-producing opportunities in industrial ecosystems"*. We identify with the above definition and in this respect, digitalization in connection with the area of corporate procurement is also viewed.

Procurement is a value-creating process that is changed by the digitalization. Srai et al. [13] identified fundamental principles of digitalization interventions in purchasing and supply management. Their grid provides two categories of digitalization technologies used by modern manufacturers in the procurement function: the first category is presented by basic technologies (Internet), the second category consists of advanced technologies (e.g., the use of IoT, artificial intelligence, machine learning, big data analytics, blockchain technologies, or advanced manufacturing technologies).

Procurement 4.0 contributes to the productivity goal of collaboration within Industry 4.0 and improves the organizational supply chain set up to meet tailored and innovative requirements as efficiently as possible. Thanks to new technologies Procurement 4.0 connects the companies with its suppliers and enables a dynamic cooperation and coordination of the procurement process [7]. Digitization contributes to increasing the transparency of information between partners in the supply chain. Process automation can help to reduce purchasing time and optimize resource utilization [14]. There are a huge number of ways to use digital procurement. Based on today's requirements, it is mainly about approving various workflows. Whether it is contracts or approving orders.

There are studies solving the implementation of technologies in procurement also to reduce costs, e.g. Klünder et al. [5] investigated the interrelation between digitization and procurement costs in German industry and confirm the hypothesis of cost reduction through digitization.

The digitization of procurement processes is still in its infancy in some companies. However, as some studies show [11, 2, 6], it brings benefits in supporting day-to-day administrative and business tasks and supporting decision-making processes. The strategic role of procurement is growing. Procurement becomes a strategic partner to support business efficiency, profitability and the creation of new business models.

We can say that digitization of certain operations and processes, when aligned with the companies's core strategies, capabilities and procedures, can bring superior competitiveness [15]. Digitization offers many opportunities for companies that succeed in adapting their business models. On the other hand, not keep pace with the changing market dynamics and rapid digital processes will lead to lower profitability and represents an existential threat to the organization [16].

3.2.2 Procurement application and IT system

When digital transformation should reach the maximum effect, digitization should include business processes as a whole part. In the area of central purchasing, this means, for example, the management of suppliers, all type of orders. Artificial intelligence can support supply chain management. Analysing the market situation, taking into account the reliability of suppliers and proposing optimal contractual conditions, better monitor the development of individual commodities. The system can even recommend what discounts to require.

The whole process is very demanding on the volume of data. It is necessary to choose a suitable Big data platform at the beginning. The deployment platform is chosen based on: speed of computation, investment cost, ease of deployment and update, etc. [17]. It is necessary to determine the scope of digitization and choose a suitable provider. The Capgemini study [2], based on 36 providers, created an overview of possible system functionalities and compares them.

In practice, the functionality of the existing system is usually extended. We also present a case study of the implementation of a new IT system in a selected multinational company, which is currently underway in the company and will be a solution for procurement.

3.2.3 Case study of IT system implementation

In Fig. 3 we see the design of key phases project of system implementation in an international company engaged in the production and distribution of alcohol. In this case study, we see the international project layout. The company currently has three manufacturing plants and several sales divisions across Europe, the company's intention is to unify processes and simply apply the model to another plant in the next probable acquisition. Due to global management in the company, some leading roles are responsible for the management of other countries, so a unified system will facilitate easier control and reporting.

The company decided to move from the SAP system to the new version of SAP S4 / HANA, so it was necessary to schedule the individual phases and set up project teams with a clear definition of powers and responsibilities. After selecting the appropriate software tool, the key phases of the project are divided in a timeline with a clear division of responsibilities. This helps to has better control in which stage of implementation we are, what the key inputs and responsibilities of individual departments will be.

The project started in 2019 and a full transition should take place in 2021. During the project, workshops are planning with participants from all participating countries. It is necessary to agree on future processes and functions that will be using, given that the company's intention is to have a unified process, often there must be compromises between countries on final solutions.

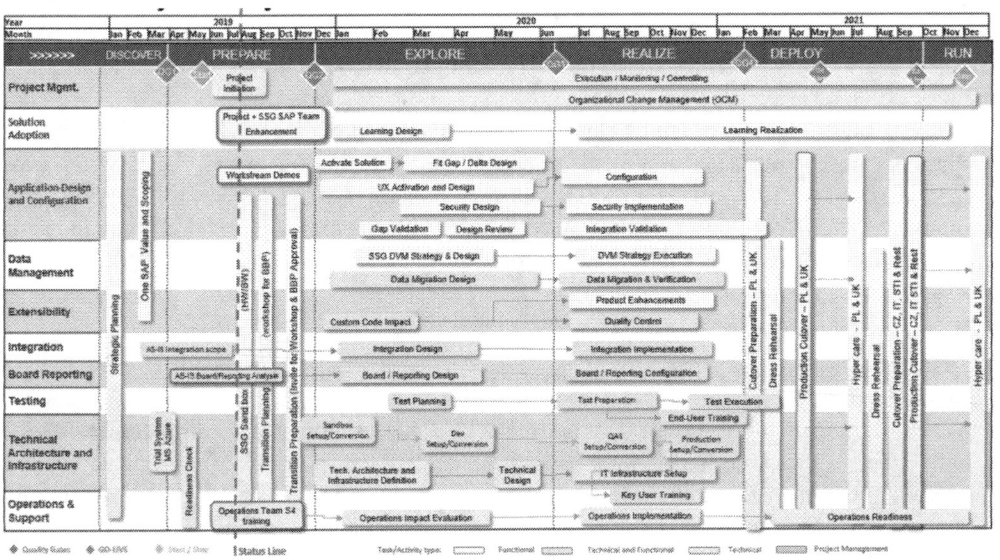

Fig. 3: Key phases of system implementation [internal source of the company]

The implementation project is divided into six basic phases. The description of individual phases is as follows:

Discover: In this section, the company decided to move to a new system. A suitable system was chosen, during a detailed analysis it was decided to expand the existing system.

Prepare: Project teams and external teams are established. Workshop dates are organized. Preparatory phases are being carried out.

Explore: December 2020 begins the first workshops for Procurement, Operations and Finance. The individual sections provide existing solutions, mostly reporting in Excel, or paper workflow, etc. During the year, the design of the future process will be set up and the first testing will take place. The technological architecture and infrastructure definition are chosen. Data migration design and operations impact evaluation will be proposed (Dec 2019 - Jun 2020).

Implementation: This phase includes setup and configuration, testing, data migration and verification, IT infrastructure design, key users training, Operations implementation (Jun 2020 - Jan 2021).

Deploy: This is a phase of the process called Go live. Processes in individual countries are ongoing, but external support is still being used, critical points are being sought and processes are being improved (Feb 2021 - Sep 2021).

Run: The last phase of full operation and official termination of the project (Oct 2021 - Dec 2021).

4 Conclusion

If a company wants to be competitive and sustainable today, it must start using modern tools to simplify processes, which can be automated and eliminate the most operational part. Part of digital transformation of companies is digitalization of procurement. This will allow people in procurement to make more strategic decisions than operations.

Digital transformation requires new technologies and software concepts that allow increasing global connectivity with cross-functional teams. Technologies like Artificial Intelligence, Big Data, and Internet-of-Things enable to automate and optimise procurement model and support the organisational digital transformation.

The effectiveness of modern tools lies in the simplification of processes and the clear definition of competencies. Process automation brings the elimination of human error and joy into everyday toil with tedious administration. Centralization of all purchasing requests into one database then allows better structuring of purchases into categories. Modern tools for automation and purchasing organization do not necessarily involve complex changes in internal regulations. In the future, this phenomenon can create competitive pressure on suppliers and procurement will therefore insist on the use of digitization on their part, it can also be an exclusionary criterion in the selection of suppliers.

We conclude that the development of new technologies brings a shift in the role of procurement. Procurement is moving from a traditional focus on cost reduction to value creation through dynamic cooperation and building transparent relationships with business partners and is beginning to play an increasingly strategic role in organizations. Of course, the digital transformation of procurement requires considerable investment and is time consuming, especially in today's globalized multinational environment. However, it can bring significant value and contribute to increasing the competitiveness of companies.

Acknowledgment

The paper was created with the support of the project No. SGS-2020-026 "Economic and financial transformation in the context of digital society" which is solved at the University of West Bohemia, Faculty of Economics.

References

1. S. Bag, L. C. Wood, S. K. Mangla, S. Luthra, Procurement 4. 0 and its implications on business process performance in a circular economy. *Resources, Conservation and Recycling*, **152** (2020)
2. Capgemini, *Digital Procurement Research 2018*, *Uncovering the solution that bring you forward,* (2018), available at: https://www.capgemini.com/nl-nl/wp-

content/uploads/sites/7/2018/10/Capgemini-Digital-Procurement-Research-2018_web-version1.pdf

3. C. L. Yang, R. J. Lin, D. Krumweide, E. Stickel, C. Sheu, Efficacy of purchasing activities and strategic involvement: An international comparison. *International Journal of Operations & Production Management*, **33**, 49-68, (2013)
4. P. Krampf, *Beschaffungsmanagement,* (München: Vahlen, 2014)
5. T. Klünder, J. N. Dörseln, M. Steven, Procurement 4.0: How the digital disruption supports cost-reduction in Procurement. *Production*, **28**, (2019)
6. Kearney, *Procurement: making digital transformation work for you,* available at: https://www.kearney.com/procurement/article?/a/procurement-riding-the-transformative-digital-wave
7. A. H. Glas, F. C. Kleemann, The Impact of Industry 4.0 on Procurement and Supply Management: A Conceptual and Qualitative Analysis. *International Journal of Business and Management Invention*, **5**, 55-66 (2016)
8. A. V. Mabert, The early road to material requirements planning. *Journal of Operations Management*, **15**, 346–356, (2007)
9. S. Snapp, *The History of MRP and DRP,* available at : https://www.brightworkresearch.com/scmhistory/2012/08/the-history-of-mrp-and-drp/ (2012)
10. C. McCue, A. V. Roman, E-Procurement: Myth or Reality. *Journal of Public Procurement*, **12**, 221–248, (2012)
11. F. Bienhaus, A. Haddud, Procurement 4.0: factors influencing the digitisation of procurement and supply chains. *Business Process Management Journal*, **24**, 965-984 (2018)
12. V. Parida, D. Sjödin, W. Reim, Reviewing Literature on Digitalization, Business Model Innovation, and Sustainable Industry: Past Achievements and Future Promises. *Sustainability*, **11**, (2019)
13. J. S. Srai, H. Lorentz, Developing design principles for the digitalisation of purchasing and supply management. *Journal of Purchasing and Supply Management*, **25**, 78-98, (2019)
14. A. Telukdarie, E. Buhulaiga, S. Bag, S. Gupta, Z. Luo, Industry 4.0 implementation for multinationals. *Process Safety and Environmental Protection*, **118**, 316-329, (2018)
15. M. Ghobakhloo, M. Fathi, Corporate survival in Industry 4.0 era: the enabling role of lean-digitized manufacturing. *Journal of Manufacturing Technology Management*, **31**, 1-30 (2019)
16. J. Bleicher, H. Stanley, Digitization as a catalyst for business model innovation a three-step approach to facilitating economic success. *Journal of Business Management*, **12**, 62-71, (2017)
17. De Lillo, F. Cecconi, G. Lacorata, A. Vulpiani, Sedimentation speed of inertial particles in laminar and turbulent flows. *EPL*, **84**, 314-381 (2008)

Impact of social media on children´s relationship with brands

Nikoleta Hutmanová[1,*] and *Peter* Dorčák[1]

[1]University of Economics in Bratislava, Faculty of Business management, Department of Business Finance, Bratislava

Abstract. The paper focuses on how social media usage by children determines their interactions with consumer brands. First it describes how and when young children develop brand awareness and which are the most important predictors of this development. Those findings are then put in connection with the impact of social media. We elaborate on a deeper level how children approach online communications with brands in the social media context. Our assumptions are supported by a research conducted on a group of New Zealand children, both boys and girls in the age group of 11-14 years. This qualitative approach was implemented using in-depth interviews and identifies three key modes of brand interaction behaviour when young consumers use social media. According to these findings we assume that there is a connection between the use of social media and children´s relationship with consumer brands.

1 Introduction

This article aims to explore based on previous research, how children interact with brands through social networks. Children are active users of social networks. According to previous findings, almost 70% of Slovak children actively use a profile on social networks or game sites. Concerning age categories, it can be stated from the data that a third of children aged 9–10 years, and up to 69% of children aged 11–12 years, use social media [1]. Therefore, understanding the nature of children's interaction with brands on social networks is also crucial for companies. Social networks offer companies many opportunities to understand children's consumer behavior and thus continuously and effectively build children's relationships with brands as relationships of potential customers.

We define children as a target group and describe the development of their brand knowledge in this article. We also analyze and elaborate on their consumer development and behavior. Furthermore, we assume the influence of social networks on the relationship of children with brands. We support our assumptions about the issue by research conducted on a small group of New Zealand children about whom we presumed that their interaction with brands through social networks would be similar to that of children in Slovakia in terms of the frequency of social network use.

* Corresponding author: nikoleta.hutmanova@euba.sk

We implemented a qualitative approach using both paired and individual in-depth interviews of New Zealand children (boys and girls) in the age group of 11–14 years. We analyzed the data by thematic transcript analysis of interviews. At the end of the research, based on the respondent's answers, it was possible to create a conceptual model that illustrates the main processes based on which children use social networks to interact with brands.

2 Literature and background

2.1 Development of children's brand knowledge

Children worldwide became an essential and unique market segment in the sixties of the twentieth century. This understanding of children as independent consumers has grown over time and continues to grow. According to this fact, marketing specialists started to deal with children and adolescents as a separate target group for several reasons. One of them is that it is not uncommon for children to receive both material gifts and money from their relatives. As a result, children begin to realize the value of money and decide what products they would like to buy at a young age. Based on this, children are already perceived as customers who have a certain power and even have the ability to influence their parents [2].

According to findings, children's consumer behavior develops at an early age. Two fundamental factors influence the process by which children become consumers: cognitive and environmental factors. Cognitive factors are children's abilities based on their age. Environmental factors apply to families (especially mothers), peers, and the media [3].

Knowing when and how children learn brand names and how they remember them in the process of becoming consumers is generally explained by Piaget's theory of cognitive development. In Piaget's theory, there are two developmental stages related to developing the faculty of understanding symbolism most closely. These phases are the preoperative and specific operative phase [3]. Research has shown that information about the brand increases in children as they cognitively develop. Ward et al. [4] reported in their study that brand information is increasing with age and that even preschoolers can remember brand names from different product categories.

The purpose of one of the studies in this area was to use selected clues to find out when children's brand awareness starts to develop. A total of 66 Turkish children aged 3 to 5 attending preschool took part in the research. Individual interviews were conducted with them in order to determine their awareness of selected brands. The results of this research show that brand awareness development takes place at the age of 3 years. Even 3-year-olds remembered specific brands and products after seeing the logo, packaging, and character of the brand. Brand awareness shows a directly proportional increase in age. Moreover, brand knowledge reaches its highest level at the age of 12 [5].

2.1.1 Impact of the theory of mind on children's brand knowledge

The cognitive perspective suggests that the individual differences between children, and their social development, can only be assessed once they have developed a so-called theory of mind. American psychologists Premack and Woodruff were the first to define the concept of the theory of mind. According to them, the theory of mind refers to the ability to think about the mental states of others. Thanks to this ability, individuals are further able to understand and predict the behavior of others. A good level of mastery of the theory of mind allows an individual to express different reactions in different social situations. Terms such as "mind-reading" or "mentalization" are also used in literature to describe this term in more detail [6].

In other words, the theory of mind explains that children need to develop the ability to understand another person's perspective because they need a mental basis to comprehend that there are different representations of the world [7]. These representations also apply to trademarks symbolizing the desired attributes of selected products. For instance, quality, prestige, warmth, comfort, and "coolness". The development of the theory of mind in this context allows children to recognize these attributes in a particular brand [3].

Other principal prerequisites for children's understanding of brands relate to their ability to develop their knowledge of market phenomena, such as what brands mean to people and how to understand their persuasive intent [7]. The theory of mind and the enhancement of powerful functional abilities form an essential developmental element for children that enables them to simultaneously understand brand symbolism and develop brand awareness to foster interaction with brands relevant to them [8].

Children's direct experience with brands can help them learn more about brand attributes and discriminate between brands later. Knowing brands helps children shape their preferences because it makes them differentiate between market incentives (Bahn, 1986) [9]. Learning from direct experience also leaves memory traces that help children with later recognition and memorization of the brand [10].

Nowadays, however, the Internet is a central medium in the whole process of consumer behavior. The use of digital technologies has shown an improvement in the memory of young teenagers, the organization of stored information, and an overall improvement in cognitive activity [11]. These benefits of actively using digital media are potentially beneficial elements in helping children build their relationships with brands.

2.1.2 Impact of social media on children's brand knowledge

The As already mentioned, children's interactions with consumer brands are in social contexts no longer limited to family relationships. Since children are active users of social networks, understanding the nature of this interaction is vital. Mainly because social networks offer companies more opportunities to connect customers with brands [12].

Previous research has examined the impact of advertising on children's brand perceptions. However, these studies have not sufficiently dealt with how children interact with consumer brands through social networks, and little is known about how children use social networks to interact with the brand or how it affects their behavior and brand relationship.

Research has suggested that children use social media for entertainment, socializing with friends, and developing their identities [13]. Changing trends in social network usage, such as the increase in media streaming, enabled children to access content through the Internet at home more [14].

The possibilities of brand interaction differ from platform to platform based on how these social media platforms are constructed. For example, children can indicate their affiliation to a brand by communicating on YouTube with a well-known vlogger. Such interactions create the perception of a two-way relationship with the celebrity brand in the form of a parasocial interaction [15]. These types of interactions create feelings of intimacy, we expect them to contribute to the positive affiliation of children with brands and help children gain market and brand knowledge.

Friendly interactions of children on social media platforms are interesting for marketers because the users, in this case, children, can potentially learn about consumer brands. Therefore, marketers strive to make children learn more about consumer brands through commercial placements or sponsored posts. Using social media, in this fashion, makes it easier for children to acquire brand knowledge and builds an affiliation with brands that are expected to determine the quality of their relationship with each other. Under these circumstances, the brand might develop various new strategies that are able to achieve involvement with a specific target group in an online environment. Advanced online branding strategies for the media allow the market to create direct, interactive, and social relationships between brands and young consumers [16].

The understanding of the persuasive side of advertising develops in children between the ages of 10 and 12. Children at this age already understand that ads and brands are able to cause changes in thinking and attitudes and can also influence their shopping intent. On the other hand, from the strategic standpoint of marketers, this age of children is perceived as the one that can be most effectively used to change the attitudes of children as potential future customers [17].

3 Investigation of the influence of social networks on the children´s relationship with brands

The qualitative research conducted through paired and individual in-depth interviews with New Zealand children offers insights on how to deepen children's understanding of their online brand communication in the context of social media. Research has shown that children's social media utilization is more active and dynamic than presented in the past. It is an important finding for the inquiry into children's relationship with brands.

For marketers, the research process can provide specific data to help children understand the brand's commercial intent and its practices on social media. The findings also clarify the extent of young consumers' interactive behavior and provide information on how children get to know the brand, how they respond to their communication on social media, and how they subsequently evaluate this communication [18].

Based on the data obtained from interviews with boys and girls in the age group of 11 to 14 years, it was concluded that in the context of children's interaction with brands on social networks, there were three main processes shown in Figure 1 that take place.

Discerning – the subprocesses of recognition are the perception of a brand, which with recognizing the brand represent a lower level of interaction. They already include a specific internalized knowledge of the brand.
Reacting – consists of a description of the brand, which covers a passive and active interaction. Therefore, it represents a more dynamic interaction resulting in the shaping of children's opinions.
Forming – may involve interaction through remote "observation" or behavior in which a more active relationship manifests, mainly if the child actively uses multiple social media platforms. Most of the children interviewed used at least two social networking platforms. Some children used up to three or more networks, and most of them had a favorite, preferred social media network [18].

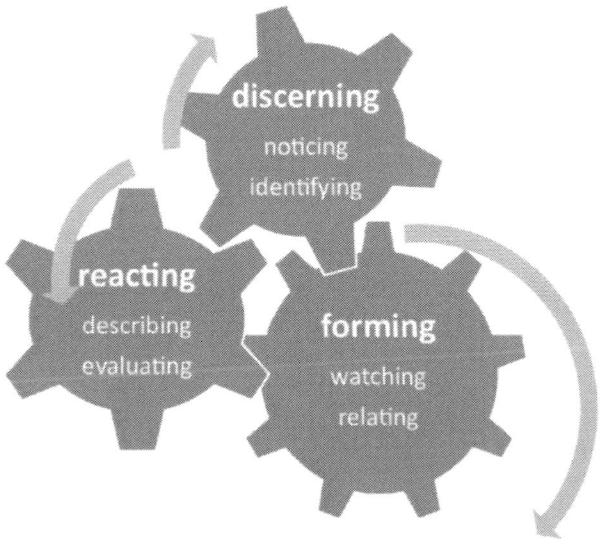

Fig. 1. Conceptual model: children using social media for brand interactions Jones & Glynn, 2019 [18].

In this study, the mentioned processes show a range of interactive behaviors, with specific behavioral manifestations on social media. They confirm that children are able to build relationships with brands through a combination of interaction processes and that some of the established relationships with brands are more dynamic than others. Marketers might find these results particularly interesting. Moreover, people offering branded products and services in a child consumer market might also find them beneficial. The findings clarify the extent to which young consumers' behavior interacts and provide information on how children

4 Conclusion

According to secondary data, we have found that children have become a fascinating and attractive market segment in recent years. Their developmental cognitive perspective suggests that children's consumer behavior appears at an early age. The advances of modern digital technology have an increasing effect on this behavior over time, mainly through social media platforms. Children were presented with a range of interactive types of behavior that clarified the extent of their interaction. Comprehending it enables marketers to create interactive social relationships between brands and young consumers more effectively.

The findings of the mentioned case study have several important implications for future practice. On the other hand, it is salient to note that the research sample of 23 respondents may not be sufficient to express a clear conclusion applicable to all or precisely most child consumers in Slovakia. Nevertheless, we believe that the respondents were able to answer questions that focused on the range of social media platforms they used and to detect what kind of marketing content they could perceive and how they interact with it. The most striking result emerging from the data is that based on this qualitative research, it was possible to compile the presented conceptual model of social network usage processes of children interacting with brands, which we believe should be a basis for further research.

The article is an output of research project VEGA (1/0240/20) "Financial Aspects of Sustainable Business - Enterprise Succession Solution for Small and Medium-sized Enterprises".

References

1. P. Izrael, J. Holdoš, M. Hasák, *Sociálne siete a excesívne používanie internetu a sociálnych médií* (Správa z výskumu EU Kids Online IV na Slovensku). Ružomberok: Katolícka univerzita v Ružomberku (2019)

2. V.M. Somesfalean, Children as target market. *Studies in Business & Economics*, **7**, 172-183. ISSN 1842-4120 (2012)

3. A.R. McAlister, T.B. Cornwell, Children's brand symbolism understanding: links to theory of mind and executive functioning. *Psychology and Marketing*, **27**, 203-228. ISSN: 1520-6793 (2010)

4. S. Ward, D. Wackman, E. Wartella, The development of consumer information processing skills: integrating cognitive development and family interaction theories. *Advances in Consumer Research*, **5**, 535-539. ISSN: 00989258 (1977)

5. Y.A. Arnas, I. Tas, I.G. Ogul, The development of Brand awareness in young children: how do young children recognize brands? *International Journal of Consumer Studies*, **40**, 36-542. ISSN: 1470-6431(2016)

6. A.I. Goldman, *Theory of mind*. Oxford Handbook of Philosophy and Cognitive Science. ISBN: 9780195309799 (2012)

7. L.J. Moses, D.A. Baldwin, What can the study of cognitive development reveal about children's ability to appreciate and cope with advertising? *Journal of Public Policy & Marketing*, **24**, 186-201. eISSN: 15477207(2005)

8. P.M. Valkenburg, Identifying determinants of young children's brand awareness: television, parents, and peers. *Journal of Applied Developmental Psychology*, **26**, ISSN: 0193-3973 (2005)

9. K.D. Bahn, How and when do brand perceptions and preferences first form? A cognitive developmental perspective, *Journal of Consumer Research*, **13**, 382-393. ISSN: 1537-5277 (1986)

10. K.L. Keller, Brand synthesis: the multidimensionality of brand knowledge. *Journal of Consumer Research*, **29**, 595-600. ISSN: 0093-5301 (2003)

11. P.Q. Brito, Pre-teens and teens conceptualization of electronic games and internet. 2012 8th *International Conference on Information Science and Digital Content Technology (ICIDT),* IEEE. Vol. 3, p. 621-625. ISSN: 0018-9219 (2012)

12. J. Ahn, What can we learn from Facebook activity? Using social learning analytics to observe new media literacy skills. *Proceedings of the Third International Conference on Learning Analytics and Knowledge* **8-12** April, Leuven,135-144 (2013)

13. L. Anderson, D.B. McCabe, A co-constructed world: adolescent self-socialization on the internet. *Journal of Public Policy & Marketing*. **31,** 240-253. eISSN: 15477207 (2012)

14. M.R. Nelson, L. Atkinson, M.A. Rademacher, R. Ahn, How media and family build children's persuasion knowledge. *Journal of Current Issues & Research in Advertising*, **38**, 165-183. ISSN: 2164-7313 (2017)

15. H.U. Khan, A. Daud, U. Ishfaq, T. Amjad, N. Aljohani, R.A. Abbasi, J.S. Alowibdi, Modeling to identify influential bloggers in the blogosphere: a survey, *Computers in Human Behavior*, **68**, 64-82. ISSN: 07475632 (2017)

16. S.C. Boerman, L.M. Willemsen, E.P. Van Der Aa, "This post is sponsored": effects of sponsorship disclosure on persuasion knowledge and electronic word of mouth in the context of Facebook. *Journal of Interactive Marketing*, **38**, 82-92. ISSN: 1094-9968 (2017)

17. P. Nuñez-Gomez, J. Sánchez-Herrera, T. Pintado-Blanco, Children's Engagement with Brands: From Social Media Consumption to Brand Preference and Loyalty. *Sustainability*, **12**, 9337. ISSN 2071-1050 (2020).

18. K. Jones, M. Glynn, How children use social media for brand interactions. *Young Consumers*, 20, 91-108. ISSN: 1747-3616 (2019)

Sustainability Concept in the Digital Age: New Opportunities for Companies

Alena Palacká[1,*], *Michaela* Krechovská[1] , and *Josef* Čiž[1]

[1]University of West Bohemia, Faculty of Eonomics, Department of Finance and Accounting, Univerzitni 22, 306 14 Pilsen, Czech Republic

Abstract. Sustainability and digitalization are two of the most powerful phenomena of recent times. The phenomena of digitalization and sustainable development are each separately massively explored, but their intersection is subject only to a few of undertaken researches. The paper aims to expand existing sustainability research by incorporating the current digital environment. In particular, the paper explores the relationship between sustainability and digitalization and maps the potential of digitalization benefits for sustainability issue of companies on the base of conducted literature review and content analysis of published researches in this field. As the results shown, digitalization brings new opportunities and challenges in the management of organizations, as well as the fulfilment of their sustainability strategy. It is time to rethink the concept of sustainability and update it in the light of the development of digitalization. Digitalization is becoming an essential part of business.

1 Introduction

The first decades of the 21st century are characterized by turbulent changes, high interconnectedness of individual economic entities (consumers, businesses, countries and international unions) due to the ongoing process of globalization, which also brings with it an increased competitive environment. For businesses being able to endure and if possible to thrive in such environment, it is vital not only to accomplish their strategic goals, but also to deal with pitfalls and vice versa to exploit the potential of current phenomena. Sustainability and digitalization are two of the most powerful phenomena of today's business environment.

Sustainability and increasing pressure on organizations to manage not only their economic but also environmental and social performance and to inform all stakeholders of the impact of their activities on the environment has risen considerably over the last few years. Several researches have shown that sustainable activities have a significant impact on business performance [1, 2, 3].

At the same time, the development of technology occurs and digital technologies have a major impact on the development of the whole society, the economy of individual countries

[*] Corresponding author: astankov@fek.zcu.cz

and individual economic subjects. We are spoken about the digital economy that is seen as a resource for maintaining competitiveness, economic growth and development. Thanks to the massive and global advent of new technologies, value chains are changing and innovative business models are arising. The newly created business models are based on the usage of resources in smart, disruptive and also sustainable ways [4, 5].

There is a significant amount of studies about how sustainability and digitalization are changing management practice, business and even the society. However, both trends are usually examined separately. Far less attention is paid to understanding how these trends combine to create new market and business conditions [6, 7]. There are some analyses of the relationship between sustainability and digitalization [8, 9, 10], but there are not many of them and they deserve to be deepened.

Therefore, the authors focus their attention on the convergence of these two trends. The paper aims to expand existing sustainability research by incorporating the current digital environment. In particular, the authors are interested in understanding the relationship between sustainability and digitalization and mapping the potential of digitalization benefits for sustainability issue of companies.

2 Methods

In order to fulfil the goal of the paper, first the definition of sustainability and digitalization is given. Furthermore, the authors use the well-known sustainability models, such as the Elkington Tripple Bottom Line model and the Epstein and Buhovac model of sustainable performance implementation and its benefits, which seek to update with regard to the era of digitalization and its challenges and opportunities. The discussion focuses on the potential benefits of implementation sustainable performance when taking into account digitalization. The findings are based on a literature review and content analysis of published researches in the field of digitalization and sustainability.

3 Results and Discussion

3.1 Sustainability

According to Mason [11] the most frequent way to define sustainability is still the definition of the UN Commission on the Environment, which states that sustainability is: *„development that meets the needs of the present generation without compromising the ability of future generations to meet their own needs."* [12, p. 43]

Sustainable development can be viewed from two perspectives. At the micro level, business sustainability can be seen as a business approach to creating long-term value for its owners, through the use of opportunities and the management of risks arising from the economic, social and environmental environment. At the macro level, the essence of sustainable development according to Garza [13] lies in the fulfilment of 3 simultaneous goals:

- maintaining a stable and high level of economic growth and employment;
- in the effective protection of the environment and the prudent use of natural resources;
- and, last but not least, social development that respects the needs of all.

3.2 Dimensions of sustainable performance

Sustainable performance according to Elkington [14] consists of 3 main pillars - economic, social and environmental. According to this world famous concept called "The Triple bottom line" the condition of sustainable business performance is met when all the pillars are in equilibrium (see Fig. 1.).

In order to achieve a sustainable balance in current business environment, Corporate Governance and Digitalization needs to be added to the core pillars [15]. Corporate Governance, on the one hand, is the way in which an enterprise is managed and, on the other, it is a system of managerial responsibility for corporate governance.

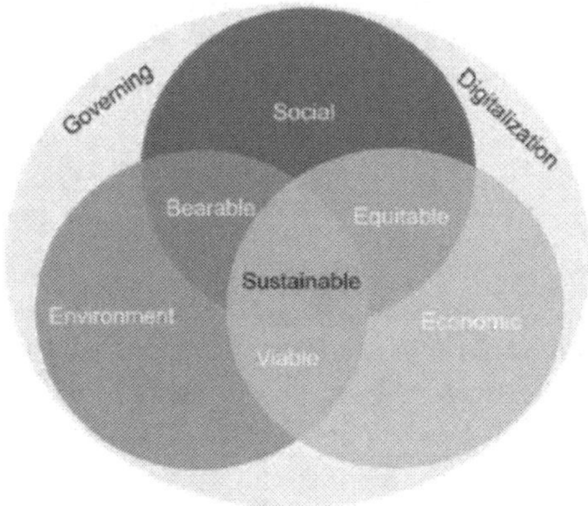

Fig. 1. Dimensions of sustainable performance – modified Elkington´s model [15]

At an equilibrium point located at the intersection of all aspects of sustainability, there is an enterprise with suitable governance that achieves excellent economic results and has achieved stable economic growth, all in combination with the environmentally friendly management of natural resources and addressing the needs of all and exploiting the potential of digitalization.

3.3 Digitalization

As well as in the theory concerning the very definition of sustainability there is no consensus in the digitalization definition. Vor dem Esche and Hennig.Thurau [16] define digitalization as: *"a major change process with enormous "disruptive power" that effects not only the area of information and communication, but also products, services and distribution channels."*

The digitalization (also referred to as "digital revolution") is already reshaping almost every part of private and corporate life thus transforming the society [15, 16]. There are several areas which are identified as the main drivers of digital revolution, for example artificial intelligence, digitalization of information, virtual reality, connectivity and synthetic biology.

In general, there are not only positive changes connected with the digital transformation, but also certain negative effects. Digitalization has positive impacts on energy, resources and carbon productivity, it can lower production costs, dematerialize production and expand access to services and goods. On the other hand, there are job losses concerning mainly the professions where the workers can be easily replaced by AI, deepening the inequality. The inequality issue reflects the current dominance of technology companies over others – the "big five" and the similar situation in other specific sectors (agricultural, food industry, etc.) which are dominated by single or few companies [16].

3.4 Digitalization and sustainability

The interdependency between sustainability and Industry 4.0 has been confirmed by research of Kamble et al. [17]. Bleicher and Stanley [18] see digitalization as a major driver of growth and sustainability. Heemsbergen [19] argue that digitization has implications for transparency and accountability, which opens up completely new ways of shaping, monitoring, communicating and managing sustainability.

According to Beier et al. [20], Industry 4.0 might offer a great chance to align the aims of digital transformation with sustainable development, however if sustainability aspects are not taken into account, it becomes a threat. One of the biggest challenges for sustainable development is the transformation of industrial production through the digitalization.

Ordieres-Mére et al. [10] note that due to the insufficient analysis of some relevant dimensions of digitalization and Industry 4.0, it is believed, that digitalization can improve corporate´s environmental performance through the consistent increase of the institutional knowledge of how to perform the sustainable development, mainly when the main aim of actions does not address sustainability.

3.4.1 Benefits of sustainable performance implementation in the digital age

Epstein and Buhovac [21] set out the reasons why a sustainable performance aspect for businesses is significant. The issue of sustainable business performance has become the subject of regulation by international organizations and, consequently, by individual states. If the company does not incorporate the concept of sustainability into its business and is mandated to do it by law, it is not only threatened by financial penalties, but also by increased costs with additional controls, potential closure of certain manufacturing operations and damage to reputation. Nowadays, the relatively close interconnection of the company with its surroundings provides many opportunities for the company as well as threats. If a company acts in a sustainable manner and takes into account its surroundings, it can, thanks not only to its positive reputation, be more prosperous, for example it can be gaining additional orders or increasing demand from stakeholders for its production. Otherwise, a company not applying sustainability elements in its business faces potential job losses or a decline in production demand. Companies, as well as individuals, have social and moral responsibility for their actions. The company influences its surroundings, whether it is a positive influence (increasing employment, increasing living standards, etc.) or negative (e.g. air pollution). The incorporation of sustainable performance into a business strategy should, when implemented effectively, promote a positive impact on the environment and eliminate or at least reduce the negative impact.

All of the above arguments usually lead to one fact - even if the company spends some financial resources to incorporate the sustainability concept into its business, it should result in greater benefits - whether by avoiding sanctions or other traffic restrictions, or by improving of the company's image and the increased demand for production.

Potential benefits of sustainable performance implementation are quite deeply explored topic, nevertheless the impact of digital revolution on achieving sustainable goals is subject to only few studies. *What could be other potential benefits, if we take into account even the challenges posed by digitalization?* The benefits classified by business area are summarized in Fig. 2.

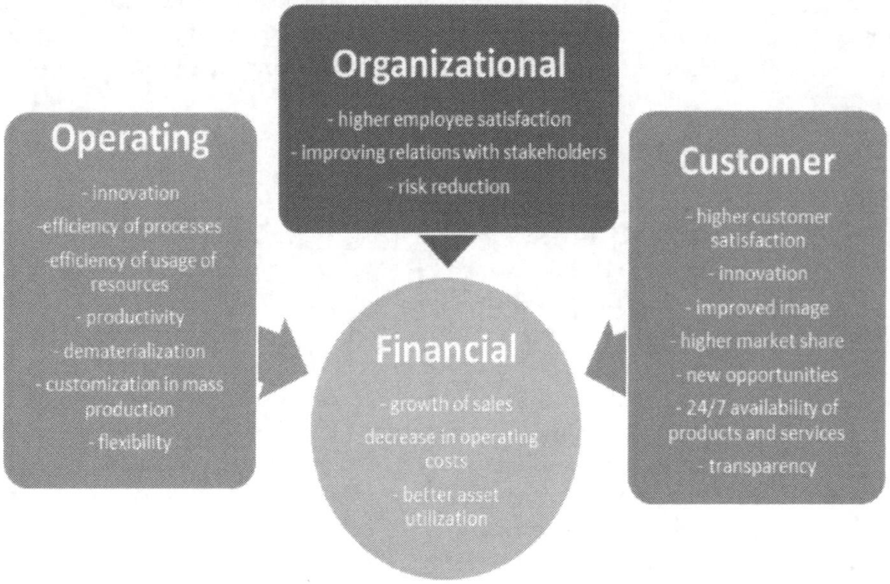

Fig. 2. Potential benefits of sustainable performance implementation combined with digitalization (own processing based on [15, 21, 22])

According to TWI2050 [15] the major transformative impacts of digitalization on production and consumption consist of **dematerialization** and **shared economy**. Dematerialization of goods and services is associated with reduction of resource needs and consumption (energy, raw materials, human capital) and lower carbon emissions. The process of dematerialization is made possible mainly by practically zero marginal costs of transactions, thus better affordability for wider spectrum of customers. The principle of shared economy lies in the real-time matching of demand and supply, thus improved quality of provided services, continuous availability of products and services, and better asset utilization.

Automation of production (using robots and AI in manufacturing process) enables to customize products in mass production and still benefit from economies of scale. Smart automation allows to create flexible manufacturing cells, operating and adjusting automatically based on configurations in production, instead of classical production lines. [22].

With the help of digitalization, the availability of services of all kinds is increasing – for example applications in smartphones or fitness trackers enables individuals to monitor their health conditions in real time and others relatively newly developed tools in telemedicine help to prevent health of employees and reduce work accidents, that might lead to increase in productivity [15].

4 Conclusion

Digitalization in the form of big data, artificial intelligence, the internet of things and more brings new opportunities and challenges in the management of organizations, as well as the fulfilment of their sustainability strategy. Companies use new technologies to communicate with stakeholders, map their environmental footprint, reduce emissions and assess the impact of environmental change on their business. New digital technologies also enable sustainable innovation [15].

As the results show, it is no longer possible to separate digital trends from sustainability trends. It is time to rethink the concept of sustainability and update it in the light of the development of digitalization and the potential of this phenomenon. Digitalization is becoming an essential part of business, the strategy of digitalization cannot be separated from the strategy of sustainability if the company wants to be successful, create value and satisfy its stakeholders in the long run.

At the moment, we are at the beginning of our research, focusing on identifying new trends and potentials that digitalization brings. We have conducted a literature review to identify directions for our future research which will focus on the impact of digitalization on sustainable performance of companies in more detail.

Acknowledgment

The paper was created with the support of the project No. SGS-2020-026 "Economic and financial transformation in the context of digital society" which is solved at the University of West Bohemia, Faculty of Economics.

References

1. B. Georgeevski, A. Alquadah, The Effect of the Volkswagen Scandal (A Comparative Case Study). *Research Journal of Finance and Accounting*, **7**, 54-57 (2016)
2. R.,Ameer, R. Othman, R., Sustainability practices and corporate financial performance: A study based on the top global corporations. *Journal of Business Ethics*, **108**, 61-89, (2012)
3. M. L. Barnett, M. L., Stakeholder influence capacity and the variability of financial returns to corporate social responsibility. *Academy of Management Review*, **37**, 794-816, (2007)
4. A. Joyce, R. L. Paquin, The triple layered business model canvas: A tool to design more sustainable business models. *Journal of Cleaner Production*, **135**, 1474–1486 (2016)
5. S. Schaltegger, F. Lüdeke-Freund, E. G. Hansen, Business models for sustainability: Origins, present research, and future avenues. *Organization & Environment*, **29**, 3–10 (2016)
6. D. Kiron, G. Unruh, The Convergence of Digitalization and Sustainability. *MIT Sloan Management Review*. (2018). Available at: https://sloanreview.mit.edu/article/the-convergence-of-digitalization-and-sustainability/
7. A. B. L. De Sousa Jabbour, C. J. C. Jabbour, M. Godinho Filho, D. Roubaud, Industry 4.0 and the circular economy: A proposed research agenda and original roadmap for sustainable operations. *Annals of Operations Research*, **270,** 273-286 (2018)
8. P. Seele, I. Lock, The game-changing potential of digitalization for sustainability: possibilities, perils, and pathways. *Sustainability Science*, **12**, 183-185, (2017)
9. M. Stuermer,G. Abu-Tayeh, T. Myrach, Digital sustainability: basic conditions for sustainable digital artifacts and their ecosystems. *Sustainability Science*, **12**, 247-262, (2017)
10. J. Ordieres-Mére, T. P. Remón, J. Rubio, Digitalization: An Opportunity for Contributing to Sustainability From Knowledge Creation, *Sustainability*, **12**, 1460 (2020)

11. M. Mason, *What is Sustainability and Why Is It important?* (2019) Available at: https://www.environmentalscience.org/sustainability
12. World Business Council of Sustainable Development, WBCSD, Corporate Social Responsibility (1998) Available at: http://www.wbcsd.org/DocRoot/hbdf19Txhmk3kDxBQDWW/CSRmeeting.pdf
13. F. A. Garza, A Framework for Strategic Sustainability in Organizations: A Three Pronged Approach. *Journal of Comparative International Management* **16**, 1, 23-36 (2013)
14. J. Elkington, *Cannibals with forks: The triple bottom line of 21st century business*. (Gabriola Island: New Society Publishers, 1998)
15. TWI2050 - The World in 2050, *The Digital Revolution and Sustainable Development: Opportunities and Challenges* (International Institute for Applied Systems Analysis, Laxenburg, Austria, 2019)
16. J. Vor dem Esche, T. Hennig-Thurau, *German Digitalization Consumer Report* (Digitalization Think:Lab, Research Report No. 2, 2014)
17. S. S. Kamble, A. Gunasekaran, S. A. Gawankar, Sustainable Industry 4.0 framework: A systematic literature review, identifying the current trends and future perspectives. *Process Safety and Environmental Protection*, **117**, 408-425, (2018)
18. J. Bleicher, H. Stanley, Digitization as a catalyst for business model innovation a three-step approach to facilitating economic success. *Journal of Business Management*, 12, 62-71, (2017)
19. L. Heemsbergen, Digital age from radical transparency to radical disclosure: reconfiguring (in) voluntary transparency through the management of visibilities. *International Journal of Communication*, **10**, 138–151, (2016)
20. G. Beier, A. Ullrich, S. Niehoff, M. Reißig, M. Habich, Industry 4.0: how it is defined from a sociotechnical perspective and how much sustainability it includes; a literature review. *Journal of Cleaner Production*, **259**, 120856, (2020)
21. M. J. Epstein, A. R. Buhovac, *Making sustainability work best practices in managing and measuring corporate social, environmental, and economic impact* (London: Routledge, 2017)
22. R. Howels, *Smart automation: How Manufacturers Strike The Right Chord With Each Customer*, (2018) Available at: https://www.digitalistmag.com/digital-supply-networks/2018/05/31/smart-automation-how-manufacturers-strike-right-chord-with-each-customer-06172797

Slovak Audit Firms and their Membership in a Network

Denisa Domaracká[1,*], *Renáta* Antalová[1]

[1]University of Economics in Bratislava, Department of Accounting and Auditing, 852 35 Bratislava, Slovakia

Abstract. Statutory audit in Slovakia is performed by statutory auditors or audit firms. Many of the audit firms are members of the globally operating audit networks. Membership in a network imposes certain rights and duties for audit firms. The existence and operation of the audit networks can be understood as one of the results of globalization that allows an increasing possibility of cooperation at the international level. Our goal is to clarify the state of audit firms operating in the audit networks. To achieve the goal, it was necessary to choose a purposeful work methodology and research methods, based primarily on the analysis of data from the list of the audit firms.

1 Introduction

As a result of historical development, accounting and auditing have been formed as complex and controversial professions [1]. Audit profession in Slovakia, focused on the performance of statutory audit (audit of financial statements), is directly connected with the process of verifying the truthfulness of information which are presented to the third parties through financial statements. The purpose of the financial statements is to provide information about the financial position, performance and changes in the financial position of the entity [2]. The information needs of individual user groups are diverse [3, 4]. Accounting entities in Slovakia prepare financial statements according Slovak legislation or IFRS [5].

Third parties rely on the auditor´s opinion to make further decisions. In this context, the audit of financial statements is a tool of ensuring that objective information is reported in the financial statements [6]. As well as a comparability of financial statements and a constant need of ensuring transparency are caused by vast of changes constantly influencing economic practices in the relevant economic environment [7]. The need to perform the statutory audit arises form the uncertainty of the third parties that information presented in the financial statements is incorrect [8]. There are many models used to detect fraud [9]. It is therefore necessary to present all business transactions in the form of financial statements [10].

* Corresponding author: denisa.domaracka@euba.sk

The fact resulting from the uncertainty of the third parties is the reason for the legislative regulation of the requirements for the auditing profession and at the same time the whole process of the statutory audit.

In Slovakia, we can meet the fact that audit firms are members of a "Network". The audit network is one of the possible forms of cooperation among the audit firms. In the supranational legislation of statutory audit, a "Network" is defined primarily in the International Standards on Auditing (ISA) and Directive 2006/43/EC of the European Parliament and of the Council of 17 May 2006 on statutory audits of annual accounts and consolidated accounts, amending Council Directives 78/660/EEC and 83/349/EEC and repealing Council Directive 84/253/EEC (Directive 2006/43), art. 2 (7).

The European Directive 2006/43 had to be transposed into national legislation – to the Act. No 423/2015 Coll. on Statutory Audit and on amendments and supplements to Act No. 431/2002 Coll. on Accounting, as amended (Act on Statutory Audit). The transposition of the Directive 2006/43 into the Act. on Statutory Audit was reflected also in the definition of a Network in art. 2 (12) of the Act on Statutory Audit, as it is identical to the definition of a Network in the Directive 2006/43. In the European Union (EU) and its member states the EU legislation is implemented [11].

Table 1. Definition of a term "Network".

	"Network"
ISA (Glossary)	A larger structure: That is aimed at cooperation, and That is clearly aimed at profit or cost-sharing or shares common ownership, control or management, common quality control policies and procedures, common business strategy, the use of a common brand name, or a significant part of professional resources
Directive 2006/43 (Art. 2 (7))	The larger structure: Which is aimed at cooperation and to which a statutory auditor or an audit firm belongs, and Which is clearly aimed at profit or cost-sharing or shares common ownership, control or management, common quality-control policies and procedures, a common business strategy, the use of a common brand-name or a significant part of professional resources
Act on Statutory Audit (Art. 2 (12))	Means a cluster aimed at cooperation and to which a statutory auditor or an audit firm belongs, and Which is clearly aimed at profit-or cost sharing, or Which shares common ownership, control or management, strategy, the use of a common brand-name or a significant part of professional resources

2 Methods

In Slovakia from its establishment in 1993 to the present, there have been several audit firms. The list of audit firms as well as statutory auditors in Slovakia is maintained in accordance with art. 11 of the Act on Statutory Audit by Auditing Oversight Authority.

Auditing Oversight Authority was established on 1 January 2008 and, as of 31 March 2008, it took over the management of the list of audit firms and the statutory auditors from the Slovak Chamber of Auditors.

2.1 Statistics of audit firms in Slovakia

Such a list, in accordance with the requirements of the European Union, is a publicly accessible register, published on the Authority´s website. As of December 2019, the number of the audit firms with an active license was 234 (see Fig. 1.).

As Fig. 1. shows, there has been no significant change in the number of audit firms in recent years. However, Auditing Oversight Authority in the framework of the management of the licenses, in addition to issuing the new licences, issues licenses again (e. g. due to a change of an audit firm name), suspends licenses (e. g. on a request or due to a breach of duties arising from the Act on Statutory Audit), activates the suspended licenses (e. g. on a request or after the expiration of the period which a license was suspended for), deletes audit firms from the list (e. g. on a request or due to an audit firm liquidation).

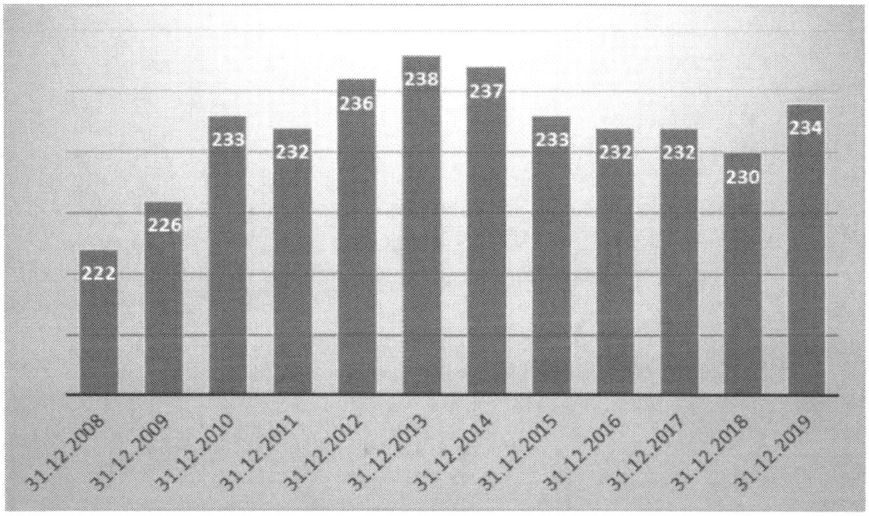

Fig. 1. Number of audit firms with an active license in individual years.

2.2 Statistics of Slovak audit firms and their membership in a network

Auditing Oversight Authority states in the list of audit firms, in accordance with art. 11 (2i), information about audit firms´ membership in a network and all entities belonging to the network and affiliates of audit firms and their registered office or reference to the place where the information is available to the public.

In the list of 234 audit firms with an active licence, 20 audit firms are members of a network. Fig. 2. shows thus, that less than 10% of all active audit firms are members of a network.

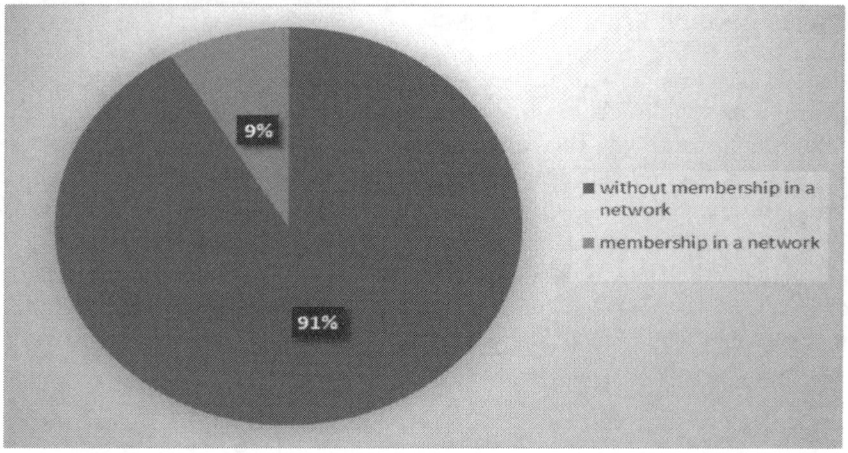

Fig. 2. Overview of audit firms and their membership in a network.

3 Results and Discussion

Based on information from the list of audit firms, most audit firms operating in the Slovak republic are members of different networks. Even some of them are subject to EU market monitoring in relation to the services provided to Public Interest Entities (PIEs). Public interest entities are understood to be accounting units that are governed by the accounting legislation as well as auditing legislation [12].
Based on the list identified by the Committee of European Auditing Oversight Bodies (CEAOB), there are 10 key audit firms operating globally:

Table 2. Key audit networks defined by CEAOB.

Key Audit Networks (by CEOB)	Slovak Audit Firms as Members
Baker Tilly	TPA AUDIT, s.r.o.
BDO	BDO Audit, spol. s r. o.
Deloitte	Deloitte Audit s.r.o.
Grant Thornton	Grant Thornton Audit, s. r. o.
KPMG	KPMG Slovensko spol. s r.o
Mazars	Mazars Slovensko, s.r.o.
Moore Stephens	BDR, spol. s r.o.
Nexia	VGD SLOVAKIA s. r. o.
PwC	PricewaterhouseCoopers Slovensko, s.r.o.
EY	Ernst & Young Slovakia, spol. s r.o.

Mentioned audit firms currently in Slovakia do not provide services only in the field of statutory audit. They extend the services mainly with other assurance services, consulting services, or risk management. It is also because every enterprise may face the risk of business default [13].

In such cases, an audit network is a network of "Professional Services", which in an interconnection of independent audit firms that choose to merge in order to reduce costs effectively and provide professional services to the clients through an organizational framework. Distribution of the costs (e. g. for trainings or technologies) among a larger number of firms is an advantage, which can effect positively the fees of services provided.

Mutual cooperation of Slovak audit firms within the networks exists in several forms. Cooperation typical for smaller national audit firms operating in the network does not focus on e. g. compliance with global quality standards of services provided.

More significant cooperation of national audit firms operating in the network is the contractual cooperation among legally and economically autonomous member audit firms that are organized under the strategic guidance of one or more member firms. It means e. g. a usage of the network´s name, exclusive representation in a specific territory, compliance with the global quality framework of services provided, etc.

Legal autonomy in this case means that each firm member in the network has its own legal status, which also results into the protection of risk of liabilities for other members of the network.

Economic autonomy in this case means that the strategic economic decisions are made independently [14]. For comparison, within a classic group of firms, e. g. based on a percentage of voting rights, one firm takes control of the other firm.

Although we state the firms within the audit network are legally and economically independent, a network membership imposes them rights and duties that should be respected. One of the typical rights of a member firm is the possibility of using a well-known brand name, common resources, manuals, databases or software, the exclusive right to provide services in specific territory. A typical duty of a member firm is to be under a constant monitoring by the network management. By monitoring and controlling, the network ensures that quality of services provided does not decrease and is constantly in accordance with legislative requirements.

In this article, we point out the activities of Slovak audit firms in the network as one of the possible forms of cooperation among audit firms, especially at global level. Many clients expand to foreign markets, so audit firms had to react. The reaction of audit firms was the establishment of the audit networks. Harmonization and standardization, as well as the overall business, represent a principal feature of present globalization [15]. Pakšiová and Loviová (2019) presented that globalisation, accompanied by rapid technological changes, has given rise to a completely new business environment [16].

In Slovakia, only 10% of all audit firms with an active licence are members of some audit network. This might be due to a fact that audit networks focus mainly on operating at a global level. Therefore, if small audit firms focus only on the national market, they do not have such a need to be integrated into a network. However, most of them belong to key audit networks, which are monitored by the EU because of services provided to PIEs.

In addition to many benefits of the network membership, there are also many disadvantages. One of the disadvantages seems to be the process of global network monitoring, which is performed in all member audit firms. However, the need to monitor the quality control system arises from the supranational legislation – International Standard on Quality Control I (ISQC I). Global monitoring should have an impact on the quality of services provided. Therefore, even in this disadvantage should be a benefit for the member audit firm.

This article is an output of the project of the Scientific Grant Agency of the Ministry of Culture of the Slovak Republic and Slovak Academy of Sciences (VEGA) no. 1/0517/20 (2020-2022) "Virtual Cryptochains as a Relevant Tool to Eliminate Economic Crime."

References

1. B. Parajka, Are Information Needs of Financial Entities served by Financial Statements in the Slovak Republic. *Financial management of firms and financial institutions*, 950-957 (2015).
2. B. Parajka, Micro Accounting Entities in the Slovak Republic – A Year After an Introduction. *Strategic management*, **21**, 45-48. (2016).
3. S. C. Green, Risks associated with corporate social media communication - Time for internal auditing to step-up. *Southern African Journal of Accountability and Auditing Research-SAJAAR*, **18**, 73-91 (2016).
4. J. Leidner, H. Lenz, Client's business risk, public-interest entities, and audit fees: The case of German credit institutions. International Journal of Auditing, **21**, 324-338 (2017).
5. Z. Kubaščíková, R. Pakšiová, Impact of accounting standards on the results of financial analysis. *IFRS: global rules & local use: proceeding of the 3nd International scientific conference*, 150-155 (2015).
6. P. Krišková, J. Užík, Professional scepticism in statutory audit under conditions of increasing requirement on the relevance of accounting information. *Účtovníctvo a audítorstvo v procese svetovej harmonizácie*, 103 – 106 (2017).
7. L. Kareš, V. Kňažková, L. Rybka, Will the identified improvements of auditor's report meet the requirements of users and provide them with the value, i.e. benefit that they seek? *Strategic Management: International Journal of Strategic Management and Decision Support Systems in Strategic Management*, 15–26 (2013).
8. F. Maděra, Audit a audítorstvo. Prvé vydanie. Bratislava : Wolters Kluwer, 2014. 124 s. ISBN 978-80-8186-147-9
9. K. Kubaščíková, M. Tumpach, Z. Juhászová, G. Almanova, Blockchain technology and verification of information from financial statements. *IFRS: global rules & local use: proceeding of the 7th International scientific conference*, 411-416 (2019).
10. N. Nabm, M. J. Ali, K. Ahmed, Corporate governance, board ethnicity and financial statement fraud: evidence from Malaysia. *Accounting Research Journal*, **32**, 514-531 (2019).
11. Z. Juhászová, P. Markovič, D. Mokošová, Fair value and its importance for financial decision-making. *IFRS: global rules & local use: proceeding of the 2nd International scientific conference*, 64-69 (2014).
12. V. Kňažková, L. Ondrušová, Public Interest Entities in the Context of Accounting and Auditing Legislation in the Wake of the Globalization Process. *Globalization and its Socio-Economic Consequences 2019*, 1-7 (2019).
13. M. Kováčová, T. Kliestik, K Valášková, P. Ďurana, Z. Juhászová, Systematic Review of Variables Applied in Bankruptcy Prediction Models of Visegrad Group Countries. *Oeconomia Copernicana*, **10**, 1-30 (2019).
14. Lenz H., James M.L. (2007) International Audit Firms as Strategic Networks — The Evolution of Global Professional Service Firms. In: Cliquet G., Tuunanen M., Hendrikse G., Windsperger J. (eds) Economics and Management of Networks.

Contributions to Management Science. Physica-Verlag HD. https://doi.org/10.1007/978-3-7908-1758-4_17

15. M. Hladika, D. Mokošová, J. Molín, Comparison of national accounting framework – the case of Croatia, Slovakia and Czechia. *Economy of Easten Croatia – vision and growth: 6 th International scientific symposium*, 1028–1044 (2017).
16. R. Pakšiová, K. Lovciová, Managerial Reporting by Food Production Companies in Slovakia in 2017. *Engineering Management in Production and Services*, **11**, no. 3, 71-85 (2019).

SHS Web of Conferences **115**, 02002 (2021) https://doi.org/10.1051/shsconf/202111502002
Current Problems of the Corporate Sector 2021

Is it Possible to Predict Earnings per Share?

Miroslav Kmeťko[1,*], *Eduard* Hyránek[1]

[1]University of Economics, Faculty of Business Management, Department of Business Finance, Dolnozemska cesta 1, 851 04 Bratislava, Slovakia

Abstract. The publication of quarterly results of publicly traded companies can have a significant impact on the valuation of their shares. This is mainly concerned with the valuation of the shares, whether it is correct, and at the same time as a prediction of the overall annual financial results. It In most of the analysed companies, we found that most of the year-on-year changes were negative. It is also not possible to draw a clear conclusion about the linear relationship between the percentage change pf surprises and the change in the market price of shares. It should also be noted that the share price in the monitored days may be affected by the current market situation. What this means in practice is that, despite the positive results and the negative mood, stock prices can end up in negative values. However, this situation was not the subject of our research. Therefore, we used a correlation coefficient for this dependence, which represent the mutual movement.

1 Introduction

The publication of quarterly results of publicly traded companies can have a significant impact on the valuation of their shares. During this period, which is called the Earning Season and is always started by company Alcoa (as the first component of the S&P 500 index) are closely watched by investors. This is mainly concerned with the valuation of the shares, whether it is correct, and at the same time as a prediction of the overall annual financial results. Outside this period, company share´s prices are mainly influenced by market sentiment and by published news about company, such as statements from top managers, change of managers, mergers and acquisitions, new products etc.

Major bank, investment firms and brokers, employ number of analysts, who publish their investment recommendations for the main areas of the expected financial situation of the analysed company. To this we may include: quarterly earnings per share (EPS), annual earnings per share, annual sales, and these estimates are prepared several periods to the future. There are also analytics opinions for the current price, whether the share price is undervalued or overvalued, in the form of recommendations to buy, hold or sell the shares.

* Corresponding author: miroslav.kmetko@euba.sk

2 The aim of the paper and methodology

The aim of this paper is how the analysts were accurate in estimating quarterly earnings per share. For every major U.S. exchange-traded company, there are several analysts from different companies. The median value (which is standard measure for expectations for financial markets) is subsequently published for the estimates. Not as is usually used arithmetic mean.

According to the information source, therefore, it also depends on sources to analysts and their views and therefore estimates of quarterly EPS can vary. For our analysis, we used data from the page www.finance.yahoo.com. We recalculate the percentage changes of surprises based on the data, to two decimal places because the real data on the page works with multiple decimal places.

In addition to comparing real data with expected data (EPS), we also used a comparison of real changes of quarterly changes of EPS by tracking the current quarter against the same quarter of the same quarter of the previous year. We also analysed the changes in the share prices in the form of the final price before the results were published to the final price of the share on the day the information was published.

Lastly, the correlation coefficient between the surprise of the current EPS value and the price change was calculated.

For the analysis, given by the current situation, we selected the largest producers of dgug makers (according to the market capitalisation) in the USA Specifically sorted from the biggest: Johnson & Johnson (JNJ), Pfizer Inc. (PFE) a Merck & Co., Inc. (MRK).

3 Estimation of EPS

The fundamental analysis of joint stock companies is devoted a lot of literature. However, it is necessary to distinguish whether we focus on ex-post or ex-ante analysis e.g. Weaver [1] or Damodaran [2]. But is it sufficient? According to Yohn [3] he highlights that financial statement analysis research has slowly evolved and has received limited attention from academics.

The analysis of expected EPS and real EPS can be based on a comparison of short selling. If the investors expect worse results than analysts expect, they will start selling shares briefly. As the authors state Hou et al. [4]: "We examine the relationship between short selling and analyst optimism bias of earnings forecasts of stocks in China. Lastly, we document that short selling information improves the accuracy of earnings forecasts. Similarly Li et al. [5]: "Further analysis of the role of short-sales constraints in the sensitivity of stock prices to bad earnings news implies that short sales can enhance informational efficiency".

Boehmer et al. [6]: "Importantly, short sellers use both public news and private information to anticipate news regarding earnings and analysts. Information that may have a significant impact on investor sentiment". As indicated by Yang and Wu [7]: "combined effect of investor sentiment and public information on the asset price, the mechanism of public information shock on investor sentiment, and how the shock impacts the asset pricing.

From the behavior point of view Murkherjee and De [8]: "the proxy for rational behavior is investor's use of private predisclosure information during earnings announcement periods, while the disposition effect they display serves as an estimate of their behavioral bias."

Also pandemic situation affect earnings. Landier and Thesmar [9]: "Forecasts over 2020 earnings have been progressively reduced by 16%."

The reaction to the currently published profit figures also depends on the concrete capital market. According to Gaston a Jarne [10] there is an evidence that the magnitude of the reward (penalty) for companies meeting (failing to meet) earnings forecasts differs on the market. Chi et. Al. [11]: "The stock price inflation is partly explained by greater earnings surprises and more positive investor reaction to the surprises."

The sensitivity of stock prices to published results is also influenced by the frequency of their publication Arif a De George [12] : "We find that the returns of semi-annual earnings announcers (i.e., low reporting frequency stocks) are almost twice as sensitive. " The time aspect of the publication was also analysed by Adams and Neururer [13]: "we find evidence that uncertainty and volatility risk premiums are higher for firms that report later in the quarter. "

It is also linked to analysts´recommendations when it is forecasting profits. Research in this area has been tested by Billing et al. [14]: "Our evidence suggests that this negative relation results from analysts revising recommendations upward for higher expected earnings growth but failing to also adjust downward for a related decrease in expected returns arising from a decrease in investor risk aversion. " Clarkson et al. [15]: "Finally, our results show that target prices are useful in predicting future stock returns beyond earnings forecasts and commonly used risk proxies. "

4 Research results and discussion

We will analyse individual companies based on selected criteria. We will start with the company with the highest capitalisation and continue with companies to lower capitalisation. But it´s important to remember that that we don´t take into account other factors impacting the shares prices.

4.1 Company JNJ

We´re starting with JNJ. First, we will analyse the relationship of changes in the current quarters compared to estimates to the current changes in EPS. This is represented in Figure 1.

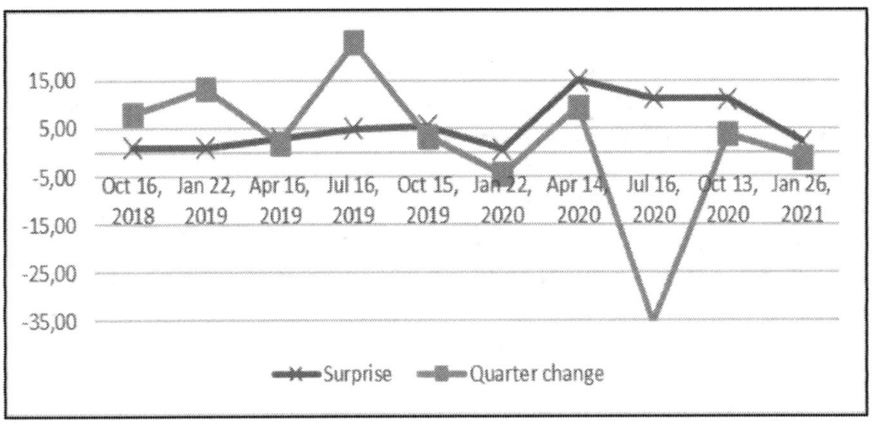

Fig 1. Earning surprise and quarterly change, own processing

Despite the high volatility of the quarterly results, the surpsises against the expected values were not so significant. Which means that analysts predicted these changes, they expected these changes to be less than they actually were.

In the next figure we present what was the real reaction of investors, in share prices, to the real published EPS data.

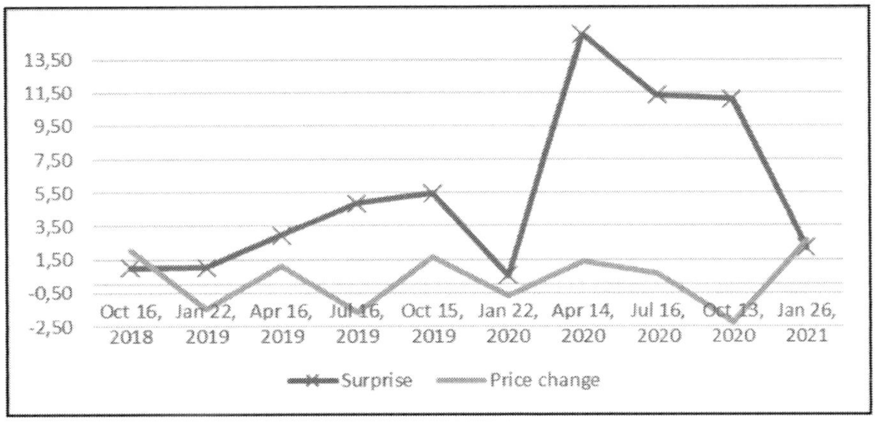

Fig. 2. Earning surprise and change in share prices, own processing

According to the figure, it is not possible to directly predict how the stock price will change with regard to the surprises in the financial resultss. Even in October 2020, despite positive information, stock prices fell.

4.2 Company PFE

PFE company is the second company in terms of market capitalization, producing drugs in the US.

As in the previous company, we will start here with an analysis of the changes in quarter changes in EPS as well as with the actually published financial results. The results are presented in figure 3.

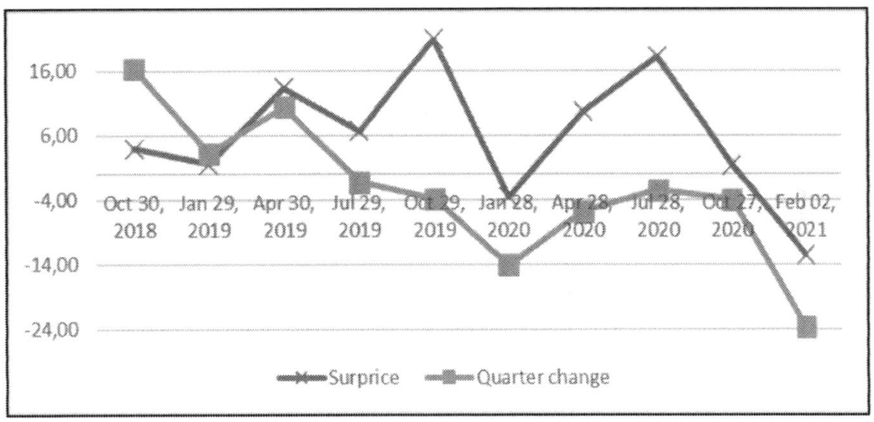

Fig. 3. Earning surprise and quarterly change, own processing

The gradual decline in quarterly results is surprising, The year-on-year declining change was expected by analysts, but it was significantly underestimated to the current values published by the company. It is looks like findings of Balashov and DeVides [16] we find that analysts can predict breaks in earnings strings by issuing less favorable earnings estimates ahead of a break announcement.

How the stock price reacted in view of the surpices of the published EPS is shown in figure 4 [17].

https://doi.org/10.1051/shsconf/202111502002

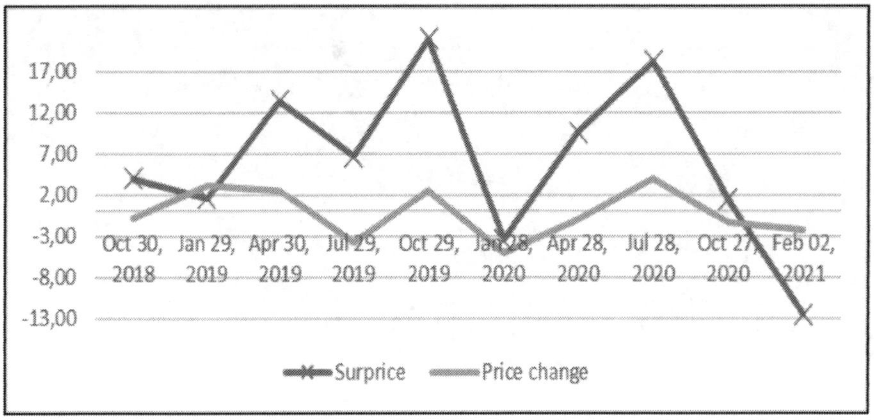

Fig. 4. Earning surprise and change in share prices, own processing.

In this case, the share price reacted in accordance with the value of the EPS surprice. For this company, if the investor can correctly predict the value of EPS, he can significantly gain from this surprise.

4.3 Company MRK

Compared to PFE, the year-to-year change in EPS is rather positive. In most quarters, teh change was grater than 10 %. The development is shown in figure 5.

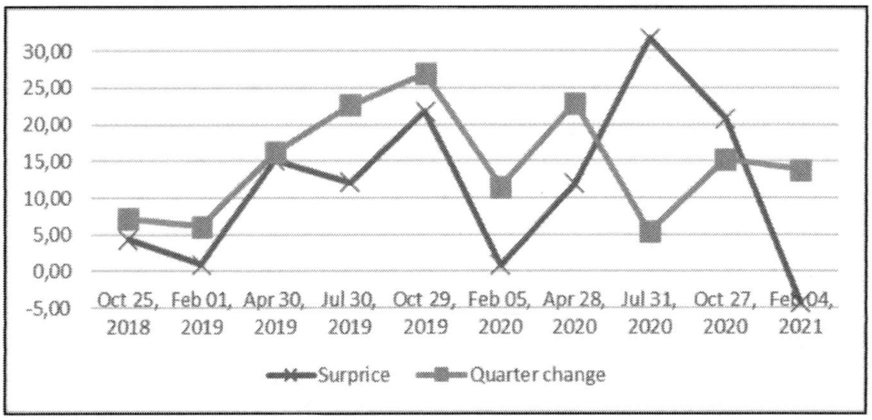

Fig. 5. Earning surprise and quarterly change, own processing

Based on the above development, it can be stated that the real were for all almost better than expected, except for the last monitored quarter.

How the share price reacted to thes published results is shown in the following figure 6.

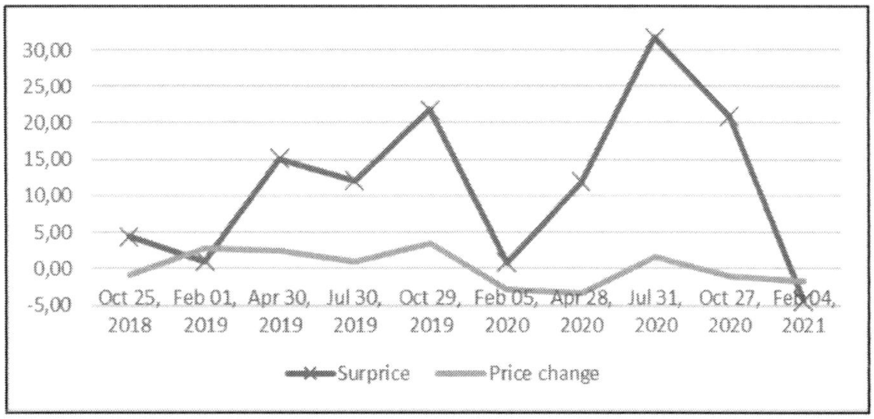

Fig. 6. Earning surprise and change in share prices, own processing.

According to the figure, it is so clear that the share price did not react to the published value of EPS as the analysts would predict. At the highest value of the surprise, the highest percentage change in price was not recorded. However, it can be stated that despite some positive changes in values in EPS, the share price decreased.

5 Conclusion

In most of the analysed companies, we found that most of the year-on-year changes were negative. It is also not possible to draw a clear conclusion about the linear relationship between the percentage change pf surprises and the change in the market price of shares. Therefore, we used a correlation coefficient for this dependence, which represent the mutual movement.

If we focus on the JNJ company, the correlation coefficient is at the level of -0,07. This value explains that it is not possible to draw any conclusions about the surprise of EPS and the change in price.

Conversely, for PFE, the value of the correlation coefficient is 0,64. Thus, there is a strong dependence between the better EPS values that have been published to analyst´s forecasts. It is therefore highly likely that if the company reports better estimates, the share price will increase. But it is necessary to stated that the EPS of this company on a long-term basis is declining in the long-term run.

The last company MRK has a correlation coefficient of 0,41. This value is in the middle among analysed companies. It is likely that stock price will rise if the value of the surprise is positive, but to the given value of the correlation coefficient, it is not possible to confirm this with high probability.

It should also be noted that the share price in the monitored days may be affected by the current market situation. What this means in practice is that, despite the positive results and the negative mood, stock prices can end up in negative values. However, this situation was not the subject of our research.

The article was elaborated within the VEGA project no. 1/0462/19 Implementing key determinants into performance models as tools financial management under the current conditions. (100 %)

References

1. S.C. Weaver, The essentials of financial analysis. McGraw-Hill Education. 480 pp. ISBN 0-07-176836-X (2012)

2. A. Damodaran, Investment valuation : Tools and techniques for determining the value of any asset. 3rd edition. Wiley. 992 pp. ISBN 1-118-01152-X (2012)
3. T.L. Yohn, Research on the use of financial statement information for forecasting profitability. *Accounting and Finance* **60**, 3 (2020)
4. D.S. Hou, Q.B. Meng, K.C. Chan, Does short selling reduce analysts′ optimism bias in earnings forecasts? *Research in International Business and Finance* **56**, April (2021)
5. Z. Li, M.T. Tian, G.D. Ouyang, F.H. Wen, Relationship between investor sentiment and earnings news in high- and low-sentiment periods. International Journal of Finance and Economics **26**, 2 (2021)
6. E. Boehmer, C.M. Jones, J. Wu, X. Zhang, What do short sellers know. *Review of Finance* **24**, 6 (2020)
7. C. Yang, H. Wu, Investor sentiment with information shock in the stock market. *Emerging Markets Finance and Trade* **57**, 2 (2021)
8. S. Murkerjee, S. De, When are investors rational? *Journal of Behavioral Finance* **20**, 1 (2019)
9. A. Landier, D. Thesmar, Earnings expectations during the COVID-19 crisis. *Review of Asset Pricing Studies* **10**, 4 (2020)
10. B.K. Billings, S. Keskek, S. Pierce, The predictability of future earnings growth and the relation between aggregate analyst recommendation changes and future returns. *Accounting Review* **96**, 1 (2021)
11. S.C. Gaston, J.I.J. Jarne, An international comparison of incentives for earnings management in order to meet analysts forecasts. *Spanish Accounting Review* **24**, 1 (2021)
12. J.D. Chi, M. Gupta, S.A. Johnson, Short-horizon incentives and stock price inflation. *Journal of Corporate Finance* **65**, December (2020)
13. S. Arif, E.T. De George, The dark side of low financial reporting frequency: Investors′reliance on alternative sources of earnings news and excessive information spillovers. *Accounting Review* **95**, 6 (2020)
14. P. Clarkson, A. Nekrasov, A. Simon, I. Tutticci, Target price forecasts: The role of the 52-week high price and recent investor sentiment. *Journal of Business Finance and Accounting* **47**, 9-10 (2020)
15. T. Adams, T. Neururer, Earnings announcement timing, uncertainty, and volatility risk premiums. *Journal of Futures Markets* **40**, 10 (2020)
16. V.S. Balashov, Z.B. DeVides, Can analysts predict breaks in earnings strings? *Review of Accounting and Finance* **18**, 4 (2019)
17. YAHOO. *Finance*, www.finance.yahoo.com (2021)

Key factors influencing family businesses: a qualitative study of selected world wineries

Vojtěch Koráb
Faculty of Business and Management, Brno University of Technology, Czech Republic

Abstract. This research study is based on the elaboration of nineteen case studies of family wineries from all over the world and one case study from the Czech Republic. All family wineries were visited by the author and semi-structured interviews were conducted with selected representatives of family wineries. The aim of the research was to look at the family winery through the lens of two systems: the family system and the business system, in order to find out the involvement of family members in a winery and the most important factors affecting a winery. Websites of family wineries were examined through contextual analysis. Based on the analysis of all case studies, four key groups of family wineries in the world were compiled. The research study identifies key family factors, i.e. familiness and the generation of the family which owns a winery. On the business side, the study identifies two key factors important for the sustainability of a family winery, namely corporate innovation and the degree of internationalization. The results of the analysis of world family wineries were compared with one winery selected from the Czech Republic.

1. Introduction

The history of family businesses is connected with the entire modern history of mankind. The oldest family business still in operation is the Japanese business hotel Hoshi Ryokan. This family business was founded as early as 718. The oldest family businesses are not concentrated in only one business sector, but their activities are spread over different business sectors. It is very gratifying that there are six companies operating in the wine sector in the top twenty oldest family businesses in the world. We can say that in most of the world's economies, the role and importance of family businesses is absolutely crucial. It can be estimated that family businesses account for approximately 70-90% of the world's wealth. The business sector in which family businesses play a key role is the agricultural and wine sectors. Family businesses in these sectors predominate, especially in the category of micro and small businesses. Family businesses are understood as synergies between two systems - the family system and the business system. In the family system, it is important to identify the family's involvement in the family business and to identify the generation that currently controls (owns and manages) the family business. In the business system, it is crucial to characterize the innovative activities of the family business and the degree of internationalization of the family business. The family wineries were analysed through this lens, and research questions are also based on this lens. This article deals with the analysis of selected wineries in the world. These family wineries are processed in the form of research case studies. These case studies were subsequently analysed and research results were discussed. The aim of the research was to perform an analysis of selected world family wineries. A partial goal was to compare the results of this analysis with selected family wineries in the Czech Republic.

1.1 Family business definition

The definition of a family business is quite widely discussed in academic literature. Family businesses are presented as entities involving family members involved in business [1], [2], [3] through ownership and control, involvement in business management and wishing to pass the business to the next generation in the future [4]. In these definitions, the business system is permeated by another system, namely the family system. These two systems are often referred to as the two-circle model. It is argued that this interconnection of the two systems imposes special requirements on the family and that the family must manage both family and business requirements so that neither the family nor the business is endangered. Most research and theoretical work on family business, which elaborates on family theory, uses systems theory [5] to assess how business and family systems relate to each other. Authors Sharma and Nordqvist [6] provide an overview of the development of this theory in family business academic literature. New frameworks include the theory of new systems. In academic papers this theory is sometimes called the "three-circle model", as well as the "open systems approach", "family ecosystems" and a "view of resources". However, these new theories will not be discussed further in the article, as they go beyond its scope. Based on the above characteristics of two systems, the family system and the enterprise system, we can give the following definition of a family enterprise: "*a family enterprise is an enterprise owned or managed by a family/ families or selected family members, it is assumed that the company will be passed on to the next generation*". The definition affects entrepreneurs, such as registered farmers and winegrowers, but also legal entities, i.e. businesses. It also preserves the basic attribute of family businesses, i.e. the transfer of the business to the next generation. This definition was used to examine global family wineries in this study. However, this definition does not cover family ties. These family ties are described by the family, which will be introduced below.

1.2 Familiness

Familiness is an important factor in understanding the functioning of family businesses [7], [8], [9], [10]. Familiness refers to the nature of the family relationship in the enterprise [11], family kinship and subsequent resources and abilities, unique to the involvement of family members in the family enterprise [12], [13]. All this characterizes the influence of the family on the family business [14]. Wilson and Tonner [15] define "familiness" as a resource that gives a business a competitive advantage. Key activities provide family members with space to negotiate their positions within the family. The study of these authors shows that family businesses benefit from wider networks and the strong and weak ties of individual family members.

1.2.1 Involvement of family members in the company

Family kinship characterizes the affinity of those family members who are involved in the control of the family business. Family kinship of family members can be quite complex. Some links can be quite distant. The academic literature lacks a deeper analysis of the differences in the structure of family relationships, i.e. the genetic links of family members holding leadership positions in the company. The study [16] captures the degree of closeness of family kinship between family managers and explains why family behaviours and motives with close relatives and distant relatives have different priorities. Research shows that various forms of family involvement in the business, for example through share ownership, board presence and management positions, have an impact on the family's influence on the family business's strategic decisions. The extent and complexity of family kinship of family members involved in a family business makes it difficult to identify family businesses and include them in a research sample.

1.3 The family generation currently holding power in family businesses

As follows from the above definition of a family business, the family strives to maintain power in the business in the future and pass on power and ownership to the next generation. When examining the family kinship of family members who are involved in the management and control of a family business, it is also necessary to examine which generation currently holds power in the family business, i.e. in which generation the family business is located. Many research studies speak of the socio-emotional wealth created by family businesses owned by the family for several generations. These studies are also focused on the wine sector. The authors Fernandez-Olmos, Diaz-Vial and Malorgio [17] state that the last generation of the family who controls the family business has a higher socio-emotional wealth and this affects the overall performance of the business. The authors Delmans and Olivier [18] point out that the focus on the future generation of the family, measured as the intention of the winery owner to pass the winery on to his children, is associated with the adoption of sustainable development of the family winery.

1.4 Innovation in family businesses

Innovation is seen as an improvement. Innovation is a complex process from the initial idea, through development to eventual implementation. Thus, innovation is more than creativity or invention, which, while a necessary part of innovation, brings new ideas, but is not in itself sufficient to fulfil innovation. Many new ideas do not lead to innovations because they are not feasible (from a technical, practical or economic point of view). The result of innovation is an improved process, product, service, or anything else. A distinction is usually made between product or service innovation, process innovation, organizational innovation and marketing innovation. It is important for wine family businesses that their innovation strategy is linked to the tradition of the family business. It is a combination that can give a family winery a competitive advantage in the wine market [19]. The literature shows that the impact of product and process innovation in the family business on consumer perception is crucial [20]. In wine consumption, cultural identity among customers is associated with a broader desire for authentic experiences. Research studies further show that customers value local products if they embody the tradition, heritage, culture, ethnic diversity and identity of a particular place [21]. A good brand is essential in the wine sector, and its importance is growing, especially in those well-known wineries operating in top markets [22].

1.5 Internationalization of family businesses

The internationalization of a company is an important factor in the sustainability of a family business. The internationalization of a family business is directed both inside the company (internationalization of products, people, systems and management of the family business) and outside the company (the ability of the family business to carry out marketing, banking and business international operations). A study by Zhou, Han and Gou [23] states that family ownership and management of a family business has a positive effect on the depth and breadth of internationalization of the family business. Furthermore, the environmental generosity of the family business, on the other hand, weakens the effect of family ownership on the depth of internationalization. The research questions that were set for the research study stem from the understanding of the family business as the intersection of two systems, namely the family system and the business system. Understanding two basic problems is key to the family system. Firstly, the involvement of family members in the management of the family business and the determination of the family generation, and secondly, which generation currently controls (owns and manages) the family business. An understanding of the degree of innovation in the family business and the degree of internationalization of the family business is crucial for the business system. The following are the research questions:
VO1: How are family members involved in the control of a family wine business?
VO2: In which generation of the family is the family wine business currently located?
VO3: What innovations does the family wine business apply in practice?
VO4: How is the family wine business internationalized?

2. Methods

2.1 Selection sample

From 2005 to 2018, the author visited 58 wineries around the world as part of his wine tours. Of these wineries, 31 wineries were identified as family wineries. Due to the limited time to conduct an interview with the winery's representatives during the visit of the winery, some of the interviews were incomplete and these wineries had to be excluded from the basic sample. Finally, 19 family wineries from around the world and one winery in the Czech Republic were included in the sample. These family wineries meet the definition of a family business presented above. All wineries included in the sample were understood as research case studies. The sample included one family winery from Africa (South Africa), two wineries from South America (Chile, Argentina), two from North America (USA, California), two from Asia and Oceania (Australia and New Zealand) and 12 wineries from Europe (France, Italy, Spain, Portugal, Germany, Austria and Switzerland). Apart from one family winery from the Czech Republic, no family wine businesses from Eastern Europe were examined.

2. 2 Methodology used

The methodology applied in this research arose from the understanding of the family business through the lens of two systems, the family system and the business system. Primary and secondary data collection methods were used. For a maximum understanding of the behaviour of individual family wineries, techniques of collecting qualitative data by the method of semi-structured interviews were applied when visiting these wineries. In identifying family wineries, the websites of family businesses were also examined. The method of contextual analysis was used in the analysis of the family wineries' websites. In the analysis of the first stage, the fullness of the definition of a family business and the nature of family viticulture were examined. In the analysis of the second stage, semi-structured interviews with selected representatives of family wineries captured on voice recorder and video camera were analysed. The interviews were converted into electronic form. Manual literal transcription and coding were used to analyse these interviews.

3 Results

3.1 First degree analysis

In this analysis, the fulfilment of the definition of a family business was assessed and the characteristics of the family winery were studied from the point of view of the size of the family business and the degree of involvement of the family winery in cooperation groups. From the point of view of the size of the family business and the degree of involvement in cooperation groups, family wineries were classified into four clusters, which are shown in Figure 1.

Wine family businesses		The size of a family winery	
		Small	Big
Degree of dependence of family wine business on other wineries	Small	Cluster I. Small, independent family winemakers	Cluster II. Big, independent family winemakers
	Big	Cluster III. Small associations of family winegrowing enterprises	Cluster IV. Big joint ventures, big associations of family wine enterprises

Fig. 1 Classification of family wineries into four clusters

Individual clusters of family wineries, including the number of wineries in the cluster, can be characterized as follows:

Cluster I. Small independent family winemakers (nine family wineries). These winemakers grow grapes and also produce wine. They produce wine mostly for the needs of their family and for sale. These are winemakers who produce wine not as a hobby, but produce it on a professional level as their main business.

Cluster II. Big, independent wine family players (six family wineries). These family businesses determine the price of the grapes and the price of the wine on the market. These family wineries also determine the global quality of wine.

Cluster III. Small, cooperative association of family wineries (one group of family wineries). These are family grape growers. These growers are then members of wine cooperatives. Family winemakers supply grapes to wine cooperatives, but also produce a small amount of wine for their own use. Family wineries included in this cluster are identifiable especially in France.

Cluster IV. Big, joint ventures, large groupings of family wineries (three large groupings of family wineries). These are joint ventures of big players from different countries, which form a joint venture in a third country, or even an alliance of large wine companies in their own country.

Although the division of family wineries into individual clusters is uneven, family wineries predominate in Cluster I and in Cluster II, these results of the first degree analysis can be understood as very useful and as a possible future direction of research in the field of family wineries.

3.2 Second degree analysis

3.2.1 Involvement of family members into the family business, determination of family kinship

The analysis of semi-structured interviews of selected members of family wineries identified three family ties which are significant:

Family kinship 1: husband-wife - this link can be called a horizontal kinship. From the point of view of the above definition of a family wine enterprise, this includes a partner-partner family kinship.

Family kinship 2: parents-children - we can call this link horizontal/vertical. It is also possible to include the ties of one parent-more than one child, or one of the parents-one of the children.

Family kinship 3: parents/children-grandparents - we can call this link a purely vertical link. One of the grandparents also works in the family winery, but mostly only as a consultant.

The results of the second-level analysis in the area of family members' involvement in the management of the family business show the involvement of the closest family members. Efforts to involve more distant relatives in the management of the family business have not been identified. These results could also be compared with the results of a research study in the Czech Republic [24], where the involvement of family members in the management of family wineries is similar.

3.2.2 The family generation currently holding power in family businesses

Clusters of family wineries show the control of a family business by the following generations of the family:

Cluster I: Here are nine small, independent family wineries. In this cluster, three family wineries were controlled by one generation of the family, one family winery controlled by three generations of the family, one family winery controlled by four generations of the family, four family wineries controlled by five generations of the family. The characteristics of family wineries in this cluster show that family wineries do not necessarily grow in their development. It is obvious that a family winery can maintain a small size and with a well-established sustainability live to longevity.

Cluster II: Six large independent family wineries are included here. In this cluster there were two family wineries controlled by five generations of the family, three family wineries controlled by six generations of the family and one family winery controlled by seven generations of the family. These wineries were not only very well managed (the family businesses grew), but the family also successfully managed to hand over power to the next generations, and the family winery was thus preserved in the family for generations.

Cluster III: Included here is one group of small family wine companies, which has the character of a cooperative. The generation of the family could not be determined during an interview with a representative of the wine group. It can only be inferred that since the wine cooperative was founded in 1930, there has now been three generations of micro-family businesses growing grapes and supplying them to the wine group. At present, 280 micro family businesses are involved in the wine group. This grouping of family businesses was observed only for family businesses in France. In other countries, these efforts to form cooperation groups have not been observed.

Cluster IV: This includes three large family wine businesses involved in international groupings. In this cluster, there was one family winery controlled by two generations of the family, one family winery controlled by three generations of the family and one family winery controlled by one family ever since its establishment in 1041. In this winery, the winery answered a question about which family generation controls the company controls: "*We do not count the family generation here due to the history of the family*". This is an Italian family winery, which is one of the oldest family wineries in the world. Family wineries, in an effort to expand into foreign wine markets, form large international alliances, are able to produce large volumes of wines of global quality and produce wines to the average consumer's tastes in global markets.

3.2.3 Innovation activities of family businesses

Innovative activities in the world's family wineries have been applied in the following areas:

Product innovation. Family wineries around the world are based on a family tradition, which is built on family values. Family wineries pay great attention to the quality of wines and the loyalty of their customers. There was not much effort in the wineries to introduce new wines, such as rosé wine, or the production of wines in clay containers called „kvevri". Efforts to innovate of this type have not been analysed in any of the clusters of family wineries. Large family businesses, included in Cluster II, produce wines from global vine varieties, such as Cabernet Sauvignon, Merlot, Pinot Noir, Malbec, Zinfandel, Tannat, Carmenere, Pinotage, Siraz, Sauvignon Blanc, Riesling, Chardonnay, or Traminer. For small family wineries included in Cluster I, the cultivation of so-called autochthonous vine varieties (vines that occur only in a given wine-growing area) and the production of wines from them were analysed.

Innovations in wine technologies. Family wineries in all clusters use advanced winemaking technology to produce wine. Pneumatic presses are commonly used for wine production, which gently press the juice from the grapes and the controlled fermentation of the juice in stainless steel tanks. Wooden barrique barrels of the French or Slavonian type are used for wine maturing. Only in South African winery were efforts analysed to use French wooden oak boards instead of wooden barrels, which were lowered in a controlled manner into stainless steel tanks. Efforts for a gentle wine production process at family businesses in all clusters were analysed. Sulphur was used to a minimal extent. The process of so-called malolactic fermentation (the process of degrading malic acid and conversion to lactic acid) was also used in small family wineries.

Innovation in vineyards. Strong efforts to invest in vineyards was analysed for all the world's family wine businesses. Efforts to protect vineyards from spring frosts have been identified in family wineries located in northern areas. The vineyards are warmed during these frosts to prevent the vines from freezing. Family businesses in all clusters were characterized by ecologically approached vineyard management. Here is a typical statement of one representative of a family winery: "*Our philosophy is based on respecting the unique conditions of the environment where we grow vines. Why not respect our soil? Why not respect natural conditions? Our approach is a biological approach to farming in our vineyards.*" Family wine businesses in all clusters were characterized by a friendly ecological approach to farming in the vineyards. A very nice example of a nice environment combined with a clever approach to vine growing was identified in California. The surprise was a chicken coop on wheels, which transported hens to the vineyard. The hens in the vineyard were picking pests between the rows of vines. Sheep were also transported to the vineyards, grazing on grass between the rows in the vineyards. So-called biodynamic management was also recorded in small family wine-growing enterprises (individual work on the vineyards was carried out with respect to the phases of the moon). Large investments were identified in large wineries, which led to the energy self-sufficiency of the family winery and the so-called zero carbon footprint.

Innovation in the wine trade. A common attribute of all family wineries analysed was a strong effort for international visibility. All the family wineries have tried to succeed in prestigious international wine competitions. The awards from these competitions were then used by wineries as a tool shop. For large family wineries included in Cluster II, a robust e-shop capable of sending wines worldwide was identified. In this context, very strong innovation activities in etiquette innovation were identified. Each bottle of wine has a label. Family wineries understand etiquette as the customer's first contact not only with wine, but also with family wineries. Great attention is paid to the design of etiquettes in family wineries. Many family wineries work with renowned designers and artists to design etiquettes and do not hesitate to invest significant resources in etiquette design.

3.2.4 Internationalization of family businesses

The internationalization of the world's wine family businesses showed a focus both inside and outside the wine family business. Wine family businesses of all clusters are open to international clients. The ability to present their wines and tour the winery in English was found everywhere. The family wineries had the capacity of their employees to prepare selected wines for international wine competitions. Family wineries have carefully maintained their websites in several languages, but at least in their home language and in English. Even the family businesses of Cluster I had an effective online e-shop, and were able to deliver wines to their own country, in selected cases throughout Europe. Family businesses of Clusters II., III. and IV were able to send wines around the world.

4 Discussion

When comparing the most important conclusions of the qualitative analysis of 19 selected global family wineries with a selected family winery from the Czech Republic, it is possible to state the following differences:

4.1 First degree analysis

After classifying selected world family wineries, it can be stated that most family wineries are located in Cluster I. - small independent family businesses and Cluster II - big independent family businesses. Also a small Czech family wine business can be included in Cluster I.

4.2 Second degree analysis

4.2.1 Involvement of family members in the family business, determination of family kinship

Family wineries around the world are built on a very strong intergenerational business of other family members. Three types of family kinships have been identified in the world's family wineries: horizontal, horizontal/vertical and vertical. Even in the case of a Czech family wine enterprise, a horizontal/vertical family connection can be identified. The family business is based on a stable family with historical traditions. The family relies on traditional Christian values. It is possible to give an excerpt from an interview with the owner of a family winery: "*Since our establishment, we have been a family business, literally and figuratively. Božena's wife takes care of almost everything and everyone. Our daughter Božena became the Czech Queen of Wine in 2001 and in 2003 she and her husband opened our store in České Budějovice. His son Josef has been involved in the running of the company since his early youth and is a capable assistant in foreign trade and travel. His brother František takes care of the vineyards with his whole family, so we have the highest quality grapes*", said its owner about his winery.

4.2.2 The family generation currently holding power in family businesses

Most of the world's family wineries have identified a higher generation of the family that currently controls the family business. Exceptions were three family wineries included in Cluster I. In addition, the Czech family winery can be included in Cluster I. The company was founded in the 1990s after the velvet revolution (1989). Based on the analysis of the Czech family winery, it is possible to state that it also relies on a deep family tradition and this tradition is often boasted of. Although the family's genealogical tree is not displayed in the wine tasting room, but the family winery wants to keep the family business in the family and pass it on to the next generation.

4.2.3 Innovation activities of family businesses

In selected global family wineries, innovation activities were analysed in four areas: Innovation in products, innovation in technology, innovation in vineyards and innovation in trade.

Innovation in products. For big global wine family businesses in Cluster II a great effort in product innovation has not been analysed. The winery grew global grape varieties. The effort of growing autochthonous vine varieties was also analysed in small family wine businesses in Cluster I. The Czech family wine business went its own way from the beginning of the business and always let the wines mature. The winemaker likes to experiment with the production of a mixture of wines or unusual processing of vine varieties. He tested new vine varieties in the vineyard, and was among the first winemakers in the Czech Republic to plant, for example, the Merlot, Cabernet Franc, Cabernet Blanc, Carmenere and Lena varieties.

Innovation in technology. Family wineries around the world are built on modern technologies. A very strong effort towards innovation was seen in the wineries. Families have a very strong relationship with the land and try to approach the land in a very gentle, ecological way. The world's wineries have also seen an effort to invest in seemingly unrelated areas. In many wineries, small wine museums were built. In larger wineries you could also see investments in art, both in paintings and in sculptures and other art. The Czech wine family business is equipped with the same modern technologies as seen throughout in the world.

Innovation in vineyards. The vineyards at the world's wineries are very heavily invested in climate protection, spring frost damage and bird damage. At the Czech wine family business, a strong relationship of family members to the land was identified, as in family wineries across the world. However, its vineyards are not as heavily invested as in wineries around the world.

Innovation in the wine trade. The world's family wineries in all clusters pay great attention to innovations in the wine trade. They have great efforts for international visibility. Strong efforts have been identified to succeed in prestigious international world wine competitions and efforts to sell wines on international markets. Also, the effort to succeed in international world wine competitions can be analysed at the Czech family winery. The Czech family winery has won a number of awards abroad. Recently we can mention, for example, big gold and gold medals from the global competitions Concours Mondial de Bruxelles, Concours International des Vins Lyon or Ice Wine du Monde. The biggest foreign success is the overall third place out of almost a thousand samples at the world competition Chardonnay du Monde 2011, supported by repeated placement in the Top 10.

4.2.4 Internationalisation of family businesses

Two dimensions of internationalisation have been identified in the world's family wineries, namely the internal dimension and the external dimension. The internal dimension was the internationalisation of the internal environment of viticulture, including the internationalisation of the product, i.e. wine. The external dimension was a particularly powerful e-shop capable of sending wines to the whole world. At the Czech winery, the wines were supplied mainly to the Czech market. Due to its wine production, the winery is not able to satisfy the demand from abroad in larger quantities. Smaller deliveries of wines abroad thus serve more to maintain the prestige of the winery and international awareness of the wines of this Czech winery.

References

1. Anderson, A.R., Jack, S.L., Dodd, S.D., The role of family members in entrepreneurial networks: beyond the boundaries of the family firm. *Family Business Review,* **18,** 135–154 (2005)
2. Upton, N., Petty, W., Venture capital investment and US family business. *Venture Capital,* **2,** 39 (2000)
3. Whiteside, M.F., Brown, F.H., Drawbacks of a dual systems approach to family firms: can we expand our thinking? *Family Business Review,* **4**, 383–395 (1991)
4. Chua, J.H., Chrisman, J.J., Sharma, P., Defining the family business by behavior. *Enterpreneurship Theory and Practice.* **23**, 19-19 (1999)
5. Jennings, J.E., Breitkreuz, R.S., James, A.E., Theories from family science: a review and roadmap for family business research. In: Melin, L., Nordqvist, M., Sharma, P. (Eds.), SAGE Handbook of Family Business. Sage, London, 25–46 (2014)
6. Sharma, P., Nordqvist, M., A classification scheme for family firms: from family values to effective governance to firm performance. In: Tapies, J., Ward, J. (Eds.), Family Values and Value Creation: the Fostering of Enduring Values within Family- Owned Businesses. Palgrave Macmillan, New York, 71–101 (2008)
7. Chrisman, J.J., Chua, J.H., Litz, R., A unified systems perspective of family firm performance: an extension and integration. *Journal of Business Venturing.* **18,** 467–472 (2003)
8. Habbershon, T.G., Williams, M.L., A resource-based framework for assessing the strategic advantages of family firms. *Family Business Review,* **12**, 1–25 (1999)
9. Nordqvist, M., Sharma, P., Chirico, F., Family firm heterogeneity and governance: a configuration approach. *Journal of Small Business Management,* **52,** 192–209 (2014)
10. Zachary, R.K., The importance of the family system in family business. *Journal of Family Business Management*, **1,** 26–36 (2011)
11. Habbershon, T.G., Williams, M., MacMillan, I.C., A unified systems perspective of family firm performance. *Journal of Business Venturing*, **18**, 451–465 (2003)
12. Pearson, A.W., Carr, J.C., Shaw, J.C., Toward a theory of familiness: a social capital perspective. *Enterpreneuship Theory and Practice,* **32,** 949–969 (2008)
13. Rautiainen, M., Pihkala, T., Ikävalko, M., Family business system models: a case study and some implications of open systems perspective. *Journal of Small Business and Enterpreneurhip,* **25**, 155–168 (2012)
14. Zellweger, T.M., Nason, R.S., Nordqvist, M., Brush, C.G., Why do family firms strive for nonfinancial goals? An organizational identity perspective. *Enterpreneurship. Theory and Practice,* **37,** 229–248 (2013)
15. Wilson, J., Tonner, A. Doing family: The constructed meanings of family in family farms. *Journal of Rural Studies,* **78,** 245-253 (2020)
16. Yu, X., Stanley, L., Li, Y.,Eddleston K.A., Kellermans, F.W., The invisible hand of evolutionary psychology: The importance kinship in first-generation family firms. *Entrepreneurship Theory and Practice.* **44,** 134-157 (2020)
17. Fernadez-Olmos, M., Diaz-Vial, I., Malorgio, G. Empirical approach to the sequential relationship between generation, relational capital and performance in family wineries in Spain. *International Journal of Wine Business Research,* **33**, 1 (2021)
18. Delmas, M.A., Gergaud, O. Sustainable Certification for Future Generations: The Case of Family Business. *Family Business Review,* **27**, 228-243 (2014)
19. Vrontis, V., Bresciani, S., Giacosa, E. Tradition and innovation in Italian wine family businesses. *British Food Journal,* **118,** 1883-1897 (2016)
20. Atkin, T., Garcia, R., Lockshin, L. A multinational study of the diffusion of a discontinuous innovation, *Australasian Marketing Journal*, **14,** 17-33, (2006)
21. Contò, F., Vrontis, D., Fiore, M. and Thrassou, A., Strengthening regional identities and culture through wine industry cross border collaboration", *British Food Journal*, **116,** 1788-1807 (2014)
22. Giacosa, E., Giachino, C., Bertoldi, B., Stupino, M., Innovativeness of Ceretto Aziende Vitivinicole: a first investigation into a wine company. *International Food and Agribusiness Management Review*, **17,** 223-236 (2014)
23. Zhou, L., Han, Y, Gou, Ch., Influence of Family Involvement on Family Firm Internationalization: The Moderating Effects of Industrial and Institutional Environments. *Sustainability,* **11**, 5721 (2019)
24. Petlina, A., Koráb, V. The Structure of Family Ties within the Family Wine Trading Business in the Czech Republic. In. Enometrics XXII, Peer-Reviewed Conference Proceedings, 52-58 (2015)

Resilience and sustainability of creative industries businesses

Helena Majdúchová[1,*], and *Daniela* Rybárová[1]

[1]University of Economics in Bratislava, Faculty of Business Management, Bratislava, Slovak Republic

Abstract. The importance of the creative economy for the development of society not only from an economic but also from a social and cultural point of view has been the subject of numerous scientific and professional studies. The connecting link between them is the observation that creativity has been considered in recent years as a new type of competitive advantage. As a generator of added value for all businesses, the creative industries have unlimited growth potential and are thus seen as a route to sustainable development. One of the problems of the creative industry is to obtain resources for business development. This paper examines the financial structure of selected areas of the creative industries in the Slovak Republic, using mean values for 2016-2019, correlation and regression analysis of financial data in 2019 (before the negative impact of COVID-19). The findings point to the absence of credit sources, which confirms the fact that these businesses are perceived as risky and unstable by banks and financial institutions in the Slovak Republic.

1 Introduction

The cultural and creative industries have an important role to play in the creative economy. The importance of the creative economy for the development of society not only from an economic but also from a social and cultural point of view has been the subject of numerous scientific and professional studies. The connecting link between them is the observation that creativity has been considered in recent years as a new type of competitive advantage. It is becoming the basis of a new direction, a new economic paradigm - the creative economy. This concept is considered to be a relatively new direction in theories of socio-economic development, which draws attention to the importance of human creativity in development. The creative economy is one of the fastest developing sectors in the global economy, providing income growth, new jobs and export earnings. In addition to the economic benefits of the cultural and creative industries, the creative economy contributes to addressing sustainable urban development issues as well as the 2030 Agenda [1]. Awareness of the potential of creativity and its proper management [2] is a value-added generator for all enterprises.

The debate around 'cultural value' has become increasingly central to policy debates on arts and creative industries policy over the past ten years and has mostly focused on the

* Corresponding author: helena.majduchova@euba.sk

articulation and measurement of 'economic value', at the expense of other forms of value-cultural, social, aesthetic [3]. There is wide consensus regarding the capacity of the creative economy as a factor in economic development and sustainability in regions in the medium term, and it is an issue that has aroused interest in measuring the creative potential of areas by estimating composite indicators [4]. Autors Lishuan and Yanli [5] shows that evaluation of cultural and creative enterprises shoud cover two parts: economic benefit and social benefit.. Hence, the benefits of the cultural and creative industries are demonstrated not only in generating more value for the region's economy but also in creating social value, such characteristics of cultural and creative industries as brand demonstration effect, social morality drive effect, education effect and social capital. Douvon [6] points out in her qualitative study on the characteristics, representation of creative and cultural entrepreneurship in a developing country like Mali the influence impact in the global economy of the country. Salder [7] proposes the Arts-based Creative Industries (ABCI) supplement resource limitations and demands for creative intensity through networks; techniques used by ABCI could be of value more widely to SMEs in an increasingly network-based economy. Similarly, these networks risk locking firms into unrealistic dependence on external sources of knowledge and funding which exert significant influence on business development.

Creative industry represent a heterodox sector with high variation levels between firms. Pressure has been placed on creative industries to adopt business practices that maximize growth and contribute to national economic objectives. However, it is important to recognize that this is a sector based on creative intensity and project-based systems. Given the prevalence of small businesses in the creative industries, many entrepreneurs find it difficult to finance creative projects due to limited access to banks or venture capital. Bedendo et al. [8] document significant differences in the financing structures of small firms with managers of diverse cultural backgrounds. To isolate the effect of culture, they exploit cultural heterogeneity within a geographical area with shared regulations, institutions, and macroeconomic cycles. Their findings suggest significant cultural differences in the preference toward debt funding and in the use of formal and informal sources of financing (bank loans and trade credit).

The financeability of small creative industry businesses can be affected by potential differences in credit constraints and in the distribution of cultural origins across sectors, by trading partners and locations too. For the same reason, we examined the financial structure in different sectors of the creative industries.

The selection of creative industry entities according to SK NACE[†] was carried out in accordance with the approaches of qualification of creative industry enterprises presented in our previous research papers [9]. The set of surveyed enterprises was divided into three thematic areas:

- **Professional, Scientific and Technical Activities**. This area includes architectural and engineering activities, design activities, photographic activities, and advertising and market research. This class covers a wide range of services that are most commonly provided to commercial clients. It includes those activities which require a higher level of professional, scientific and technical expertise, but does not include ongoing, routine business services which are mostly of a short-term nature. In terms of SK NACE Rev. 2, this is section 'M' and divisions 69-75.
- **Information and communication**. The information and communication sector includes enterprises engaged in publishing, film production, video and television

[†] SK NACE Rev. 2 is fully compatible with the European classification NACE Revision 2, which is laid down for the countries of the European Community by Regulation (EC) No 1893/2006 of the European Parliament and of the Council of 20 December 2006.

programme production, preparation and publication of sound recordings, radio and television broadcasting, telecommunications, computer programming, consultancy and related services. It is an activity that deals with a wide range of copyright content (information products) and the creation of that content which is made available to the general public by engaging in (or preparing for) the reproduction and distribution of that content in various forms. In terms of the SK NACE, these are section 'M' and divisions 69-75, and also section 'J' and divisions 59-63.

- **Arts, entertainment and recreation**. This industry comprises the operation of establishments and the provision of services to satisfy the interests of customers in the fields of culture and entertainment. It includes the production, promotion, and participation in live performances, events, or exhibitions intended for the public; the provision of artistic, creative, and technical expertise to produce artistic products and live performances. Also included are the activities of libraries and archives; the operation of museums, botanical and zoological gardens. For the sake of completeness, it should be added that some entities providing cultural, entertainment and recreational facilities and services are classified in other divisions, such as: film and video production and distribution, (59.11, 59.12, 59.13), film exhibition (59.14), radio and television broadcasting, (60.1, 60.2). Within the meaning of SK NACE Rev. 2, this is section 'R' and divisions 90-93.

2 Metods

In this paper we used the analysis of means and the method of regression and correlation analysis.

The source of data for our mean value analyses are the financial statements of creative industry entities for the years 2016-2019, which are aggregated into the so-called mean values of financial indicators of economic activities in the Slovak Republic, which are published by the Slovak Credit Bureau, s.r.o. as its eponymous publication [10]. The median values of financial indicators are divided into four quartiles with values located at the level of the lower quartile (DK), the median (Me) as the mean value of a series of values arranged according to size, and the upper quartile (HK).

Within the structure of financial ratios, we focused on the ratios that determine the financial structure of the creative industry enterprises namely: total debt to total assets ratio, credit debt assets. Total debt to total assets ratio tells about the share of foreign resources (both interest-bearing and non-interest-bearing) in financing the needs of the enterprise. The credit leverage ratio indicates the extent to which the enterprise's needs are financed by interest-only sources (e.g. bank loans, short-term financial loans, etc.). We have also included the return on equity ratio (ROE) in the set of assessed indicators, which indicates the level of financial performance of the enterprise. The indicator is mainly monitored by the owners of the enterprise and demonstrates the appreciation of the capital invested by them. Its amount depends not only on the size of the profit generated, but also on the size of own resources.

Further, the method of regression and correlation analysis was used to know and mathematically describe the statistical dependence between quantitative statistical features for the selected group of economic activities. Regression analysis is used to capture the nature of correlation dependence and correlation analysis is used to measure the tightness of correlation dependence. The object of the regression and correlation analysis were enterprises of the group 90.0 Creative, artistic and entertainment activities, belonging to the division 90 Creative, artistic and entertainment activities, according to SK NACE. The database of economic results of Slovak companies of FinStat, s.r.o. [11] was used for the analysis.

3 Results

In the following subsections we present the results of the analysis of means for the selected creative industries and the results of the regression and correlation analysis for one of the selected industries, i.e. the group of the group 90.0 Creative, artistic and entertainment activities.

3.1 Analysis of means

The mean values of financial indicators of economic activities in the Slovak Republic (Credit Bureau, s.r.o.) are processed in the following table for selected sectors of the creative industries according to SK NACE 71, 74, 73, 58, 59, 60,90 and 91.

Table 1. Setting Word's margins

		Architectural and engineering activities -71				Specialised design activities - 74.1				Photographic activities -74,2			
Business entities		4682	5109	5358	4343	200	293	349	290	347	443	492	363
Indicators		**2016**	**2017**	**2018**	**2019**	**2016**	**2017**	**2018**	**2019**	**2016**	**2017**	**2018**	**2019**
Total debt to assets ratio (%)	DK	15,74	14,96	12,75	10,32	21,31	12,13	9,51	8,86	14,43	10,77	8,19	6,16
	Me	44,46	41,87	39,86	35,04	68,83	51,84	44,31	39,86	47,66	36,54	36,47	37,07
	HK	83,08	78,86	76,72	72,35	97,91	93,15	92,12	91,82	90,42	85,52	83,37	83,28
Credit debt to assets ratio (%)	DK	0,00	0,00	0,00	0,00	0,00	0,00	0,00	0,00	0,00	0,00	0,00	0,00
	Me	0,00	0,00	0,00	0,00	0,00	0,00	0,00	0,00	0,00	0,00	0,00	0,00
	HK	0,00	0,00	0,00	0,00	0,00	0,00	0,00	0,00	0,00	0,00	0,00	0,00
ROE (%)	DK	-7,33	-5,11	-1,73	-2,19	-21,73	-10,71	-24,95	-16,58	-26,27	-10,54	-15,37	-6,87
	Me	13,45	13,42	10,43	8,93	9,93	8,12	2,80	2,16	10,88	9,85	3,64	3,53
	HK	38,19	38,37	36,89	32,59	43,09	38,21	37,12	28,46	33,93	32,73	27,09	29,04
		Advertising and market research -73				**Publishing activities -58**				**Motion picture, video and television programme product., ... - 59**			
Business entities		4670	5227	5605	4147	860	896	889	676	804	942	1023	762
Indicators		**2016**	**2017**	**2018**	**2019**	**2016**	**2017**	**2018**	**2019**	**2016**	**2017**	**2018**	**2019**
Total debt to assets ratio (%)	DK	17,99	14,25	10,96	9,47	15,59	15,07	11,91	10,63	12,86	11,62	9,30	7,54
	Me	57,84	51,74	49,94	45,51	52,21	50,26	44,66	44,72	41,69	38,12	34,61	31,28
	HK	95,61	93,41	91,88	90,27	94,82	92,74	88,60	86,69	82,15	77,86	77,48	73,17
Credit debt to assets ratio (%)	DK	0,00	0,00	0,00	0,00	0,00	0,00	0,00	0,00	0,00	0,00	0,00	0,00
	Me	0,00	0,00	0,00	0,00	0,00	0,00	0,00	0,00	0,00	0,00	0,00	0,00
	HK	0,00	0,00	0,00	0,00	0,00	0,00	0,00	0,00	0,00	0,00	0,00	0,00
ROE (%)	DK	-28,85	-21,46	-9,06	-10,88	-29,11	-22,78	-5,58	-13,62	-6,07	-6,63	-4,23	-3,64
	Me	6,61	4,25	3,50	2,18	7,52	6,06	5,41	3,40	13,90	12,10	6,37	6,72
	HK	32,31	29,57	28,74	25,96	31,59	29,92	29,79	26,31	33,04	29,56	26,09	25,91
		Programming and broadcasting activities -60				**Creative, arts and entertainment activities -90**				**Libraries, archives, museums and other cult. activities -91**			
Business entities		43	45	47	29	558	721	828	584	58	64	67	47
Indicators		**2016**	**2017**	**2018**	**2019**	**2016**	**2017**	**2018**	**2019**	**2016**	**2017**	**2018**	**2019**
Total debt to assets ratio (%)	DK	32,30	31,23	30,66	22,70	13,06	12,20	10,16	10,05	27,31	22,56	16,60	13,43
	Me	69,63	60,42	71,01	61,59	39,49	46,04	44,74	44,80	66,32	79,33	67,81	55,30
	HK	110,11	99,72	105,37	97,97	79,93	87,72	87,05	86,65	130,05	115,62	109,95	108,51
Credit debt to assets ratio (%)	DK	0,00	0,00	0,00	0,00	0,00	0,00	0,00	0,00	0,00	0,00	0,00	0,00
	Me	0,00	0,00	0,00	0,00	0,00	0,00	0,00	0,00	0,00	0,00	0,00	0,00
	HK	0,00	0,00	0,00	0,00	0,00	0,00	0,00	0,00	0,00	0,00	0,00	0,00
ROE (%)	DK	-40,64	-127,60	-65,01	-16,55	-20,34	-28,47	-16,73	-17,65	-33,62	-70,13	-19,09	-3,20
	Me	0,36	0,13	7,50	2,86	9,77	5,82	2,86	1,69	5,02	1,08	3,37	1,83
	HK	57,45	32,87	40,13	31,73	31,17	28,16	27,39	25,57	55,47	36,33	47,59	27,27

Source: Own processing according to mean values of financial indicators of economic activities in the Slovak Republic, which are published by the Slovak Credit Bureau, s.r.o.

Total debt to assets ratio across the set of thematic areas under review showed a declining trend in virtually all quartiles. For example, while in 2016 the median for **Professional, Scientific and Technical Activities** was approximately 44.46% to 68.83%, in 2019 it ranged from 35.04% to 45.51%. The highest indebtedness of assets was reported by the Libraries, archives, museums and other cultural activities **sector**. In this sector, the ratios in the upper quartile were above 100%, which means that the amount of foreign funds was higher than the amount of assets, indicating the over-extension of the enterprise, i.e. negative equity. The situation was similar in **the** the Programming and broadcasting activities. By the nature of their activities, these two sectors are partly public utilities and partly owned by municipalities, towns or the State. The other, much more significant part is made up of privately owned entities, such as private radio and television stations and private archives.

Credit debt to assets ratio in the assessed set of thematic areas and over the whole analysis period shows zero values. This means that the enterprises in the analysis set do not operate with interest-bearing capital. This may indicate the fact that, given the size of these entities and the size of their assets and revenues, they do not have the possibility to obtain interest-bearing, especially long-term, bank funding. The size of their assets limits their ability to secure loans, while the ability to hold components of their assets that can serve as collateral has an impact primarily on long-term interest-bearing resources. Thus, the zero share of interest-bearing resources in the financial structure of enterprises is also an indication of the structure of the assets at their disposal. It can therefore be assumed that creative industry enterprises have a higher proportion of current assets than non-current assets.

On the one hand, this increases their independence and autonomy; on the other hand, the limitations of their own resources may be a barrier to their growth. At the same time, these enterprises do not achieve the leverage effect of the tax shield

The return on equity (ROE) showed a variation. In the assessed thematic set, for the whole period analysed, enterprises in the bottom quartile achieved a negative return on equity. This means that these enterprises had either negative equity or a negative economic result. For the entities classified in the Programming and broadcasting activities **and** Libraries, archives, museums and other cultural activities , the negative return on equity values correspond to the total indebtedness of assets indicator, i.e. these enterprises had negative equity. In other areas, negative ROE values in the bottom quartile are due to negative economic performance. In the upper quartile, where the best performers in the set are located, the highest return on equity was achieved by the enterprises classified in **the** Programming and broadcasting activities , but with a decreasing trend. Comparable values were also achieved by enterprises in Libraries, archives, museums and other cultural activities . Other areas of the creative industries had relatively balanced ROE ratios , e.g. in 2019 they ranged from 25.7%-32.59%.

It can be concluded that the financial statements in all areas follow a similar pattern. This is also the case for both the asset structure and the financial structure, which may be due to the fact that the creative industry is a people-based industry and not an asset-based industry. It may also be a reflection of external environmental influences on micro and small enterprises within the creative industries.

3.2 Regression and correlation analysis

The correlation analysis revealed the existence of a direct correlation between the variables EAT, EBITDA, Equity, Liabilities, Sales and Assets. As the values of one variable increase, the values of the other variable increase. The strongest correlation was observed between Sales and Liabilities, Assets and Liabilities, Assets and Sales.

Table 2. Results of correlation analysis

	EAT	EBITDA	Equity	Liabilities	Sales	Assets
EAT	1					
EBITDA	0,8945	1				
Equity	0,5789	0,8098	1			
Laibilities	0,5321	0,7711	0,7821	1		
Sales	0,6024	0,8377	0,8238	0,9438	1	
Assets	0,5902	0,8387	0,9529	0,9343	0,9308	1

The aim of the regression analysis was to estimate the functional relationship between the variables and to estimate the parameters of the regression function. Simple regression was used to describe the relationship.

The null hypotheses that are tested in this analysis relate to the significance of the location constant and the regression coefficient, with the null hypothesis asserting the insignificance of the coefficient in question and the alternative hypothesis asserting its significance. P-values are used to evaluate these claims. The table shows the functional relationships for pairs of variables for which the P-values are statistically significant.

Table 3. Results of regression analysis

Indicators	Regression function	Adjusted R Square	*P-value*
y = Assets x = EBITDA	y = 4,6132x + 20406	$R^2 = 0,7035$	Intercept = 0,0089 EBITDA = 7,8E-213
y = Assets x = Liabilities	y = 1,9192x + 16132	$R^2 = 0,8729$	Intercept = 0,00157 Liabilities = 0
y = Sales x = Liabilities	y = 2,203x + 16648	$R^2 = 0,8908$	Intercept = 0,00195 Liabilities = 0
y = Sales x = EAT	y = 6,5751x + 52200	$R^2 = 0,3628$	Intercept = 5,56E-05 EAT = 3,66E-80
y = Equity x = Laibilities	y = 0,9192x + 16132	$R^2 = 0,6116$	Intercept = 0,001568 Liabilities = 4,6E-166
y = Equity x = Sales	y = 0,4148x + 9422,7	$R^2 = 0,6786$	Intercept = 0,04301 Sales = 6,7E-199

For other combinations of variables, the chosen models were not appropriate. According to Adjusted R-squared, which also takes into account the number of estimated parameters and the number of measurements, a high degree of tightness is achieved by the relationship between Sales and Liabilities, as well as between Assets and Liabilities. In both cases, the growth of Sales is affected by the growth of Liabilities.

3.3 Suggestions and recommendations

Based on our analysis, it can be concluded that most creative industry businesses operate with short-term, non-interest-bearing capital. This fact practically excludes the possibility of long-term financial planning. This means that these enterprises operate in an unstable financial environment and work with short-term financial plans, but this is also partly due to the nature of their activity (many enterprises have a very short life cycle, e.g. film production). The capital structure is also affected by the use of intellectual property rights. These enterprises are dependent on intangible assets. They work with copyrights, licences, trademarks, but these are not reflected in the assets, despite the fact that without their active exploitation these enterprises often cannot operate at all. Consequently, the value of their assets is lower and the accounting structure of their assets does not include those assets that can serve as hedging instruments. It would be possible to value these assets at their fair value, but here we are faced with the problem of the variability of the value of these assets, the relatively short life cycle of many businesses (e.g. film production), the relatively high costs of registering intellectual property rights and also the tendency to value these rights, e.g. by means of an expert valuation.

Banks are thus unable to recognise the true economic value of these assets and, as a result, the possibilities for creative industries to obtain bank loans are very limited. Creative industries are thus dependent on self-financing, which is the most important but insufficient source of financing for their needs.

In view of the above, we are therefore formulating a proposal for measures that should contribute to improving the financial situation of creative industry enterprises.

1. Motivate the banking sector to set up financing programmes for creative industries. An example is the bridging loans prepared by Slovak Investment Holding in cooperation with BKS Bank, Všeobecná úverová banka, Slovenská sporiteľna and UniCredit Bank. The purpose of these bridging loans was to help entrepreneurs whose operations had been affected by a pandemic or quarantine measures. However, these are rather one-off measures which will not solve the problems of creative industry enterprises in the long term. The solution could be the real launch of a so-called guarantee mechanism under the Creative Europe programme, as proposed by the European Parliament in its Report on a coherent EU policy for the cultural and creative industries. It is also important to broaden the range of financial instruments available to businesses in the creative industries, such as micro-credits, repayable grants and contributions, venture capital funding, various models of collective financing and the creation of public-private partnerships. The EFSI should focus on financing projects with a higher risk profile as a priority, rather than financing projects that are normally viable through bank loans.
2. The creative industries are mainly funded by public subsidies and grants. The amount of these depends on their focus and on the individual regions. The main source of funding is the European Structural and Investment Funds. There are many instruments to support the creative and cultural industries, an overview of which can be found on the websites of the relevant line ministries and the Government Office.
   The cultural and creative industries are supported by a number of grants. Regular calls are launched to support entrepreneurship in these sectors. The following areas are supported by grants:
   - Creative vouchers,
   - Worth-partnerships,
   - Integrated Regional Operational Programme,
   - Creative Europe 2014-2020,
   - Creative Europe - Creative Europe 2020, Creative Europe 2020, Creative Europe 2020, Creative Europe 2020, Creative Europe 2020,
   - Audiovisual Fund.
3. It is clear that public sector subsidies can help with start-up and funding for creative industries businesses, often being more accessible to key cultural sectors than to the cultural and creative industries as a whole. In addition, the views of CCI representatives (presented at various academic workshops and seminars) suggest that public initiatives are not able to meet the expectations and needs of CCI entrepreneurs because they are difficult to understand, often remote from the territory in which the company operates, administratively demanding and do not necessarily represent long-term financial support.

   The European Parliament also stresses the idea of tax exemptions and the use of tax credits (lower VAT rates and other tax incentives for both online and offline creative services and labour-intensive products). It points to successful projects that have been implemented in this area in some Member States. Tax breaks should not only be targeted at businesses, but also, for example, at the taxation of labour, which affects the mobility of artists between countries.

4 Discussion

Perhaps the biggest obstacle facing entrepreneurs and businesses in the creative industries is getting enough funding to finance their needs. The lack of finance for the creative industries has been exacerbated by the pandemic crisis of COVID-19. Many scientific articles are currently devoted to this topic. Eikhof [12] demonstrates how considering the cultural economy's typical business models and resultant work and employment practices allows understanding that the inclusion and diversity impacts of the COVID-19 pandemic will be driven by more than workers' differing abilities to buffer short-term income insecurity. Betzler et al. [13] highlight that early policy measures have been developed against the backdrop of a complex interplay of public policy priorities over the last years. In these priorities the intensity of state involvement, the economic situation before the pandemic and the propensity of self-employment in society are key to understanding the adoption of specific policy measures across countries.

The uneven impact of Covid-19, coupled with the various government measures in place, will very likely result in a widening of existing inequalities, which is more worrying in the long term because of the diversity and inclusiveness of the sector.

The resilience of the creative economy as an important sector of the national economy should not only be ensured in the short term. Securing the long term requires critical reflection on business models and the inner workings of the sector, and perhaps a move away from discussions of economic growth to new agendas for creative social economies [14] and environmental economies [15, 16].

Acknowledgment

This paper is an output of the science project VEGA č 1/0340/19 "*The Entrepreneurial Dimension of the Creative Industries in the Context of Innovation and Smart Growth*" - project share is 50% and project VEGA č. 1/0708/20 „*Socio-economic determinants of sustainable consumption and production in terms of impact on performance and competitiveness of enterprises*" - project share is 50%;

References

1. D. Streimikiane, T. Kacerauskas, The creative economy and sustainable development : The Baltic states. *Sustainable development* **28,** 6 (2020) DOI: 10.1002/sd.2111
2. J. Zhou, X.M. Wang, D. Bavato, S. Tasselli, J. Wu, Understanding the Receiving Side of Creativity: A Multidisciplinary Review and Implications for Management Research. *Journal of Management* **45**, 6, 2570-2595 (2019)
3. E. Belfiore, Whose cultural value ? Representation, power and creative industries. *International journal of cultural policy* **26**, 3, 383-397 (2020) DOI: 10.1080/10286632.2018.1495713
4. M.I. Boal-San, L.C. Herrero-Prieto, Reliability of creative composite indicators with territorial specification in the EU. Sustainability **12**, 8 (2020) DOI: 10.3390/su12083070
5. Q. Lishuan, G. Yanli, An Exploratory Study on Benefit Evaluation of Cultural Creative Enterprises. *International Journal of Economics, Finance and Management Sciences,* **8**, 6, 214-218 (2020) DOI: 10.11648/j.ijefm.20200806.11

6. R. Douvon, The cahracteristics of cultural and creative entrepreneurship the Malian context. *Revue internationale PME* **33**, 3-4, 81-103 (2020)
7. J. Salder, The creative business : enterprise development in the arts-based reative industries. *Creative Industries Journal* **14**, 1, 63-80 (2021) DOI : 10.1080/17510694.2020.1789414
8. M. Bedendo, E. Garcia-Appendini, L. Siming, Cultural Preferences and Firm Financing Choices. *Journal of Financial and Quantitative Analsis* **55**, 3, 897-930 (2020) DOI:10.1017/S0022109019000103
9. S. Šagátová, Development of Cultural and Creative Industry in EU Countries. *Proceedings of Scientific Papers*, 539-544 (2019) ISBN 978-80-225-4680-5
10. CRIF – Slovak Credit Bureau, *Stredné hodnoty finančných ukazovateľov ekonomických činnosti v Slovenskej republike za rok 2019.* (CRIF, Bratislava, 2020) ISBN 978-80-971109-8-7
11. FinStat, s.r.o., Company database with financial data [online]. Available at : https://finstat.sk/databaza-financnych-udajov
12. D. R. Eikhof, COVID-19, inclusion and workforce diversity in the cultural economy : What now, what next? *Cutural Trends* **29**, 3, 234-250 (2020) DOI : 10.1080/09548963.2020.1802202
13. D. Betzler, E. Loots, M. Prokupek, L. Marques, P. Grafenauer, COVID-19 and the arts and cultural sectors : investigating countries´ contextual factors and early policy measures. *International journal of cultural policy* Early Access (2020) DOI: 10.1080/10286632.2020.1842383
14. R. Comunian, L. England, Creative and cultural work without filters : Covid-19 and exposed precarity in the creative economy. *Cultural Trends* **29**, 2, 112-128 (2020) DOI : 10.1080/09548963.2020.1770577
15. I. Kufelová, N. Grisáková, P. Štetka, *Hodnotový rozmer ekoinovačnej stratégie podniku* (VŠERS, České Budejovice, 2020) ISBN 978-80-225-4656-0
16. R. Lacko, P. Markovič, S. Šagátová, Z. Hajduová, Evaluation and Development of Environmental and Health Efficiency: Case of V4 Countries. *Ekonomický časopis* **68**, 10, 1040-1056 (2020)

Return on Education Using the Concept of Opportunity Cost

Peter Marinič[1,*]

[1]Masaryk University, Faculty of Education, 603 00 Brno, Czech Republic

Abstract. Education is an important part of individual life. Everyone is learning whether realizing it or not. There is pressure for a longer period of formal education due to the growing volume of knowledge and the need to use it in working life nowadays. In connection with new technologies, there is a need for growing population with tertiary education, but there is also a need for a skilled workforce with appropriate vocational training achieved by upper-secondary education. Formal education takes place mainly in the full-time form, which makes it necessary to give up the possibility of earning income from classic employment relationship. Improving employability at labour market and possibility to get higher net income, forces individuals to stay at education for longer time and to sacrifice possible incomes during the time. Therefore, education is analysed as opportunity costs and the payback period of education is assessed with respect to the increase in the net mean income. The results of the analysis carried out according to data from the Member States of European Union show significant differences in the opportunity cost and in the return on education as investment, both, for tertiary education and for upper-secondary education.

1 Introduction

Education is important activity influencing everyone, influencing ability of perceive the environment adequately to the stimuli from the environment, to make decisions in everyday situations, but also to create long-term plans, achieve work and personal success. It also shapes the society. Education is a prerequisite for development of society, whether in the field of innovation, technology, production of wide variety of products or services, thus educational society, but also in the development of civil society, due to creating preconditions for development of democratic institutions, freedom, environment protection, social responsibility, solidarity and belonging [1, 2].

As the amount of knowledge increases during the time, there is also need to ensure the transmission of ever-increasing amount of curriculum in teaching process. This creates pressure on the intensity of educational process at each educational level, but also increases demands on the length of the educational process. This tendency pushes to differentiate and specialize education, education institutions and the education system, and creates wide range of different types of schools, especially at upper secondary and tertiary level.

* Corresponding author: marinic@ped.muni.cz

The changes in the educational structure of active population, thus people from 15 to 64 years of age, according to the completed educational level is in figure 1. There is proportion of primary, upper-secondary, and tertiary education displayed for analysed countries, mostly the Member States of European Union. The educational structure of countries differs, but there is clear tendency of increasing proportion of people with completed higher than primary education when comparing the situation in 2008 to 2018.

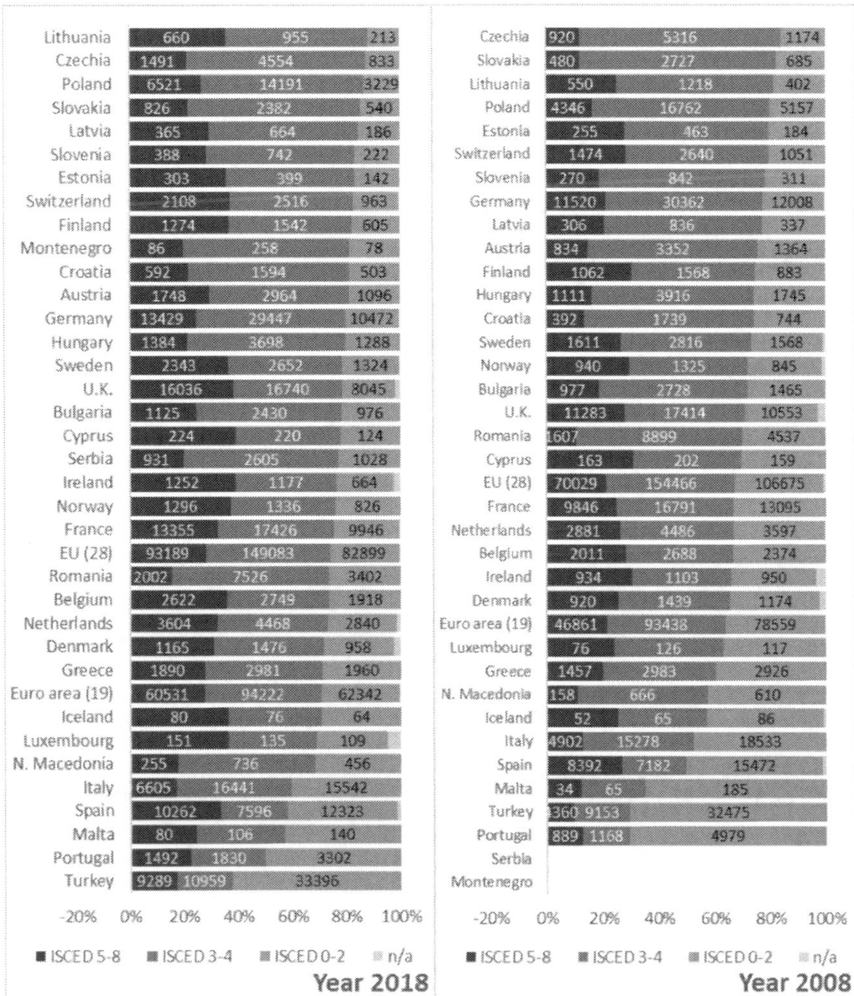

Fig. 1. Structure of the population (15-64 years) according to completed education level.

2 Methodology

All these aspects are considered by individuals when choosing their educational paths. This choice usually takes place at the end of primary education level, when choosing the next upper secondary education level. However, it is not uncommon for future direction of educational path to be decided in early childhood, or, conversely, that even at the end of primary education level or even at the end of upper secondary education level, in case of choosing the general education orientation, it is not definitely decided.

Many variables affect the decision-making process about educational path. It could be family traditions and preferences of certain professions, the possibility of achieving social status associated with certain professions, or perspective of promising earnings associated

https://doi.org/10.1051/shsconf/202111502005

with the chosen profession. The analysis of the justification of the last-mentioned motive is the aim of the paper [3].

There are several approaches to the analysis of the economic effects of the education. The classical approach to assessing the impact of education on individual's economic situation typically focuses on the completion of tertiary education level. The approach to these impacts has been developing for several decades and there are studies from international organizations as well [4-10]. The consider particularly the effect of education on the potential of higher earnings and lower unemployment, respectively higher employability, of graduates of tertiary education, in various lengths of evaluations. Thus, there is a certain volume of educational benefits expressed by money. The studies also consider the need to devote certain time to the education process, considered as costs associated with education, even in relation to the loss of earnings from employment in the labour market due to devoting to education.

The paper uses this approach of opportunity costs in a simple way, but analysis not only the tertiary education but also the upper-secondary education is provided. The time spend in the upper-secondary and tertiary education is taken as the loss of opportunity to earn mean net income from the employment with adequate educational level. Thus, for the upper-secondary education the net mean income of the primary educated employee is considered for the 4 years period assumed as the proper time for upper-secondary education process. And for the tertiary education the net mean income of upper-secondary educated employee for the 5 years period was assumed.

The opportunity costs were then compared to the difference of net mean incomes of the appropriate educated employees, thus the difference between the primary educated and upper-secondary educated employee and the difference between the tertiary and upper-secondary educated employee. The return rate is the result of this approach which can be marked as the return on investment in education.

3 Economic effects of education – Data and its analysis

In the analysis the net mean income according to completed education level is used. The first part of table 1 presents the results of this variables in the year 2008 and 2018. The typical structure of the net mean incomes is increasing with the increase of completed education level, which is understandable. Interestingly, although net mean income is rising in many countries, it is possible to identify EU Member States where the net mean incomes has fallen. The impact of economic crisis after 2008 has probably manifested itself, and it has affected some EU countries quite significantly.

Second part of the table 1 shows the results of the analysis of opportunity costs and return rate for upper-secondary education. It shows that there is higher net mean income of the upper-secondary educated employee that that with the completed primary education. The opportunity costs of upper-secondary education vary in the unit tens of thousands and the return rate in units of per cent. The average opportunity costs of upper-secondary education for European union are €61,048 with the return rate 4.99 % in 2018. The highest return rate is for Bulgaria (20.85 % with OC €10,028) and Romania (16.73 % with OC €9,312), the lowest return rate is for Iceland (1.41 % with OC €164.444) or Finland (1.77 % with OC €97,460).

The third part of the table shows the results for tertiary education using the same approach. There is also positive difference in net mean income for all countries. The average opportunity costs in European Union are €91,545 with the return rate 9.25 % in 2018. The highest return rate is in Romania (17.07 % with OC €19,430) and Bulgaria (15.8 % with OC €22,990), the lowest for Sweden (2.17 % with OC €239,655).

Table 1. Mean net income and chosen indicators of economic effects of education for upper secondary education and tertiary education.

	Mean Net Income					
	2018			2008		
	ISCED 0-2	ISCED 3-4	ISCED 5-8	ISCED 0-2	ISCED 3-4	ISCED 5-8
EU (28)	15,262	18,309	26,778	n/a	n/a	n/a
EA (19)	16,149	21,228	28,404	15,568	19,381	25,820
Belgium	19,418	24,855	32,448	16,849	20,138	26,832
Bulgaria	2,507	4,598	8,230	2,023	2,995	4,083
Czechia	7,924	10,400	13,429	5,577	7,082	9,637
Denmark	30,684	33,095	39,862	24,660	26,871	31,664
Germany	19,276	23,869	32,485	17,556	20,661	27,087
Estonia	9,370	11,164	15,515	5,396	6,240	8,762
Ireland	21,948	27,719	35,305	21,817	26,932	37,783
Greece	6,896	8,524	12,320	9,971	12,787	19,278
Spain	12,833	16,233	22,886	13,862	17,221	22,653
France	21,580	23,399	31,144	19,382	22,108	28,607
Croatia	5,433	7,381	11,008	n/a	n/a	n/a
Italy	15,184	20,040	27,117	15,354	19,534	26,500
Cyprus	13,279	16,305	22,894	15,294	18,803	25,859
Latvia	6,022	8,077	13,032	4,473	5,910	9,229
Lithuania	5,903	7,602	12,242	3,461	4,714	7,376
Luxembourg	31,028	39,447	50,352	28,534	34,475	51,107
Hungary	4,329	6,166	8,360	3,763	4,817	7,086
Malta	14,248	18,351	25,238	10,573	13,672	17,281
Netherlands	23,271	25,597	33,107	19,721	21,827	28,837
Austria	22,412	27,750	34,051	18,002	22,793	28,108
Poland	5,333	6,776	10,193	3,535	4,692	8,127
Portugal	8,843	11,276	16,399	8,897	12,259	20,149
Romania	2,328	3,886	7,202	1,624	2,490	5,117
Slovenia	11,455	13,455	17,911	9,652	11,584	16,665
Slovakia	5,647	8,212	9,414	4,227	5,308	7,003
Finland	24,365	26,089	35,161	20,349	21,125	29,490
Sweden	23,182	29,730	32,959	20,351	22,300	26,310
Iceland	41,111	43,431	49,559	35,464	37,380	46,735
Norway	36,117	42,608	49,769	31,284	35,055	42,387
Switzerland	38,207	47,931	61,333	26,364	31,451	43,105
U.K.	20,007	23,837	32,406	16,885	21,988	31,994
Montenegro	2,632	4,147	6,300	n/a	n/a	n/a
Macedonia	2,076	3,003	4,072	n/a	n/a	n/a
Serbia	1,941	2,950	4,502	n/a	n/a	n/a
Turkey	3,985	5,475	9,392	3,575	5,495	9,470

	Upper Secondary Education					
	2018			2008		
	Mean Difference	Opportunity Costs	Return Rate	Mean Difference	Opportunity Costs	Return Rate
EU (28)	3,047	61,048	4.99 %	n/a	n/a	n/a
EA (19)	5,079	64,596	7.86 %	3,813	62,272	6.12 %
Belgium	5,437	77,672	7.00 %	3,289	67,396	4.88 %
Bulgaria	2,091	10,028	20.85 %	972	8,092	12.01 %
Czechia	2,476	31,696	7.81 %	1,505	22,308	6.75 %
Denmark	2,411	122,736	1.96 %	2,211	98,640	2.24 %
Germany	4,593	77,104	5.96 %	3,105	70,224	4.42 %
Estonia	1,794	37,480	4.79 %	844	21,584	3.91 %
Ireland	5,771	87,792	6.57 %	5,115	87,268	5.86 %
Greece	1,628	27,584	5.90 %	2,816	39,884	7.06 %
Spain	3,400	51,332	6.62 %	3,359	55,448	6.06 %
France	1,819	86,320	2.11 %	2,726	77,528	3.52 %
Croatia	1,948	21,732	8.96 %	n/a	n/a	n/a
Italy	4,856	60,736	8.00 %	4,180	61,416	6.81 %
Cyprus	3,026	53,116	5.70 %	3,509	61,176	5.74 %
Latvia	2,055	24,088	8.53 %	1,437	17,892	8.03 %
Lithuania	1,699	23,612	7.20 %	1,253	13,844	9.05 %
Luxembourg	8,419	124,112	6.78 %	5,941	114,136	5.21 %
Hungary	1,837	17,316	10.61 %	1,054	15,052	7.00 %
Malta	4,103	56,992	7.20 %	3,099	42,292	7.33 %
Netherlands	2,326	93,084	2.50 %	2,106	78,884	2.67 %
Austria	5,338	89,648	5.95 %	4,791	72,008	6.65 %
Poland	1,443	21,332	6.76 %	1,157	14,140	8.18 %
Portugal	2,433	35,372	6.88 %	3,362	35,588	9.45 %
Romania	1,558	9,312	16.73 %	866	6,496	13.33 %
Slovenia	2,000	45,820	4.36 %	1,932	38,608	5.00 %
Slovakia	2,565	22,588	11.36 %	1,081	16,908	6.39 %
Finland	1,724	97,460	1.77 %	776	81,396	0.95 %
Sweden	6,548	92,728	7.06 %	1,949	81,404	2.39 %
Iceland	2,320	164,444	1.41 %	1,916	141,856	1.35 %
Norway	6,491	144,468	4.49 %	3,771	125,136	3.01 %
Switzerland	9,724	152,828	6.36 %	5,087	105,456	4.82 %
U.K.	3,830	80,028	4.79 %	5,103	67,540	7.56 %
Montenegro	1,515	10,528	14.39 %	n/a	n/a	n/a
Macedonia	927	8,304	11.16 %	n/a	n/a	n/a
Serbia	1,009	7,764	13.00 %	n/a	n/a	n/a
Turkey	1,490	15,940	9.35 %	1,920	14,300	13.43 %

	Tertiary Education					
	2018			2008		
	Mean Difference	Opportunity Costs	Return Rate	Mean Difference	Opportunity Costs	Return Rate
EU (28)	8,469	91,545	9.25 %	n/a	n/a	n/a
EA (19)	7,176	106,140	6.76 %	6,439	96,905	6.64 %
Belgium	7,593	124,275	6.11 %	6,694	100,690	6.65 %
Bulgaria	3,632	22,990	15.80 %	1,088	14,975	7.27 %
Czechia	3,029	52,000	5.83 %	2,555	35,410	7.22 %
Denmark	6,767	165,475	4.09 %	4,793	134,355	3.57 %
Germany	8,616	119,345	7.22 %	6,426	103,305	6.22 %
Estonia	4,351	55,820	7.79 %	2,522	31,200	8.08 %
Ireland	7,586	138,595	5.47 %	10,851	134,660	8.06 %
Greece	3,796	42,620	8.91 %	6,491	63,935	10.15 %
Spain	6,653	81,165	8.20 %	5,432	86,105	6.31 %
France	7,745	116,995	6.62 %	6,499	110,540	5.88 %
Croatia	3,627	36,905	9.83 %	n/a	n/a	n/a
Italy	7,077	100,200	7.06 %	6,966	97,670	7.13 %
Cyprus	6,589	81,525	8.08 %	7,056	94,015	7.51 %
Latvia	4,955	40,385	12.27 %	3,319	29,550	11.23 %
Lithuania	4,640	38,010	12.21 %	2,662	23,570	11.29 %
Luxembourg	10,905	197,235	5.53 %	16,632	172,375	9.65 %
Hungary	2,194	30,830	7.12 %	2,269	24,085	9.42 %
Malta	6,887	91,755	7.51 %	3,609	68,360	5.28 %
Netherlands	7,510	127,985	5.87 %	7,010	109,135	6.42 %
Austria	6,301	138,750	4.54 %	5,315	113,965	4.66 %
Poland	3,417	33,880	10.09 %	3,435	23,460	14.64 %
Portugal	5,123	56,380	9.09 %	7,890	61,295	12.87 %
Romania	3,316	19,430	17.07 %	2,627	12,450	21.10 %
Slovenia	4,456	67,275	6.62 %	5,081	57,920	8.77 %
Slovakia	1,202	41,060	2.93 %	1,695	26,540	6.39 %
Finland	9,072	130,445	6.95 %	8,365	105,625	7.92 %
Sweden	3,229	148,650	2.17 %	4,010	111,500	3.60 %
Iceland	6,128	217,155	2.82 %	9,355	186,900	5.01 %
Norway	7,161	213,040	3.36 %	7,332	175,275	4.18 %
Switzerland	13,402	239,665	5.59 %	11,654	157,255	7.41 %
U.K.	8,569	119,185	7.19 %	10,006	109,940	9.10 %
Montenegro	2,153	20,735	10.38 %	n/a	n/a	n/a
Macedonia	1,069	15,015	7.12 %	n/a	n/a	n/a
Serbia	1,552	14,750	10.52 %	n/a	n/a	n/a
Turkey	3,917	27,375	14.31 %	3,975	27,475	14.47 %

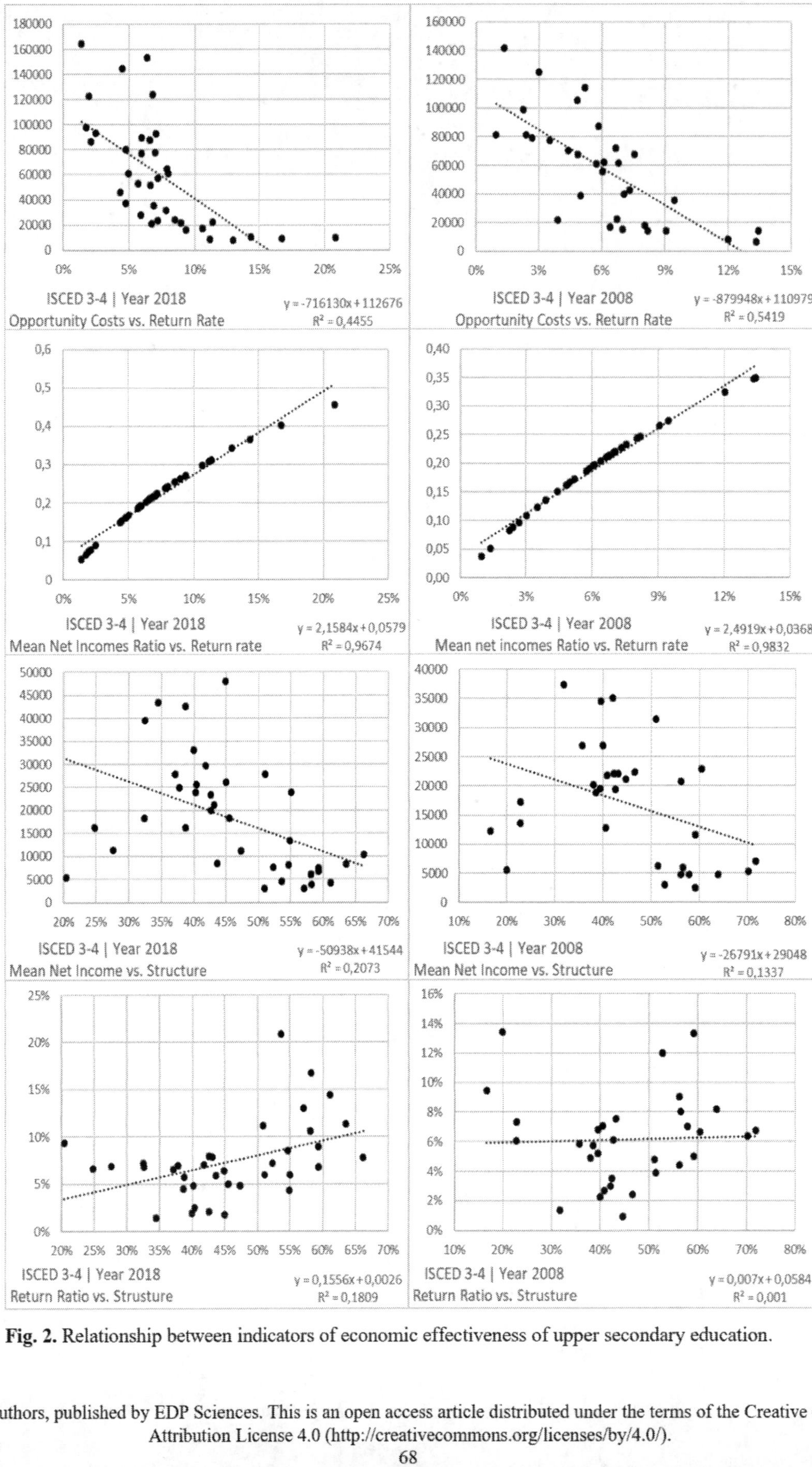

Fig. 2. Relationship between indicators of economic effectiveness of upper secondary education.

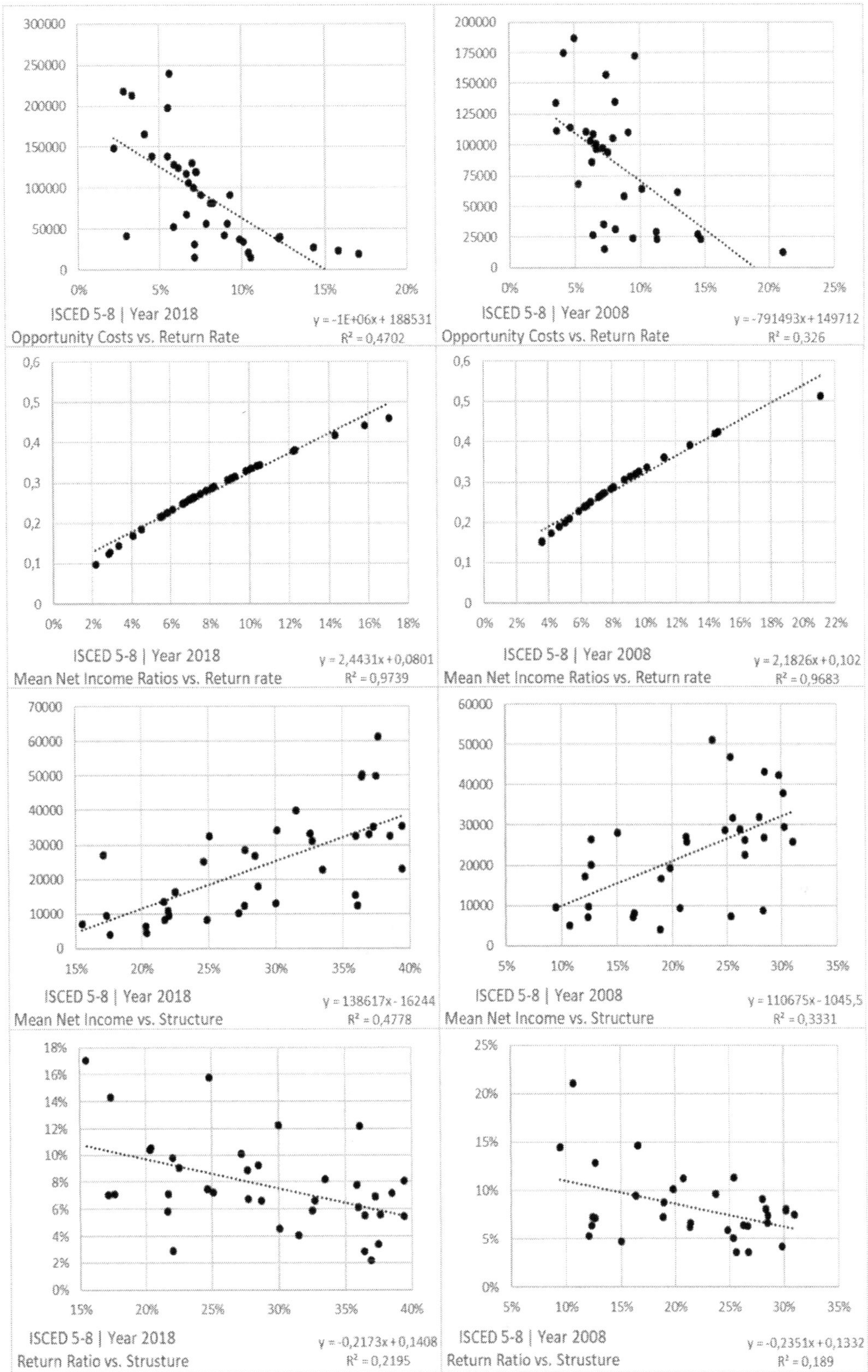

Fig. 3. Relationship between indicators of economic effectiveness of tertiary education.

Since the return rate is related to opportunity costs and net mean income difference by its construction, figures 2 and 3 provide a closer analysis of this relationship. In addition to the above relations, the relation to the educational structure and the volume if net mean income of the examined group is also assessed. The relation between return rate and opportunity costs, both for upper-secondary education and for tertiary education, is inversely proportional, where logically the lower the opportunity costs the higher the return rate and vice versa. There seems to be no relationship between the return rate and the educational structure in terms of the share of the educational level group and the total active population in both educational levels.

Quite interesting could be the relationship between the net mean income and the education structure. Although no direct causality can be deuced from the analysis, is can be assumed, that the higher net mean income attracts the individuals to tertiary education and as the proportion of tertiary educated employees rise, the proportion of the upper-secondary educated employees stagnates or decreases, it pushes up the net mean income.

4 Discussion of results – limitations and implications

The analysis is based on some simplifications. The time needed for obtaining a certain level of education has not been differentiate between countries and has not consider the wide range of options for passing through the education system. The whole period of education was taken as opportunity costs, although the study load of individual education field and levels differ, and thus the possibilities of performing paid employment during studies, which were not considered at all, differ. It was also abstracted from inflation and the changes in the net mean income structure over time, as some other analysis consider [4-9].

However, the aim of the analysis was to point out the possibility of a simple assessment of the economic effectivity of education. The benefit is in the effort to implement not only tertiary education analysis but also upper-secondary education approach. Given the initiative Industry 4.0, which often mention the need for highly educated employees, the importance of tertiary education is emphasized, but the need for skilled workers with upper-secondary education is unfortunately neglected [11-14].

5 Conclusion

Education is important part of each of our lives. Its importance is in transfer of information, knowledge, and skills from generation to generation, in developing innovation and creativity, or formation of civil society. Education also contributes to development of individuals, to creation and maintenance of social status and to increase of potential future earnings.

Education is also connected with the need to make efforts to achieve and increase a certain level of education. Thus, it is associated with certain costs that must be expend. These costs can be understood either as direct costs associated with education, such as school fees or the need to purchase of school supplies, or as indirect costs for sacrificed opportunities, such as transformation of leisure time into learning time, or scarification of job opportunities.

Analysis of economic effectiveness of education used in paper is based on such opportunity costs of education. The net mean income of primary educated employees represents the opportunity costs of upper-secondary education and the net mean income of upper-secondary educated employees represents the opportunity costs of tertiary education. On the other hand, the increase in earning due to higher education, thus the net mean income difference between education levels, should replace those opportunity costs.

The analysis shows big differences in opportunity costs of education in Member States of European Union and in return rate as well. The values vary from 1.77 % for Finland with opportunity costs €97,460 to 20.85 % with opportunity costs €10,028 for Bulgaria in the case of upper-secondary education. In the case of tertiary education values vary from 2.17 % for Sweden with opportunity costs €239,655 to 17.07 % with opportunity costs €19,430. These values imply the payback period of less than 5 years to more than 50 years.

The relationship of the return rate to other variables was also analysed. The results show, besides directly proportional relationship between return rate and net mean income ratio, ratio between two considered education levels, also the inversely proportional relationship between the opportunity costs and return rate. Relationship between return rate and educational structure has not been demonstrated. Last relationship analysed did not include the opportunity costs or return rate, the relationship between the net mean income and the educational structure has been examined instead. Different proportional relationship has been identified for upper-secondary education ad for tertiary education.

References

1. Santiago, P., Tremblay, K., Basri, E., & Arnal, E. *Tertiary education for the knowledge society : volume 1.* OECD Publishing (2008).
2. Santiago, P., Tremblay, K., Basri, E., & Arnal, E. *Tertiary education for the knowledge society : volume 2.* OECD Publishing (2008).
3. Schomburg, H.,Teichler, U. *The educational path and attainments*. Springer (2006).
4. Becker, G. S. *Human Capital*. Columbia University Press (1964).
5. Urbánek, V. *Vzdělání a lidský kapitál* (2005).
6. Boarini, R., & Strauss, H. *What is the private return to tertiary education : New ecidence from 21 OECD countries*. OECD Journal : Economic Studies. http://ssrn.com/abstract=1860961 (2010).
7. OECD. Education indicators in focus : What are the returns on higher education for individuals and countries ? doi : 10.1787/22267077 (2012).
8. Beblavý, M., Lehouelleur, S., Maselli, I. *How returns from tertiary education differ by field of study :Implication for policy-makers and students*. CEPS, 2015.
9. Krajnakova, E., Pilinkiene, V., Bulko, P. Determinants of economic development and employability of higher education institutions graduates. *Inzinerine Ekonomika-Engineering Economics*. **31**, *2020*. doi : 10.5755/j01.ee.31.2.24751.
10. Marinič, P., & Pecina, P. Is tertiary education worth it ? *INTED2021, (*2021) doi : 10.21125/inted.2021.1607.
11. Pecina, P., & Sládek, P. Fourth industrial revolution and technical education. *INTED2017*, (2017) doi : 10.21125/inted.2017.0621.
12. Záthurecký, V., Marinič, P. (2019). Industry 4.0 – Analysis of the economic development in the chosen European countries. *ISCSS2019*, (2019) doi : 10.21125/inted.2021.1607.
13. Grenčiková, A., Kordoš, M., Navickas, V. The impact of Industry 4.0 on education contents. *Verslas Teorija ir Praktika*. *22*, (2021). 10.3846/btp.2021.13166.
14. Marinič, P., Pecina, P. Industry 4.0 – Relationship between capital equipment and labour productivity. *HED2021*. doi : 10.21125/inted.2021.1607 (2021).

Pension Funds: Administrative and Investment Cost

Jan Pokorný[1*], and *Pavlína* Hejduková[1]

[1]University of West Bohemia, Faculty of Economics, Department of Finance and Accounting, Univerzitní 8, 301 00 Pilsen, Czech Republic

Abstract. The pension funds show us the alternative to the unfunded systems. This is the reason for long-terming attention. Disadvantages of funds financing include problems as indexation, inflation, or administrative and investment costs, which are decreasing the future profit for pensionary and efficiency of pension funds. This study aims to describe and analyse the administrative and investment costs and the determinants that affect these costs. The method of literary research and search of professional studies focused on empirical analysis of administrative and investment costs is used. Furthermore, the method of analysis and synthesis is used. The result of this study is an indication of the content and significance of costs and the definition of the main determinants influencing these costs. A prerequisite for the quality functioning of pension funds is the organization of the labor market and a high share of active participants. Determinants that increase costs include the number of participants in pension funds and total assets, the guarantee of income by the pension fund, and some investment instruments. Based on a search of empirical studies, the existence of economies of scale was confirmed. The limit of the study is the comparison of several countries where different legislation operates.

1 Introduction

In 1981 the pension reformation was conducted in Chile, which became an inspiration for the rest of the countries while making the pension systems. The reformation consisted of the transition from public unfunded system (known as PAYG) to fully funded. It means a mandatory savings program, where the residents have individual pension accounts. Other countries were inspired by this reformation of pension systems and decided on the privatization of part of the pension systems or implementation of these pension schemes [1].

Considering that, the pension systems became the main interest of politics in many countries because the ability to reformation the pension systems is a very important question due to social, financial, and fiscal sustainability.

The transition to fully funded schemes is of further importance because of the aging population. Limited population growth, together with an aging population, causes public finance revenues allocated to pension systems to fall, while pension expenditures increase.

* Corresponding author: pokorny@kfu.zcu.cz

As a result, deficits are created in the imaginary public pension accounts of individual states. The way to fix this is, according to some countries, is the involvement of the private sector.

The reform of Chile has become an example of how the change can be taken a place. However, Chile was able to take advantage of its undemocratic establishment at the time of the introduction of the funded pension scheme. In the case of countries with democratic governance, it is necessary to perceive the possible problems with acceptance of the reform regarding the individual interests of agents, who may influence, for example, the rules for private entities that set up individual pension accounts.

In the 1990s, the privatization of the pension system was an idea that was reflected in the recommendations of the World Bank and was also used in some post-communist countries, such as Poland and Hungary. However, the result of this attempt did not turn out satisfactorily, also due to political changes [2]. Nor does the current Chilean pension system escape criticism, which was also evident in 2016, as there were nationwide protests [3].

Fund schemes, therefore, have their advantages and disadvantages for the pension system. The ideal pension system is a combination of pension schemes so that creates a robust system that will balance individual risks, which is not entirely simple, also due to the need for political permeability of this kind of solution. Withal that, an appropriate combination can address the current problems associated with public spending and its growth in pension areas [4].

Both systems, the PAYG and the fund systems have their costs, which are closely linked to the need to ensure the functioning of the institutions that take care of the operation. We call these costs administrative costs. There are other costs in fund schemes, which are called investment costs. These costs are related to what distinguishes between a public pay-as-you-go system and a privately funded system. Notwithstanding this, even the public sector can manage public fund schemes.

Administrative and investment costs are the subject of this study. The following chapter presents the scientific goal of the study and the chosen methods that serve to fulfil the goal. The third chapter focuses on administrative and investment costs and gives reasons why it is necessary to address this issue. The fourth chapter presents an analysis of scientific studies and compares the results of this work. The last chapter of this study is the Conclusion, which summarizes the results of the research.

2 Research aim and methodology

As it was mentioned in the previous chapter, the objects of this study are administrative and investment costs. This study aims to define the importance of administrative and investment costs and assessment of determinants of administrative and investment costs. This view is very crucial because the pension funds introduce the alternative to a public unfunded scheme.

This study serves as a literature search for current knowledge of administrative and investment costs. The authors deal with this issue mainly on the examples of Australia, the USA, the Netherlands, and Chile. The reason for choosing these countries is their focus on pension fund schemes [5].

The study analyzes the determinants that affect administrative and investment costs. For practical application we have chosen countries, which are called developing countries. In these countries the labor market is still not organised rightly [6]. The second group of selected countries is developed countries, where the labour market is organised and have quality assumption for pension funds. Developed countries in this study have long-term experience with pension funds [5, 7, 8, 9]. This literature search is still relevant as it defines variables that can affect costs. At the same time, determinants can be influenced by a country's legal framework.

This study is about to be the literal research, which is focused on the issue of costs in pension costs. Further is used the method of description for specification of the theoretical background. The analysis is used in the parts Results and Discussion, where are introduced the costs due to different studies.

Between the resources for making this study were chosen the research papers, which were dealing with models of administrative and investment costs and they offer the solution of relationship between variabilities and the impact on the costs at all. In the same part is also used the method of deduction. In Conclusion of this work is practiced the method of synthesis for summarizing the results of this study.

3 Administrative and investment cost

The existence of public and private institutions, which established the pension schemes or pension funds, is connected to the existence of incomes and making of costs. The incomes are primarily represented by contributions from members or residents, so the people, who compulsorily or voluntarily participate in the given pension scheme.

On the other hand, there are costs. The initial costs that need to be mentioned are again related to the members of the pension schemes or funds. These are benefits that they usually receive based on previous participation and/or after meeting the conditions. Another important item is administrative and investment costs.

Administrative expenses represent items such as salaries of employees who take care of the operation of the institution and administrative activities associated with the existence of the pension system or fund, costs of management and marketing of the company, rent, depreciation, and other items. Thus, these are items that are necessary to ensure the functioning of the institution. Investment costs are associated with fund financing. These include items such as investment research, management, and consulting fees [9, 10].

As early as 1998, James [11] pointed out that the administrative costs of financial distortions and distributional effects posed new problems that future research should focus on. And as will be shown in the next chapter, this research focused mainly on Australia, the USA, the Netherlands, and Chile [10].

Another who dealt with the problem of pension funds and costs was Stiglitz [12]. Stiglitz [12] has generally defined several disadvantages of pension funds. These problems include indexation, the risk of inflation, difficulty to predict life expectancy (therefore it is necessary to use general prognoses that do not respect the health status of the individual). Another disadvantage, according to Stiglitz [12], is the higher administrative costs of pension funds compared to a public social security scheme.

This disadvantage is associated with what is called system decentralization. The existence of smaller pension funds creates a competitive environment. Such an environment can be advantageous in the market, but it brings marketing costs. Marketing costs aim to increase interest in a given pension fund, which should lead to an increase in the size of the pension fund. However, it would be necessary to gain ample dominance to realize cost savings from scale [13]. As James [11] points out, marketing costs can lead to the consumer choosing a pension fund that is not one of the best options.

The main reason to deal with administrative and investment costs, resp. in general, costs, are important within fund schemes. Costs represent a reduction in potential profits for clients and managing institutions. Reducing administrative and investment costs will increase the funds that can be invested. The funds invested will thus increase and may bring greater benefits, specifically mention the increase in return for a member, replacement ratio, and performance of pension fund [6].

Fig. 1 shows the issue of allocation of financial means on costs. Bikker [10] pays attention to the relationship between annual costs and erosion of retirement income. The model on Fig. 1 presumes annual wage growth 3 %, annual inflation 2 %, nominal return 7 %. The time of contribution is 40 years and pensioner gains benefit for 20 years.

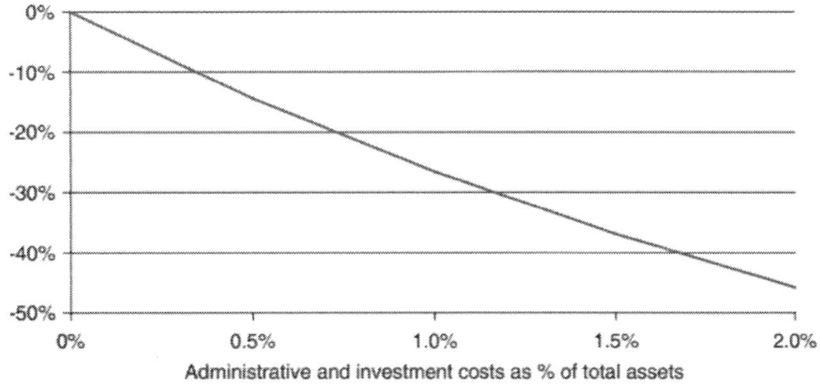

Fig. 1. Erosion of retirement benefit based on costs [10]

Another common question, that is inflected in this area, is the optimal size of pension funds. That is, whether pension funds achieve economies of scale based on the size of the pension fund and whether a U-shaped curve is generated [9]. The next chapter will be devoted to this question.

4 Results and Discussion

Recently, several studies have been conducted to empirically understand administrative and investment costs in different parts of the world. There are different times in different countries when pension funds were created and assets were acquired into these funds. We can take the period of existence as one of the important factors, which is essential in the existence of pension funds and the evaluation of administrative and investment costs.

Another factor is the development of the labor market itself, as evidenced by studies from Southeast Asia. While Singapore and Malaysia have more developed markets, the situation is the opposite for Thailand and Vietnam. For a larger and better-formed labor market, the organization of funded pension schemes is also better and the involvement of active participants is higher. This is also evidenced by the data in Table 1, which shows the cost concerning gross income, contribution, and assets.

Thailand and the Philippines are countries where the availability of fund schemes is shorter. At the same time, the system in the Philippines is based on a combination of investing and paying benefits to retirees in part of benefits and income. Thus, the current good demographic structure of the population is being used [6].

As mentioned, in the Philippines, part of the contributions to the current payment to pensioners is allocated, which negatively affects the performance of the pension fund and the future return of the members of the pension fund. A comparison of operating expenses and assets shows that the funds in Singapore and Malaysia have the best ratio, which is also evident in other categories. The Philippines, on the other hand, is the worst in all categories. Thailand moves between these countries [6].

Table 1. Estimated administrative and compliance costs of select pension and provident fund [6]

Variable	Employees Provident Fund (Malaysia)	Social Security System (Philippines)	Central Provident Fund (Singapore)	Social Security Organization (Thailand)

Operating expenses as share of gross income	1.77	6.03	2.48	3.65
Operating expenses as share of contributions	2.92	8.20	0.77	4.77
Operating expenses as share of assets	0.25	2.13	0.09	0.63

Asher & Bali [6] see the reason for a large number of inactive members in the Philippines (about one-third). Another reason is the low level of compliance, ie the already mentioned poor organization of the labor market and law enforcement, which leads to the collection of lower contributions than they should be.

The above-mentioned problems are mainly related to developing countries and the least developed countries, which are still shaping the labor market. Other countries that will be described here belong to the developed countries [14], and therefore their labor market can be considered as creation, including the payment of basic components of wages. An overview of countries and determinants that affect administrative and investment costs is provided in Table 2.

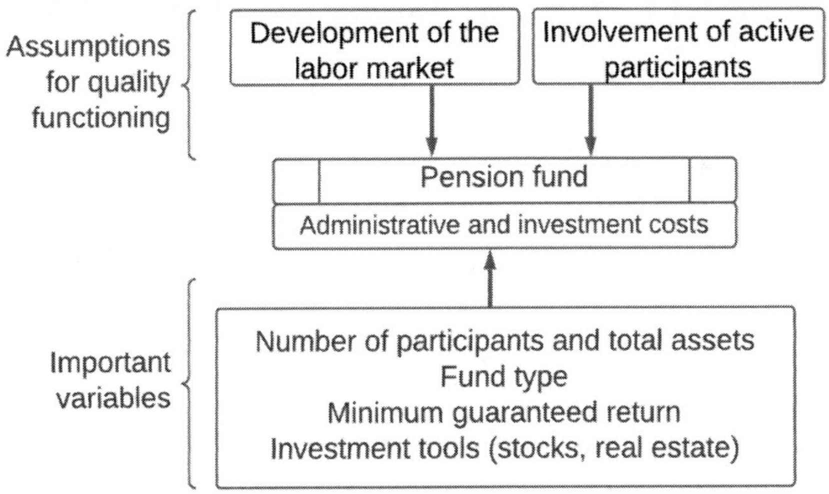

Fig. 2. Assumptions for pension fund and important variables affected administrative and investment costs

Assumptions for quality functioning of pension fund we can see on Fig. 2, where are displayed important variables. These variables have an impact on the level of administrative and investment costs.

The first of the studies focus on Italian pension funds. In this study, the relationship between cost and fund type is found. If the fund is closed for a sector or region, then its operation is more efficient than for open ones (universal funds). On the contrary, with a minimum yield guarantee, costs increase, albeit insignificant. A similar result is for share equities in total financial assets. Guaranteed schemes' net asset value has a positive effect on costs, however, this determinant in the natural logarithm harms costs. As a result, the relationship between dimension and cost is a U-shaped curve and it is possible to find savings from the range [7].

Campani & Riddier [8] focus on the Netherlands, as does Bikker [5]. In the first study on the Netherlands, Campani & Riddier [8] conclude that the number of participants that affects administrative costs is primary. However, costs per participant show economies of scale with increasing membership to a certain point, from which costs per participant increase again.

Another study [5] uses three different cost functions, here for simplification the results will be mentioned, which confirm all three functions. Furthermore, administrative and investment costs are divided here.

Table 2. Important variable of administrative and investment

Authors	Country	Important variable
De Vincentiis, Isaia & Zocchi [7]	Italy	Fund type (open/close) - Minimum guaranteed return + Guaranteed schemes' net asset value - Squared natural logarithm of the guaranteed schemes' net asset value + Share of the total financial assets invested in equities +
Campani & Riddier [8]	the Netherlands	Number of participants + Ex-contributors (ratio) + Real coverage ratio + Logarithm of the assets per participants +
Bikker [5]	the Netherlands	<u>Administrative costs:</u> Participants+ Type of funds + Pensioners + Inactive participants - Assets per 1,000 participants + Report invest cost - Outsourcing + <u>Investment costs</u>: Total assets + Type of funds + Stocks + Real estate +
Bikker, Steenbeek & Torracchi [9]	Australia, Canada, the Netherlands, the US	Number of participants + Standardized CEM service quality score and complexity score + Share of deferred participants - Public sector (state or provincial government and municipality) +

Administrative costs will be mentioned first. It is essential to confirm the achievement of administrative cost savings per participant. Other important elements include the type of fund (company, industry, professional group, or other funds). And industrial funds offer the least expensive type. Bikker [5] assumes that the reason is the universality of this fund compared to others, which offer pension plans more appropriate to the individual, but this increases administrative expenses more.

The investment cost model shows that there will be an increase in investment costs for total assets, which is around 1 according to the model chosen. Other factors that lead to growth are again the type of fund and sub-types of assets (stocks and real estate), which are more expensive, for example, against bonds, which, however, are not significant.

The last of the presented studies [9] deals with a comparison of four countries (Australia, Canada, the Netherlands, and the US) and their costs in selected pension funds. In this study, cost savings are found concerning the relationship between costs and activities performed. Table 2 shows the determinants that primarily affect growth. The increase in costs is associated with determinants such as the number of members and what needs to be highlighted is the focus of the pension fund (similar to the fund type from [5], here it is already focused on specific sectors). Funds for public sector employees show higher costs than funds for private-sector employees. State and provincial government funds show 70% higher administrative costs than average. There is therefore room for consolidating these pension schemes so that they are more effective [9].

The presented studies indicate which variables need to be taken into the account when examining administrative and investment costs. Based on these studies, it has been confirmed that there are economies of scale in pension funds, due to the growing number of participants and the cost per capital. However, the point that again increases costs per capita is different for each pension fund and country.

Other important variables include the type of fund that can be examined as open or closed. If the pension fund is closed, then it is possible to further examine for which group the pension fund is intended.

Guaranteeing the yield cannot be neglected either, which may be important for some states given their history and the need for security for citizens or political decisions. The Italian example shows that a guaranteed return increases costs, but the increase is not high. Of course, it depends on the specific situation of the pension fund and its performance. It is also worth mentioning the individual investment instruments, but here it is necessary to register the restrictions that can be imposed on pension funds and the allocation of assets to individual investment instruments, as stated, for example, by Kritzer [15].

5 Conclusion

Administrative and investment costs are part of the money market funds spent. The issue of these costs is very important, as pension funds are a possible way to implement the reforms of many countries' pension systems. At the same time, the implementation of fund schemes is also recommended by international organizations.

Administrative and investment costs, together with inflation, indexation, and life expectancy, thus form one of the key issues in the creation of pension funds. As well as the issue of financial distortions, and distributional effects. In addition, the issue of costs is linked to the performance of the funds and the subsequent payment of benefits to pensioners. By reducing costs, a higher return can be achieved for members of pension funds, and thus a consequent increase in benefits, which should be the goal of members. For pension funds, cost reductions lead to better fund management and performance.

As part of the examination of empirical studies, the preconditions for the good functioning of pension funds were defined. These assumptions include a well-functioning and organized labor market and a high proportion of active participants.

An examination of several countries has confirmed the existence of economies of scale in pension funds. This confirmation is important due to the creation of pension funds and their optimal size. The main determinants that affect administrative costs include the number of participants and the fund type, i.e. whether the fund is universal (open) or selective (closed - for a defined geographical area or sector of the economy).

Some pension funds guarantee a return, which again affects the growth of administrative costs. This link is important concerning the political permeability of the implementation of pension funds into the pension systems of states.

Investment costs are affected by total assets, i.e. it is a similar determinant as the number of participants in administrative costs. Investment tools that increase costs include stocks and real estate. However, stocks are important investment vehicles and therefore cannot be rejected as part of investment instruments. In general, for investment instruments, it is important to compare their return, which subsequently affects the return of the entire fund as well as the benefit paid to fund participants.

A limitation of research in the area of administrative and investment costs is the differing legislation in individual states, which may limit the number of costs or may affect the allocation of assets and the distribution of risk. At the same time, there may be other limitations that relate to the length of development of individual pension schemes. This limitation means that international comparisons may not be meaningful to some extent.

This research focused on the theoretical background and empirical research to date. Further research can thus focus on empirical evidence based on the findings of this study. As a next possibility we can mention deeper analysis of investment cost and tools due to their performance.

This paper was created within the project SGS-2020-026 'Economic and financial transformation in the context of digital society' at the University of West Bohemia, Faculty of Economics.

References

1. K. McKiernan. Social Security reform in the presence of informality. *Review of Economic Dynamics* **40**, 228-251 (2021).
2. J. Vostatek. *Penzijní teorie a politika.* (Praha: C. H. Beck, 2016)
3. M. J. Azocar. Policy Debates on Pension Reform in Chile: Economists, Masculinity and the Mobilization of Strategic Ignorance. *SOCIAL POLITICS* **27**, 648-669 (2020).
4. B. Heer, V. Polito, M. R. Wickens. Population aging, social security and fiscal limits. *Journal of Economic Dynamics & Control* 116, 103913 (2020).
5. J. A. Bikker. Is There an Optimal Pension Fund Size? A Scale-Economy Analysis of Administrative Costs. *Journal of Risk and Insurance* **84**, 739-769 (2017).
6. M. Asher, A. S. Bali. Public Pension Programs in Southeast Asia: An Assessment. *Asian Economic Policy Review* **10**, 225-245 (2015).
7. P. DeVincentiis, E. Isaia, P. Zocchi. Institutional disparities and asset allocation homologation in Italian defined contribution pension funds. How do they affect the guarantee commitment? *Journal of Pension Economics and Finance* **16**, 205-232 (2017).
8. C. H. Campani, I. De Ridder. Administrative costs of Dutch pension funds: the impact of fund characteristics. *Economics Bulletin* **39**, 1361-1370 (2019).
9. J. A. Bikker, O. W. Steenbeek, F. Torracchi. The Impact of Scale, Complexity, and Service Quality on the Administrative Costs of Pension Funds: A Cross-Country Comparison. *Journal of Risk and Insurance* **79**, 477-514 (2012).
10. J. A. Bikker. Operating costs of pension funds: the impact of scale, governance, and plan design. *Journal of Pension Economics and Finance* **8**, 63-89 (2009).
11. E. James. New Models for Old-Age Security: Experiments, Evidence, and Unanswered Question. *The World Bank Research Observer* **13**, 271-301 (1998).
12. J. E. Stiglitz. An Agenda for Reforming Economic Theory. Frontiers of Economics in China, **14**, 149-167 (2019).
13. M. Festić, J. Mencinger. The Perspective of Pension System Reformsin the New Meber States. *PRAGUE ECONOMIC PAPERS* **18**, 291-308 (2009).
14. UN DESA. World Economics Situation and Prospects 2014: Country classification. [accessed 02.05.2021] https://www.un.org/en/development/desa/policy/wesp/wesp_current/2014Chap1_en.pdf
15. B. E. Kritzer. Social security privatization in Latin America. *Social Bulletin* **63**(2), 17-37 (2000).

The Role of Quality Improvement in the Innovation practices in SMEs in Latvia

Ketevan Rizhamadze[1,*] and *Anna* Ābeltiņa[2,**]

[1]MBus (Global Business Analysis), Doctoral student, Turiba University, Rīga
[2]D.Sc. (Economics), Associate Professor, EKA University of Applied Science, Rīga

Abstract. The hasty pace of globalization depicted by fundamental changes in the production technologies is utterly altering international trade. Furthermore, the rapid technological transformation requires organizations to refer to quality improvement while introducing innovative services and products. The following study scrutinizes the role of quality improvement in the innovation within the SMEs operating throughout Latvia. The nature of the study is quantitative and the views of SMEs representatives were captured via an online survey. SMEs operating in the field of manufacturing and retail in Latvia were selected for the participation in the study. As the research results reveal, quality improvement is considered as an important factor upon which innovative activities are implemented. Besides, in the rapidly changing consumer needs, product quality improvement is seen as a process of modifying and redesigning the products and services to satisfy well swiftly fluctuating demands for SMEs. In addition, our research discloses that the major sectors in Latvia within which quality management systems certification was requested were: construction, transportation, communication, wholesale and retail trade. Furthermore, research findings show that financing of productive capacity is central to any development strategy, but access to finances still remains a major obstacle for SMEs.

Introduction

In this rapidly changing world, customer demands evolve and proposing innovative products and services tend to be pivotal for organizations. It is empirically evidenced that companies that exploit their creativity and innovation tend to be profitable and their market share is increased [1]. Innovation is a drive to prosperity and it is profoundly reliant on management and not just a technological innovation. Innovation management implies that a firm's structure, norms and practices might evolve for the purpose of establishing technological and innovative base upon which competitive advantage is gained [2]. Managers should ensure that relationships are well managed and conflicts within the organization are properly resolved. Firms succeeding in managing complex relationship patterns would reap the benefits of fierce competition [3]. Moreover, the more advanced the

*Corresponding author: katerizhamadze@gmail.com
**Corresponding author: anna.abeltina@gmail.com

company is, the more ability it possesses to make alterations and improve itself [4]. As there exist diverse quality management programs, the organizations are facing the dilemma of selecting the program that efficiently and effectively corresponds to their needs [5]. The process of quality improvement in the organization can cover various actions like benchmarking, reengineering, upgrading, etc. [6]. As a matter of fact, the business process implies phases upon which an input is transformed into a viable output. In addition, a continuous process improvement forms a particular course of enhancement that is based on business strategy [7]. The other challenge that the organizations have to overcome is differentiating innovation and improvement. Continuous improvement relies on the enhancement of existing products and services. Whereas, innovation requires breakthroughs and process disruptions [4]. SMEs, known as a heterogeneous population of firms have an impact upon process of innovation and to economies. The contribution and engrossment could be expressed in various methods. Innovative products could be created by using research and development (R&D) inputs or without it. The SMEs that are present in high technological industries rely on science based knowledge to practice innovation. However, SMEs working in low technological settings, don't refer to R&D for innovation purposes [8]. In developing countries, the role of SMEs is meaningful in job creation and in economic development. They account for more than 90% of businesses and above 50% of employment globally. One of the main obstacles for SMEs to grow and to extend their operations, appears accessibility to financial means [9]. Struggles faced by SMEs in accessing bank finance are hindering economic development in all sectors in Latvia. Therefore, putting emphasis on the area of cross-border, e-commerce and focusing on productivity, growth and export tend to be fundamental to the development of Latvian economy [10].

1 SMEs and innovation in Latvia

In 2016 Latvia became member of the OECD (The Organization for Economic Co-operation and Development) and in 2014 joined the Eurozone. Agriculture, chemicals, logistics and woodworking are the cornerstones of the Latvian economy [11]. SMEs contribute hugely in non-financial business economy of Latvia. SMEs and large firms created employment opportunities and added value at an analogues degree during the period of 2014-2018 [10]. Even though the principal source of funding for SMEs tend to be EU funds, Latvia currently undergoes difficulties as numerous EU funding lines are expiring and there is a harsh decrease of economic relations with Russia [11]. Moreover, country faces a challenge to improve the skills and innovation via vocational training. As a result, it would be beneficial for Latvian companies to reap the benefits of digital skills derived from life-long learning programs. Even though the total valued added to the total employment in Latvia is above the respective EU averages and is 79.4%, the average productivity is half than the EU average as illustrated in Table 1.

 https://doi.org/10.1051/shsconf/202111502007

Table 1: Value added of SMEs

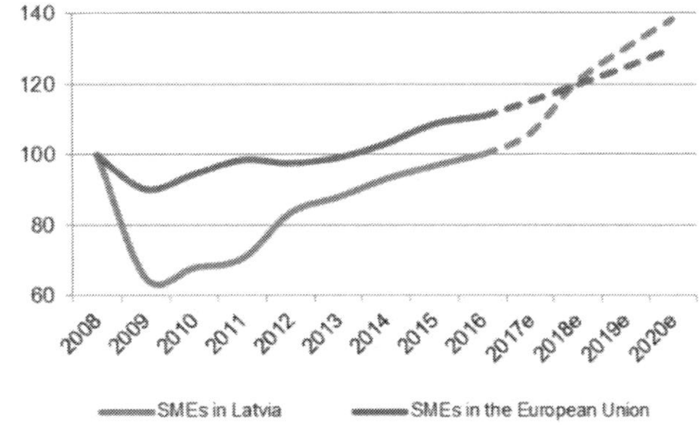

Source: European Commission 2021, [5]

Manufacturing, wholesale & retail trade are chief sectors that add value and create employment in EU countries and Latvia is no exception. As illustrated in Table 2, SMEs in Latvia employ around 4.6 people that tends to be the above EU average of 3.9.

Table 2: Number of persons employed in SMEs

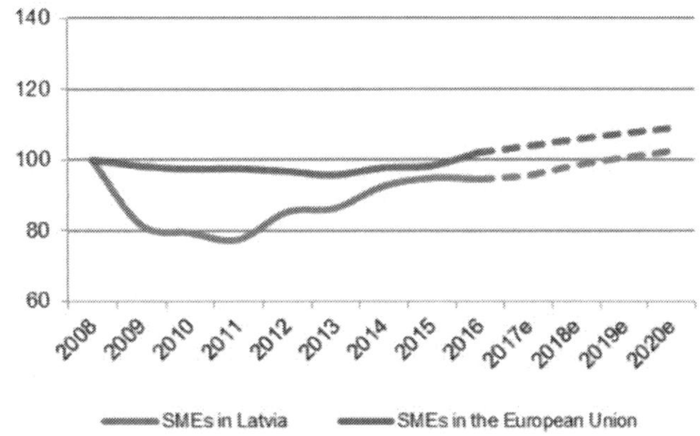

Source: European Commission 2021, [5]

Micro firms are defined as organizations that hire up to nine persons, small firms employ from 10 to 49 persons and firms that employee from 50 up to 249 persons are considered as medium sized enterprises. As illustrated in Table 3, small firms contribute greatly in adding the value. Micro companies were the major contributors in terms of employment growth. The increase in manufacturing production and e-commerce were evident and it had effect upon growth.

Table 3: SMEs basic figures, Number of persons employed in SMEs

Class size	Number of enterprises			Number of persons employed			Value added		
	Latvia		EU-28	Latvia		EU-28	Latvia		EU-28
	Number	Share	Share	Number	Share	Share	Number	Share	Share
Micro	104,705	91.6%	93.0%	217,364	33.2%	29.7%	2.7	21.0%	20.8%
Small	7,976	7.0%	5.9%	158,454	24.2%	20.1%	3.1	23.9%	17.6%
Medium-Sized	1,450	1.3%	0.9%	143,879	22.0%	16.8%	3.4	26.3%	18.0%
SMEs	114,131	99.8%	99.8%	519,697	79.4%	66.6%	9.3	71.1%	56.4%
Large	195	0.2%	0.2%	135,144	20.6%	33.4%	3.8	28.9%	43.6%
Total	114,326	100.0%	100.0%	654,841	100.0%	100.0%	13.1	100.0%	100.0%

Source: European Commission 2021, [5]

According to small business act of European Commission, the weakest performing area for Latvia is skills and innovation. In addition, compared to other EU countries, Latvia is lagging behind on digitalization of activities [10]. In addition, the growth is hampered by weak innovation and inefficient resource allocation. Compared to other OECD countries, Latvia is behind other EU countries, in terms of implementing innovation by the business field. In addition, the amount dedicated to R&D activities is low within the OECD [12]. There is also lack of talent in this field and the scarcity of skilled professionals hampers the development and proper implementation of innovative activities as illustrated in table 4:

Table 4: Skills and innovation

Area	Latvia	EU average
Introduction of new products and processes by SMEs	18.68%	33.32%
Introduction of organizational and marketing innovations by SMEs	21.41%	33.43%
In-house innovation by SMEs	15.2%	28.95%
Cooperation of SMEs with other companies	5.6%	12.23%
SMEs selling online	10.33%	16.57%
Turnover from e-commerce	5.36%	10.13%
Percentage of enterprises employing ICT specialists	14.04%	18%
% of enterprises ensuring CT skills learning to staff	9.56%	21.22%

Source: Elaborated by authors based on Small Business Act, Latvia 2019, [5]

As the problem persists in the area of skills and innovation, since 2014 policy interventions focus on supporting the policies that improve and strengthen these activities. Policy measures and support programs are developed to ensure and to improve innovation capacity of SMEs. The models developed by European commission sets the rules and guidelines for cooperation among research institutes and organizations [10]. Moreover, by ensuring the quality cooperation between scientific institutions and companies innovation ecosystem will be reinforced and digital technologies shall be disseminated broadly throughout the economy [12]. Following the ranking of European innovation scoreboard, Latvia falls in a moderate category. The differences between EU and Latvia are illustrated in the table below.

Table 5: European innovation scoreboard, 2020

Description	Latvia	EU
Turnover share SMEs (%)	52.8	38.3
Launching new firms (>10 staff) (%)	1.9	1.1
Entrepreneurship (%)	14.8	6.7
R&D expenditures (per 10 mln population)	0.0	16.2
Easiness of commencing a business (max.100)	80.2	76.5

Source: Elaborated by authors based on European Innovation Scoreboard 2020, [6]

As the figures demonstrate the revenues generated by SMEs tend to exceed the average amount among European Union members. The firm's birth is 1.1% in EU and in Latvia it's higher and tends to be 1.9%. In entrepreneurship activities, Latvia has also higher percentage and sets the percentage at 14.8%. However, there is a different situation when it comes to R&D expenditures. The R&D spending is negative and is lower than the average. When it comes to ease of starting a business, the figure is positive and tends to be above the EU average. As the figures in table 6 illustrate, the weakness is remarkable in innovation capacity for star-ups [13]:

Table 6: The Community Innovation Survey (CIS), 2018

Description	Indicator scores Latvia	Indicator scores EU
Product innovation presented by SMEs	134	123
Business process innovation presented by SMEs	106	109
Cooperation of SMEs with other companies	112	141
Non-R&D innovation expenses	61	103

Source: Elaborated by authors based on community Innovation Survey (CIS). [6]

2 Previous researches

The research conducted by Kotane and Kuzmina-Merlino concludes that generally the performance of the SMEs is evaluated based on the financial figures, however it tends to be insufficient and other indicators that demonstrates the revenues should be scrutinized [14]. According to research conducted by Riekstina and Zvirgzdina, reveals that micro and small firms are presented in overwhelming numbers in the total number of Latvian companies. In addition, following main sectors were identified: wholesale, retail trade, financial & insurance activities and construction [15]. The research conducted by Vasilevska and Rivza identified that problems related to the operation of quality management systems in SMEs were: financial expenditures necessary for staff training and certifications, as well as long term implementation of the system and sustaining and maintaining the system in place [16]. In order to examine the quality management system, it is advisable to scrutinize the results of the Survey for 2019 which shows an estimation of the number of valid certificates issued by the International Organization for Standardization (ISO). The ISO survey results consists of three sets of the data: the number of valid certificates, the number of sites covered by the certificates and the number of sectors. The number of valid certificates issued in Latvia were 889 within 1031 sites as illustrated in Table 7 below:

Table 7: ISO 9001:2015 Quality management systems

Latvia	
Certificates	**889**
Sites	**1031**

Source: Elaborated by authors based on the ISO Survey, [7]

The major sectors in Latvia within which ISO 9001 certification was requested were: construction, transport, storage and communication and wholesale & retail trade. The data is illustrated in the Table 8:

Table 8: ISO 9001:2015 Quality management systems

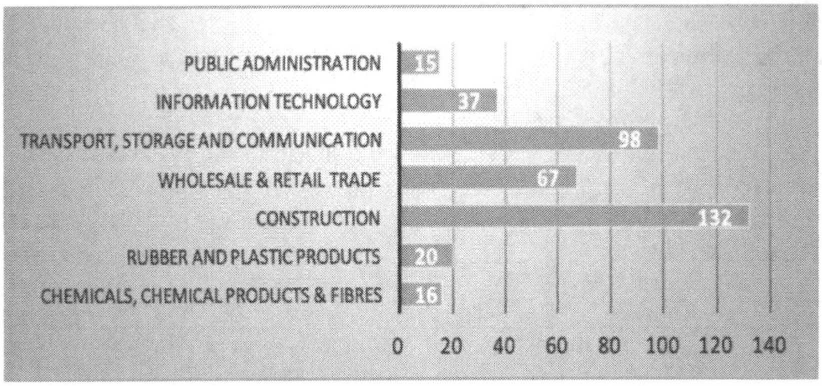

Source: Elaborated by authors based on the ISO Survey, [7]

Previous researchers have identified four broad strategic types of SMEs:

- Firms that are concerned with progression;
- Firms that are concerned with customer demands;
- Firms that are concerned with product development;
- Group of firms without having clear strategic direction

The categorization of strategic types, gives an opportunity to managers and entrepreneurs to comprehend well strategic course and their attitudes towards global functioning of firms [17].

3 Overall aim of the research and methodology

The main aim of the study to identify a significant platform upon which innovation practices are implemented. In addition, to address the relationship between quality improvement and innovation in SMEs. The target of the study is SMEs working in the manufacturing and retail. In addition, we posit that quality improvement is considered as an important factor upon which innovation is build and developed. Furthermore, this paper empirically considers that internal factors that exist in the organization, contribute greatly in regulating the link among innovative activities and quality management.

The nature of the study is primarily quantitative. The views of SMEs representatives were captured through an online survey. SMEs operating in the field of manufacturing and retail in Latvia were selected. The participants of the survey were chosen based on their scope of activities. Contact information was obtained through official registry of companies. The criterion of selection was the following: firms with more than ten staff members up to 250 employees were eligible to participate in the survey. Nevertheless, participation in the survey was voluntary and questionnaire was used as a research instrument. The questionnaire was administered online and survey was distributed among SMEs via company's official e-mails. Survey included both closed and open questions. Closed questions gave the respondents a chance to select the answer from pre-defined Reponses. In addition, the questionnaire included 'yes' or 'no questions along with multiple choice questions. The summary of research methodology is illustrated in table 9:

Table 9: Research Methodology

Research methodology	**Quantitative**
Data	**Secondary/ Primary**
Paradigm	**Positivism**
Research type	**Analytical(Exploratory)**
Research outcome	**Pure**
Logic of the research	**Deductive**
Sample size	**110 SMEs**
The margin of error	**5%**
Uncertainty margin	**95%**
Survey completion rate	**86.30%**

Source: Elaborated by the authors

Following research limitations were identified: the number of the units of analysis used in the study is dictated by the type of specific research problem. As sample size is not large, it will be difficult to ensure a representative distribution of the population. In addition to sample size, there are lack of prior research studies on the topic and study will be conducted in Latvia only. Samples readiness in participating and prone to bias can also be added limitations to the research.

4 Research findings

The research findings reveal that within internal factors, innovation is considered to be an essential feature upon which the firm's growth is achieved. In addition, the results reveal that 95% of participants deem that innovative activities do have an influence upon firm's growth and development. Study finding demonstrate that innovation is considered as an important practice that has effect upon growth and firm's development. In addition, per participants responses, human, technological resources, organizational structures and systems endorse firm growth (See Figure. 1).

Figure 1: Internal factor for growth

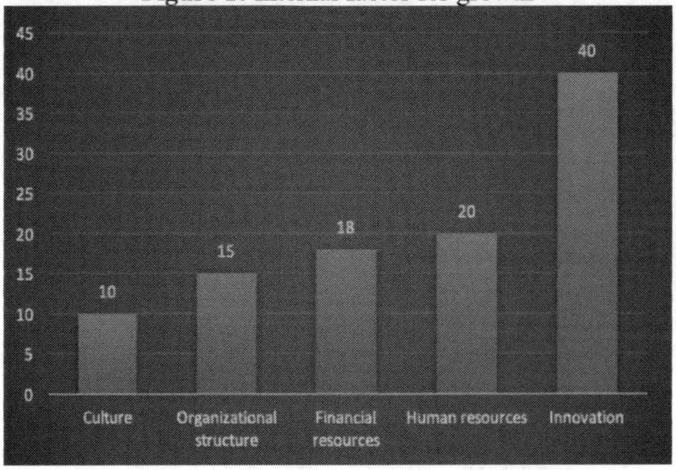

Source: Elaborated by the authors

In the survey, there were more participants from medium-sized enterprises than there were from small organizations, 52% and 48% respectively (See Figure 2).

Figure 2: The breakdown of participants of the survey

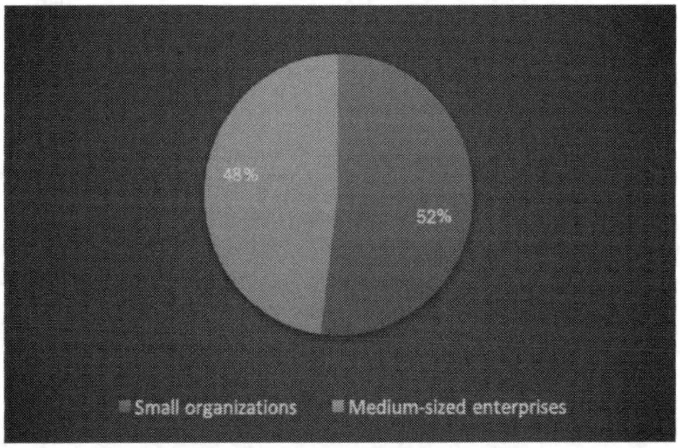

Source: Elaborated by the authors

Pursuant to study findings, the platform that SMEs generally rely on for implementing innovation practices is quality improvement. For them, improving quality or adding value to existing products and services tend to be the platform upon which innovation practices are implemented. Alongside quality improvement, market competition and added value to customers is also considered to be important platform for innovation practice drive 26% and 21% respectively. As illustrated in below table, the majority of respondents, 35% consider quality improvement as an essential platform for implementing innovation (see Figure 3).

Figure 3: Base for innovation

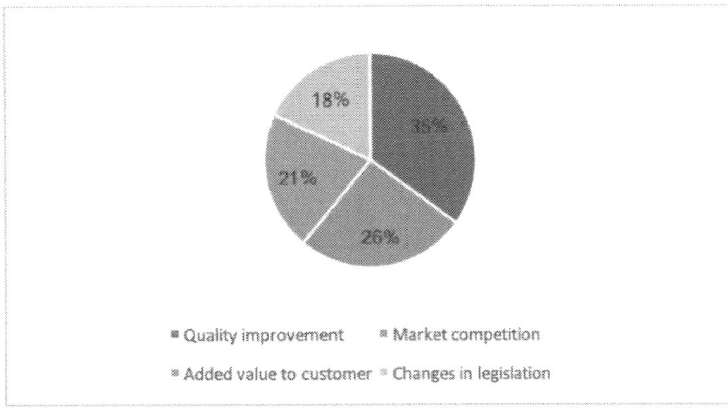

Source: Elaborated by the authors

Even though, quality improvement was cited as an important platform that drives innovation, implementing it in practice is complicated for SMEs. As survey results show, the major challenge for it is access to finances [18]. Majority of survey respondents not only regard access to finance as an impediment for product development, but also as a huge hindrance to growth, productivity and form's development. Competitor's response to market changes and legal environment have also a major role in firm's growth and development (See Figure 4).

Figure 4: Outer factor that endorse growth

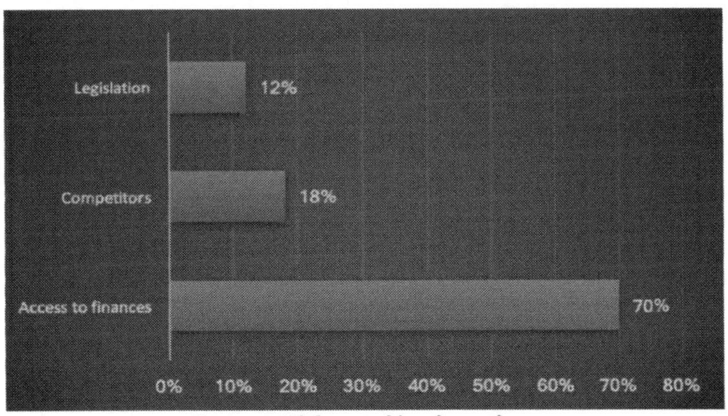

Source: Elaborated by the authors

The research findings demonstrate that, innovation process that is linked with quality improvement is present in up to 45% of SMEs. However, small firms are hesitant to conduct research and development activities as it is associated with additional finances. However, while redesigning product and adding new features to existing products small firms are referring to innovation for quality improvement.

Conclusion

The SME sector is a backbone for a number of developed and emerging economies thus assisting and complementing larger firms. SMEs play a crucial role in the world economy and contribute substantially to income, output and employment. In today's changing world innovation and quality management is becoming one of the key drivers for change. However, global financial crisis and increasingly fast technological changes created a particularly tough climate for SMEs. The lack of innovation strategy, well defined and structured quality improvement process, and performance indicators are rarely assessed due to lack of structures and systems in place. The issues of innovation and quality improvement is pivotal, especially when competitive advantage of SME's is highly dependent on it. From management side of view, managing these processes is even complicated and requires special attention and tactics. Present research proposal has been planned with a view to discuss the various issues related to the SME's. In order to meet a wide scope of international development goals, it is necessary to ensure and guarantee that SME's are given possibilities and opportunities where they can exhibit their full potential and skills especially in developing countries. Following major findings emerge from the different analyses presented in the study:

- SME tend to be the main creators of jobs in Latvia and major contributors to national economy;
- Quality improvement is considered as a significant base upon which company implements its innovation activities;
- The major sectors in Latvia within which quality management systems certification was demanded were: construction, transport, communication, wholesale and retail trade;
- Quality improvement is considered to be an important factor to implement innovative activities broadly;
- SMEs are less likely to conduct comprehensive research and development activities due to lack of accessibility to finances

Based on the study and survey findings, pursuing innovation is a key to productivity and creativity. Continuous innovation will make business more competitive and profitable. For quality enhancement and improvement, it is vital to conduct researches. If small and medium sized enterprises wish to grow, to become innovative and to add features to existing products it is suggested to refer to researches frequently. However, R&D is associated with finances and SMEs usually lack additional financial resources, it is less probable that they will conduct research activities.

References

1. Maital S., and Seshadri D.V.R. Innovation management: Strategies, concepts and tools for growth and profit. New Delhi: Sage. P38 (2017).
2. Volberda, H. W., Van Den Bosch, F. A., Heij, C. V. Management innovation: Management as fertile ground for innovation. *European Management Review*, **10**, 1-15 (2013). [Online]. Retrieved on February 15, 2021 from https://doi.org/10.1111/emre.12007
3. Ābeltiņa, A., Rizhamadze, K. The complexity of relationships within business and the importance of common ground. *SHS Web of Conferences*, 90, 01001 (2021). https://doi.org/10.1051/shsconf/20219001001.
4. Boonstra, J. J. Dynamics of organizational change and learning. Chichester: Wiley (2008).
5. Backström, T., Fundin, A., Johansson, P. E. Innovative Quality Improvements in Operations Introducing Emergent Quality Management. Cham: Springer International Publishing (2017).
6. Møller, C. and Chaudhry, S. S. Advances in enterprise information system II proceedings of the fifth international conference on recearch and pratical issues of

entreprise information systems (CONFENIS 2011), Aalborg, denmark, October 16-18, 2011. New York: CRC Press (2012).
7. Plenert, G. J. Strategic continuous process improvement. New York: McGraw-Hill Professional (2012).
8. Organisation for Economic Co-operation and Development, OECD. Enhancing the competitiveness of SMEs in the global economy: strategies and policies. Conference for Ministers responsible for SMEs and Industry Ministers Bologna, Italy, 14-15 June 2000. [Online]. Retrieved on February 07, 2021 from http://www.oecd.org/cfe/smes/2010176.pdf
9. World Bank. Official web-site. [Online]. Retrieved on February 20, 2021 from https://www.worldbank.org/en/topic/smefinance
10. European Commission, EC. SME Performance Review, Small Business Act (SBA) 2019, Latvia. [Online]. Retrieved on 21 February, 2021 from https://ec.europa.eu/growth/smes/sme-strategy
11. The Innovation Policy Platform, IPP. official website. [Online]. Retrieved on February 21, 2021 from https://www.innovationpolicyplatform.org/www.innovationpolicyplatform.org/content/latvia/index.html
12. Organisation for Economic Co-operation and Development, OECD (May, 2019) Economic Surveys, Latvia. [Online]. Retrieved on February 21, 2021 from http://www.oecd.org/economy/surveys/latvia-2019-OECD-economic-survey-overview.pdf
13. European Union, EU. European Innovation Scoreboard 2020 [Online]. Retrieved on February 2, 2021 from https://ec.europa.eu/info/index_en
14. Kotane, I., and Kuzmina-Merlino, I. Analysis of small and medium sized Enterprises' business performance Evaluation practice at transportation and storage services sector in Latvia. Procedia Engineering, 178, 182-191 (2017). [Online]. Retrieved on February 19, 2021 from https://doi.org/10.1016/j.proeng.2017.01.093
15. Riekstina, I., Zvirgzdina, R. Small business in Latvia – trends and facts in five years. Economic Science for Rural Development. (2019) https://llufb.llu.lv/conference/economic_science_rural/2019/Latvia_ESRD_50_2019-169-176.pdf
16. Vasilevska, D., and Rivza, B. Implementation of quality management in small and medium enterprises: Problems and solutions. Engineering for Rural Development. (2018)[Online]. Retrieved on February 21, 2021 from http://www.tf.llu.lv/conference/proceedings2018/Papers/N116.pdf
17. Birgit Hagena, Antonella Zucchella, Paola Cerchiellob, Nicolò De Giovannic (June 2012). International strategy and performance - Clustering strategic types of SMEs. *International Business Review*, **21**, June 2012, Pages 369-382. [Online]. Retrieved on February 6, 2021 https://doi.org/10.1016/j.ibusrev.2011.04.002
18. International Organization for Standardization, ISO. The ISO Survey. [Online]. Retrieved on February 21, 2021 from https://www.iso.org/the-iso-survey.html

Dimensions of the creative industry in individual regions of the Slovak Republic

Slavka Šagátová[1,*]

[1]FPM EU in Bratislava, Department of Business Economy, Dolnozemská cesta1, 852 35 Bratislava, Slovakia

Abstract. The creative industries are an area of the economy that uses human creativity for their production and thus becomes an important sector in promoting the ideas of sustainable development. However, differences of opinion in the definition of the term keat industry complicate the study of its potential. The present article therefore focuses on identifying the basic areas of the creative industries and revealing their links to other elements of the economy. It also examines the representation of the creative industry in individual regions of the Slovak Republic in terms of their number and achieved revenues. The survey confirms the presence of the creative industry in all regions of the Slovak Republic. Its greatest concentration is in large cities and the most represented industry is the advertising industry.

1 Definition of the creative industry

The creative industry is a sector with growth potential, not only in the sector, but its activities can be a determinant of the development of other sectors of the economy of individual countries. The benefit of the creative industry is also its cross-sectoral dimension with minimal territorial constraints. [1] The volume of the creative industry in the world is constantly growing and employs more and more employees. The development of the creative industry can also make a significant contribution to globalization, as it creates a larger volume of tradable services applicable in the broader foreign market. [2] The importance of this issue is underlined by increased publishing activity in this area. Sinhof identified in his survey the keyword "creative industry" as one of the two most common in the sample of 131 scholarly articles examined. [3]

As a sector based on human creativity, the creative industry is closely linked to the cultural industry. In some publications, the combination and overlapping of these terms occur. An even more complicated fact is the phenomenon of understanding the practice of creative companies, which represent a heterodox sector with high levels of variation between companies. [4, 5] KEA characterizes the creative industry as an area where culture becomes a creative input in producing non-creative goods. [6] These are activities that are not necessary of an industrial nature. However, prototypes can also be products. From a legal point of view, the outputs of the creative industry may be based on copyright, while

* Corresponding author: slavka.sagatova@euba.sk

they may include other inputs in the field of intellectual property rights (e.g., trademarks) too. The essence of the production of this industry is the use of various forms of creativity, creative skills, and creative professionals (for example, in the field of art). [6, 7] The specific types of enterprises included in the creative industry sector may differ slightly depending on the authors (experts or institutions) or the establishment time (given the current state of knowledge).

In the conditions of the Slovak Republic, the issue of the creative industry is covered by the Ministry of Culture, which perceives this branch as a part of the Cultural and Creative Industry. In order to define its content, in 2019 it adopted the ESSnet-Culture methodology, according to which the CCP was divided into three areas in accordance with the definitions of the KEA: Cultural core, Cultural industry, Creative industry. Three basic types of companies were included in the creative industry.

Table 1. Areas of creative industry according to the Ministry of Culture of the Slovak Republic

Sector according to KEA	Industry name according to ESSnet	Proposal of the Slovak name of the area	Code NACE	NACE nomenclature
Creative industry	Visual arts	Design	7410	Specialized design activities
Creative industry	Architecture	Architecture	7111	Architectural activities
Creative industry	Advertising	Advertising	7311	Advertising agencies

Source: Podrobná definícia odvetví kultúrneho a kreatívneho priemyslu [8]

The content of particular areas is based on the sector specification in NACE. However, specialized design activities include fashion design of personal or household goods, industrial design, activities of graphic designers and interior decorators. Architectural activities include architectural consultancy in the field of constructions as well as agglomeration and landscape. Finally, advertising agencies are perceived in the context of a whole range of advertising services, from consulting through the creation, production, and placement of advertising materials, space management up to the implementation of entire advertising campaigns.

2 Key areas of the creative industry

Defining individual areas of the creative industry is an essential tool for analyzing and evaluating their development and finding opportunities for their effective development. The product of subjects in the field of design is the design of an object intended for mass reproduction or consumption (cars, furniture, clothing, etc.). The design represents added value not only from an aesthetic point of view but also has an impact on the entire production process. Production in the field of design can be divided into three primary areas: industrial design, interior and architectural design, and graphic design. The industrial design focuses on specific production to design a completely new product or innovate the original product. Innovations are focused on improving appearance, performance, increasing user comfort, etc. They can also be a link between smart technologies and innovations, the development of which is linked to industry 4.0. [9] Design proposals may relate to consumer products, packaging, industrial goods, fashion brands. In the case of interior and architectural design, it is a production to improve the performance of interiors or architectural elements, whether in terms of their functionality or aesthetics. As the design industry thus identified may overlap within the creative industry with the field of architecture, only the activity of interior decorators is included in the design as an area of the creative industry. Graphic design focuses primarily on the production of designs such as the company or product logos, the appearance of corporate documents, promotional materials, and manuals, multimedia presentations, and websites. The impact of design on the production capabilities of businesses makes it an essential factor in competitiveness. [10] The use of the services of the design industry could help to balance the competitive forces between technology leaders and leaders (leaders and followers), the distance (increases) of which increases with the competition. [11]

Architecture creates an environment that provides space for human activities but also shapes and influences people. [12] The field of architecture consists of both independent architects and architectural firms. They provide architectural and engineering services and related technical advice. Their product can be an architectural work, delivery, and coordination of product documentation, or the processor of its parts, cooperation in the proper implementation of the project, coordination of individual actors involved in the project from its completion to implementation. The results of the architect's work in the form of building designs, public spaces, urban complexes, restoration, and works of landscape architecture have a significant impact on the health, lives, and property of citizens and the natural and cultural environment. [13] The final consumers of their activities can be private persons, companies or public institutions.

The product of advertising is an advertising work. The process of its production itself includes two aspects of production and process. Production is the embodiment of ideas, which on the one hand, are formed outside the production environment; for example, in society, in demand as an expression of the development of society, or on the other hand, represent an artificial influence on behavior. The whole process is relatively long and involves a large number of active participants and professions. According to the above facts, the subjects of advertising as a branch of the creative industry are advertising agencies and similar companies within, e.g., communication and brand building. [14] These companies sell services to other companies and organizations. At the same time, they buy services and products from other industries, including the cultural and creative industries. Consumption of production factors in the creative industry can be identified in the production process as well as distribution. In its production activities, the advertising industry consumes creative products from industries such as film, fine arts, music, etc. The distribution process involves the consumption of media space, for example, in TV, radio, periodicals. According to Stephan Loerke, CEO of WFA, advertising promotes competition and business innovation, provides information for consumers' purchasing decisions, and contributes to the financing of media services by purchasing media space. At present, online advertising plays an important role, which represents a substantial part of all advertising expenses. [15] In this context, however, it is necessary to draw attention to new paradigms in advertising. Some of the creative tasks traditionally provided by advertising agencies are now being taken over by media platforms. [16]

Creative industry business can be carried out by independent individuals (especially in the field of design and architecture) or companies providing design, architectural, and advertising services. They can provide their production to final consumers but also companies, organizations, or public institutions. Due to the creative side of this production, it has an impact not only on the direct satisfaction of a certain customer need but also influences the general public with its aesthetic value. The connection of the creative industry to the urban and regional economy is also identified through creative clusters. [17] It is also important to mention that the final impact expected from a creative product should not be achieved by the contracting authority but by its customers. For example, in graphic design, a graphic designer becomes a facilitator of the end-user's creative input. [18] A similar link can be observed in the advertising sector. Thus, businesses in the creative industries can have a direct and indirect impact on trends in demand and consumption. The link between these sectors and the sustainability trend is also important. For example, ecodesign can have a positive impact on all stages of a product's life cycle. [19] The impact of the creative industry is also significant in the field of innovations and increasing of the economic value of products. [20] Creativity, together with an adequate organizational culture, can even be considered a necessary prerequisite for the creation of innovations. [21]

3 Methodology and methodology

This article aims to map the state of the creative industry in individual regions of the Slovak Republic. The companies registered in the Finstat database [22] were subject to review. In terms of the above theoretical background, we have filtered from the database active companies classified according to NACE codes between: Advertising agencies, Specialized design activities, and Architectural activities. To determine the number of companies in individual regions, we used a database of 8326 active companies identified in the above branches of the creative industry. In addition to finding out the structure of creative industries in individual regions, the revenues of these companies were also examined. As the relevant information was available mainly for 2019, companies with information for a different period were excluded from the sample examined in the first step. In the second step, we also excluded from the examined sample companies that had all recorded data for 2019 zero. Thus, a total of 4959 companies were surveyed. We also want to draw attention to certain disadvantages of the database used, which does not provide data on the subjects of the creative industry, which carry out their activities on the basis of trade licenses or as freelancers.

4 The creative industry in individual regions of the Slovak Republic

The most represented branch of the creative industry in Slovakia is Reklama, in which 2,5 times more companies operate than the Architecture and Design sector together. In terms of regional distribution, more than 50 % of companies in the creative industries are located in the Bratislava region. On the contrary, the smallest creative companies are located in the Trenčín Region.

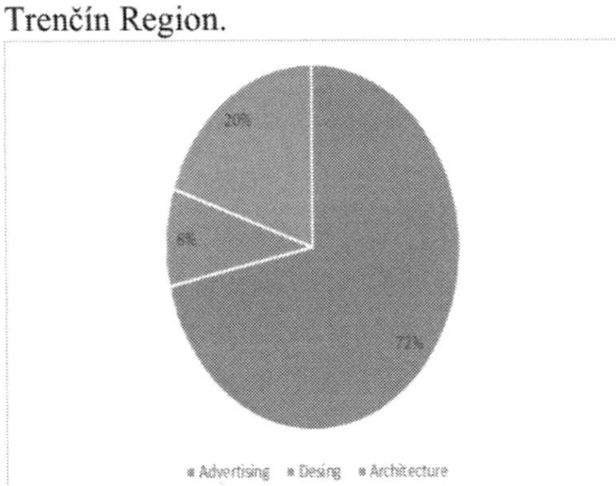

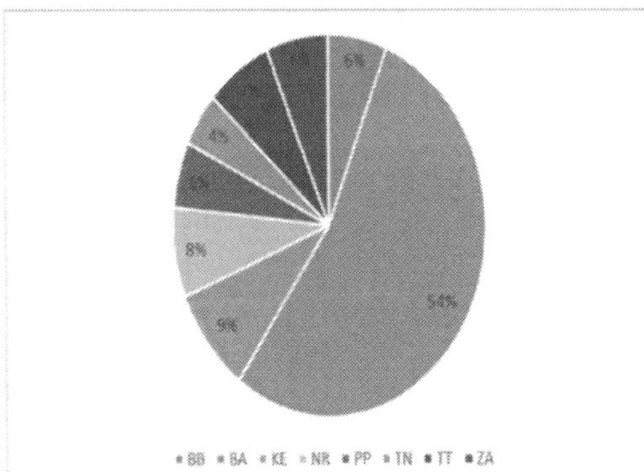

Fig. 1. Representation of the creative industry regions of the SR

Fig. 2. Representation of KP in individual in the SR

Source: own processing

4.1 Advertising agencies

Advertising agencies make up the largest share of the creative industry in each region. The number of companies copies the overall representation of individual regions in the creative industry. Most companies are located in Bratislava and the least in the Trenčín region.

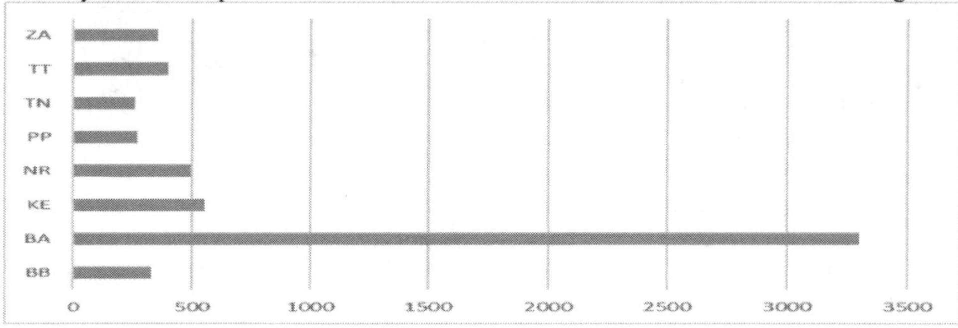

Fig. 3. Representation of Advertising Agencies in individual regions in the SR
Source: own processing

The average volume of revenues of one company in the field of advertising is 408 219 €. From the analyzed sample, 42 % of companies achieved revenuesin the range of 10 001 € to 100 000 € and others just over 20 % of revenuesin the range of 100 001 € to 1 000 000 €. At the same time, 17 % of companies did not report any sales, which indicates that they did not actively carry out their business activities. It is also interesting to note that 6% of companies had revenuesof more than 1 mil. €.

Table 2. Extent of revenues in individual regions of the SR in 2019

Revenues	BB	BA	KE	NR	PP	TN	TT	ZA
to 0 €	17%	17%	18%	18%	14%	19%	13%	14%
from 1 € to10 000 €	19%	13%	16%	9%	26%	17%	16%	16%
from 10 001 € to100 000 €	46%	39%	44%	50%	38%	44%	45%	51%
from 100 001 € to 1000 000 €	16%	22%	19%	22%	20%	19%	25%	17%
from 1000 001 € to 10 000 000 €	2%	8%	3%	1%	3%	2%	1%	2%
above 10 000 000 €	0%	1%	0%	0%	0%	0%	1%	0%
Average revenues per company in €	127 075	603 161	127 211	91 737	119 323	97 776	167 362	129 527
Median revenues in €	23 875	36 516	23 044	29 871	21 600	19 798	36 871	32 876

Source: own processing

The highest revenues were achieved in the Bratislava Region, where the average value of revenues per company exceeded 3,5 times the value of average revenues of other regions. However, the highest average value of revenueswas achieved in the Trnava region, where companies achieve revenues mainly in the range from 10 001 € to 1 mi. €. In third place, we could include companies in the Žilina region, followed by the Košice and Banská Bystrica regions. Finally, the company achieved the lowest revenues in the Trenčín Region, where most companies did not achieve any revenues.

4.2 Architectural activities

Enterprises carrying out architectural activities are the second largest group of enterprises in the field of creative industries. They have the most significant representation in the Bratislava region, where almost 50 % of them operate. The least of these companies are established in the Trenčín and Nitra regions.

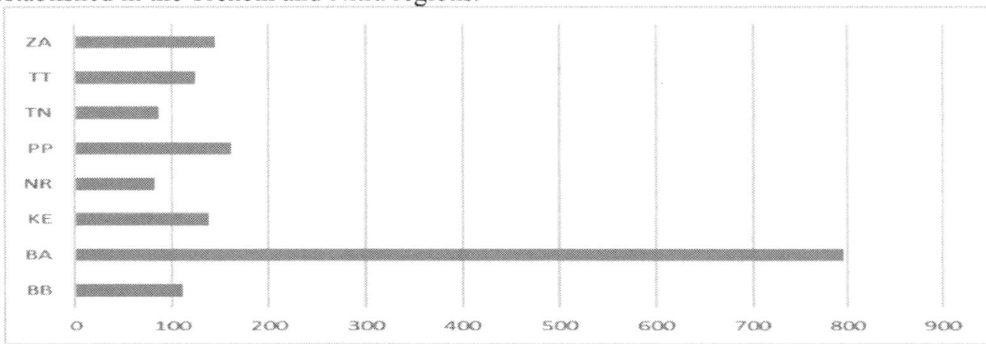

Fig. 4. Representation of companies in the field of Architectural Activities in individual regions in the SR
Source: own processing

The company's average revenues in the field of architecture were 128 867 €, which represents 68 % lower revenues than in advertising. 51 % of companies achieved revenues in the range of 10 001 € to 100 000 €. In the range of 100 001 € to 1 000 000 €, 20 % of companies reported revenues. 13 % of companies did not report any revenues and the same number reached revenues in the range of up to 10 000 €. Revenues over 1 000 000 € reached only 2 % of companies.

Table 3. Extent of revenues in Architectural activities of individual regions of the SR in 2019

Revenues	BB	BA	KE	NR	PP	TN	TT	ZA
to 0 €	16%	15%	13%	6%	13%	8%	8%	8%
from 1 € to 10 000 €	19%	11%	18%	13%	11%	16%	22%	17%
from 10 001 € to 100 000 €	33%	49%	58%	57%	52%	53%	42%	63%
from 100 001 € to 1 000 000 €	33%	21%	10%	22%	22%	21%	28%	10%
from 1 000 001 € to 10 000 000 €	0%	3%	0%	1%	2%	2%	0%	2%
Average revenues per company in €	68 034	154 677	49 437	83 515	111 270	122 643	62 066	99 657
Median revenues in €	38 761	35 479	26 260	38 675	35 122	38 465	31 919	31 653

Source: own processing

The highest revenues were reported in the Bratislava Region, while the highest number of companies achieving revenues in the range of 10 001 € to 100 000 €, which copies the national parameters. To the second place, we could rank companies in the Trenčín region and then in the Poprad region, while in both these regions, the average revenues exceeded the value of 100 001 €. The smallest revenues in this sector were achieved in the Košice Region, which was caused mainly by the fact that only 10 % of companies achieved revenues above 100 000 €.

4.3 Specialized design activities

As we have already mentioned above, the area of design has the smallest representation in Slovakia within the creative industries. Most companies focused on specialized design activities are traditionally located in Bratislava, while their number exceeds the sum of design companies in other regions. The second place in the number of companies was achieved by the Banská Bystrica region, in which there are five times fewer companies focused on design than in the Bratislava region. The fewest companies can be observed in the Trenčín.

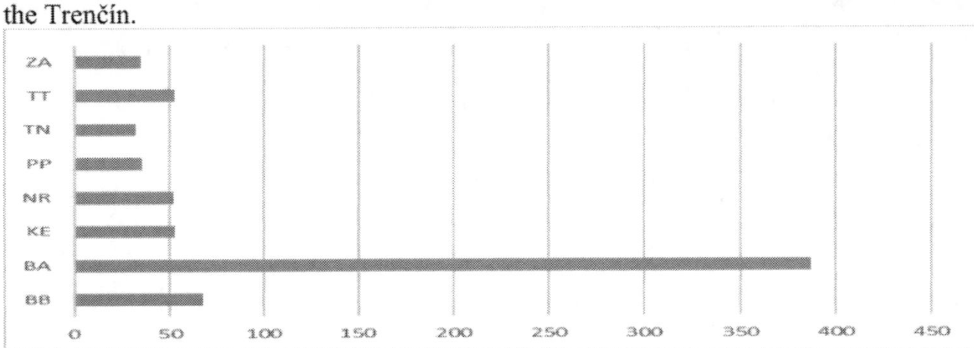

Fig. 5. Representation of companies in the field of specialized design activities in the Slovak Republic
Source: own processing

Average revenues in design were even lower than in architecture. At the same time, most companies achieved revenues in the range of 10 001 € to 100 000 €, up to a total of 47 % of companies, and this trend was visible in all regions. Revenues over € 1 million were achieved by only 1 % of companies and over € 10 million. no business. The average revenues were at the level of 84 040 € per company.

Table 4. Extent of revenues in Design activities of individual regions of the Slovak Republic in 2019

Revenues	BB	BA	KE	NR	PP	TN	TT	ZA
to 0€	17%	18%	20%	16%	13%	13%	18%	11%
from 1 € to 10 000 €	19%	13%	25%	25%	25%	23%	24%	31%
from 10 001 € to 100 000 €	51%	50%	45%	43%	47%	52%	40%	50%
from 100 001 € to 1 000 000 €	10%	18%	10%	15%	13%	13%	18%	8%
from 1 000 001 € to 10 000 000 €	3%	2%	0%	1%	3%	0%	0%	0%
Average revenues per company in €	65 097	105 468	32 650	69 908	153 222	77 467	48 316	39 447
Median revenues in €	22 162	21 900	14 375	18 455	19 245	36 768	14 240	16 368

Source: own processing

In terms of regions, the largest average revenues were recorded in the Bratislava region. To its primacy contributed the fact that 20 % of companies achieved revenues of more than 100 000 € of which 3 % of companies achieved revenues in the range of 1 million € to 3,5 million €. The second region in terms of average revenues is the Prešov region. However, its location was mainly due to the fact that there are very few design companies operating in this region, but one achieved revenues of over 2 million €. The third highest average revenues were achieved in the Trenčín Region. This result was influenced by the fact, that the highest number of company in this region (52 %) reported revenues in the range of 10 001 € to 100 000 €.

5 Conclusion

The region with the strongest branch of the creative industry is the Bratislava Region, where there are the highest number of companies involved in the creative industry, and which also at the same time achieve the highest revenues. Within the creative industry in this region, the first place belongs to the Advertising Agencies, which accounted for up to 74 % of creative companies in this region. The fact that several companies in this area achieved revenues of more than 20 mil. € in 2019 also contributed to this matter of fact. In this region, 18 % of creative companies and 9 % of design were dedicated to architecture. The trend in the structure of representation of individual branches of the creative industry in the regions, which can be observed in the Bratislava Region, is also confirmed in all other regions. Their order according to the number of identified creative enterprises is: Košice, Nitra, Trnava, Žilina, Banská Bystrica, Prešov and Trenčín regions. In terms of the share of individual industries in the creative industry within the region, the highest share of the advertising industry can be observed in the Nitra region, the architecture industry in the Prešov region and the design industry in the Banská Bystrica region. Our research confirms the idea of concentrating creative potential in large cities. On the other hand, it can be observed that the creative industry is also alive in regions outside Bratislava.

This paper is an output of the science project VEGA č. 1/0340/19, *„Podnikateľský rozmer subjektov kreatívneho priemyslu v kontexte inovácií a inteligentného rastu“*, project share is 100%;

References

1. D. Rybárová, Creative industry as a key creative component of the Slovak economy. *Globalization and its Socio-Economic Consequences 2019 : Sustainability in the Global-Knowledge Economy.* - Zilina : University of Zilina, Faculty of Operation and Economics Transport and Communications, 2020. ISSN 2261-2424, pp. 1-8 online.
2. H. J. R. Z. Lawrence, E. E. Leamer, M. J. Slaughter. Globalization and U.S. Wages: Modifying Classic Theory to Explain Recent Facts. *Journal of Economic Perspectives*, **26**, 119-140 (2019).
3. P. S. Santana, F. F. Silve, Entrepreneurship in the creative industries: A bibliometric study, *Brazilian Journal of Management,* **12,** 125-141 (2019)
4. A. Kovalev, J. Kintler, M. Uhliar. Prognóza reliability legislatívnej kategorizácie podnikov kreatívneho priemyslu v komparácii na dizajn matematických predikčných modelov. *Ekonomika a manažment : vedecký časopis Fakulty podnikového manažmentu Ekonomickej univerzity v Bratislave.* - Bratislava : Fakulta podnikového manažmentu Ekonomickej univerzity v Bratislave, **16**, 9 - 26 (2019)
5. J. Salder, The creative business: enterprise development in the arts-based creative industries. *Creative Industries Journal*, **14**, 63 – 80, (2021)
6. KEA, *The economy of culture in Europe* [online]. Brusel: KEA, Avaliable at: https://ec.europa.eu/assets/eac/culture/library/studies/cultural-economy_en.pdf (2006)
7. G. Soda, D. Stea, T. Pedersen, Network Structure, Collaborative Context, and Individual Creativity. *Journal of Management* **45,** 1739-1765 (2019)
8. Podrobná definícia odvetví kultúrneho a kreatívneho priemyslu. Avaliable at: https://www.culture.gov.sk/wp-content/uploads/2019/12/Priloha_9_Definicia_ KKP1.pdf (2019)
9. Z. Závadská, J. Závadský, *Industry 4.0 and intelligent technologies in the production management innovation.* Banská Bystrica : Belianum, 158 p. (2020)

10. D. Koshy Importance of Design as a Factor of Competitiveness Address at the WIPO International Symposium on Design Santiago, Chile, Avaliable at: https://www.wipo.int/edocs/mdocs/hague/en/ompi_di_san_11/ompi_di_san_11_2.pdf (2011)
11. P. Aghion, N. Bloom, R. Blundell, R. Griffith, P. Howitt, Competition and innovation: an Inverted-U Relationship. *Quarterly journal of Economics,* **100** (2005)
12. E. Žáková a kol. Kulturní a kreativní průmysly v České republice. Avaliable at: https://www.idu.cz/cs/publikace/726-kulturni-a-kreativni-prumysly-v-ceske-republice (2011)
13. Slovenská komora architektov, Štandardy služieb architekta a postupy stanovenia honoráru. Avaliable at: file:///C:/Users/EU/Downloads/tandardy_sluieb_architekta_a_postupy_stanovenia_honorru.pdf (2021)
14. The Danish Government, Denmark at work plan for growth in the creative industries · design. Avaliable at: https://eng.em.dk/media/10605/07-02-13-summary-vaekstplan-kreative-erhverv-english.pd (2013)
15. Y. L – Thompkins, A Decade of Online Advertising Research: What We Learned and What We Need to Know. *Journal of Advertising*, **48**, 1-13 (2019)
16. I. Derda, At the intersection of interests: evolving creative processes and new agency-media dynamics. *Creative Industries Journal*, **14**, 3-22 (2021)
17. S. Yum, The relationship between creative industries and the urban economy in the USA. *Creative Industries Journal*, **13**, 95-116 (2020)
18. S. Taffe,Tensions Facilitating Codesign in Graphic Design: Working with Asthma Educators in Australia. *International Journal of Desing*, **15** (1) (2021)
19. T. Bhamra, R. J. Hernandez, Thirty years of design for sustainability: an evolution of research, policy and practice. *Design Science*, **7**, 1 – 17, (2021)
20. H. Majdúchová, M. Kmety Barteková, Innovations in the Creative Industry Entities. - International Scientific Conference. *Globalization and its Socio-Economic Consequences 2019 : Sustainability in the Global-Knowledge Economy.* - Zilina : University of Zilina, Faculty of Operation and Economics Transport and Communications, 2020. ISSN 2261-2424, pp. 1-8 online.
21. P. Aghion, U. Akcigit, A. Bergeaud, R. Blundell, D. Hemous, Innovation and Top Income Inequality. *The Review of Economic Studies,* **86**, 1– 45 (2019)
22. Finstat, Online platforma na prácu s dátami o slovenských firmách. Avaliable at: https://www.finstat.sk/ (2021)

The relevance of educational fairs in the 21st century – decision matrix for exhibitors and visitors to participate

Susanne Schulze[1] and *Peter* Markovič[1,*]

[1]Department of Corporate Finance, Faculty of Business Management, University of Economics in Bratislava, Dolnozemská cesta 1/b, 852 35 Bratislava, Slovakia

Abstract. The aim of this article is to consider why exhibitors and visitors participate in educational fairs and whether these kinds of trade fairs are important for the professional and study orientation of young professionals. The main content of the article is to demonstrate the relevance of educational fairs and their opportunities and to show a basis for decision-making for exhibitors on how to determine whether a cost-intensive participation in an educational fair is worthwhile.
In principle, the analysis of current literature has shown that education fairs take place in large numbers, especially in Germany, and there is a desire for such presence and direct contact with target groups, especially under the current limitations of the Corona pandemic.
Based on research using expert interviews and an online survey, results were generated as to what information provides, why exhibitors and visitors use educational fairs. In conjunction with the literature research, criteria of exhibitors and visitors were compared for participation in the educational fair and then the decision criteria were merged into a pestel analysis. Finally, necessary decisions were summarised on educational fair planning, exhibition execution and educational fair analysis.
Educational fairs give exhibitors the possibility of direct contact with target groups and feedback. For visitors, educational fairs provide an overview of offers, possibly new impetus for educational offers or by practical testing the consolidation of the career wish.
It is to be expected that educational fairs will remain an important marketing tool in the future, possibly establishing hybrid formats.

1 Introduction

As part of professional and study orientation, students of the graduates are looking for a professional perspective after school. This can be study, training, year abroad or voluntary service. Education fairs are an important guide in the jungle of education. Visitors can get an overview of the variety of offers and, if necessary, get to know unknown possibilities.

* Corresponding author: peter.markovic@euba.sk

Educational fairs usually only show part of the overall offer, but give orientation and can awaken interests. The educational fairs are among the trade fairs with a regional catchment area. This includes an investigation by the ifo Institute published by AUMA (Exhibition and Exhibition Committee of the German Business Association) consisting of six million visitors and 52,000 stand exhibitors in Germany in 2018. The effects on the site are also highlighted in particular, as exhibitors and visitors make expenses on the spot, among other things. [1-3]

The research question in this article is as follows: Why do exhibitors and visitors use educational fairs and what criteria should a decision be taken to participate in the trade fair?

2 State of the problem in literature

Major events in the field of education are mainly fairs and, if necessary, congresses, symposia or specialist congresses. According to Gabler Wirtschaftslexikon, a trade fair is an event or event with a market character. The fair can be divided/classified as follows (trade fair typology): time (once a year or several times), spatial classification (regional, national, international as well as location), mobility (stationary, mobile), organisation by trade fair organisers, both in the pre- and post-processing and implementation influence on marketing.

The European Institute for ConferencesWirtschaft GmbH (EITW) at the Harz University of Applied Sciences in Germany annually examines the event market in Germany with the Meeting and Event Barometer, including the trade fairs. The types of events are distinguished according to professionally motivated events, events, international events or sports and cultural events, with only a few mentioned here, as the concentration is related to. Trade fairs and major events in this article are based on professionally motivated events and events. In the context of the above-mentioned study, 4,000 venues (more than 100 seats) were included in Germany for the year 2017 and 8,000 organisers were interviewed worldwide, with a return rate of 7.1 %. [4] A total of 405 million participants participated in the events (up 2.8 % year-on-year), of which 2.97 million attended professionally motivated events and events. [4] It is interesting that the events of conference hotels are shifting more and more towards event centers (such as exhibition halls) or event locations. The total number of participants grows by 2.8 percent. The proportion of professionally motivated events is very high at 66.9 percent. Among the professionally motivated events, the medium-sized events in particular show increases in participants (51-100 participants), a decline in large events up to 5,000 participants. [4]

As education fairs in Germany, fairs dealing with the topics of education, education and occupation are summarised. [5, 6] Over 150 educational fairs are recorded for 2019 (without claim to completeness and missing data from two Länder). The regional difference between the federal states or city states is striking. North Rhine-Westphalia and Brandenburg have about the same territory. In North Rhine-Westphalia, however, there are 38 trade fairs in the field of education and only six in Brandenburg. [6] With more than 150 study and training fairs, this type of information on educational offers is of high importance in Germany. In comparison, the search for the same trade fairs only includes an offer from a trade fair organiser in the Slovak Republic: Job Forum at the Expo Center Trenčín (October 24-25, 2019). [7] It can be assumed that further decentralised trade fairs will take place at schools or universities. [8, 9] The Slovak Republic is certainly the smaller country with about 5.4 million inhabitants compared to Germany with 83.5 million inhabitants. On the other hand, there is a clear focus on the marketing instrument Education Fair in Germany and the use as a business model.

3 Research design

The analysis of current literature has shown that in Germany there are numerous offers in the field of educational fairs, which are also attended in large numbers. Educational fairs are organised either by private trade fair organisers or by municipalities/schools etc. But why is this marketing instrument perceived by exhibitors and visitors?

Therefore, primary data are collected inresearch activities and associated with secondary data from existing studies or literature. The focus is on the primary survey of qualitative (Research Phase I: Expert interviews) and quantitative data (Research Phase II: Questioning). The main players involved in education fairs will be involved: Exhibitors or visitors. The focus in the collection of primary data lies with the trade fair organisers and exhibitors.

The primary data were evaluated and then connected to the secondary data. The aim is to answer the research question (see introduction) in order to gain insights into how the marketing instrument education fairs will be used in the future by exhibitors and visitors or how a decision can be made for or against an educational fair.

4 Results of research

Five expert interviews with key players for German education fairs were conducted to obtain the following results: Trade fair organisers/economic promotions, municipal or institutional representatives of the location marketing, such as the Chamber of Crafts or Chamber of Commerce and Industry as well as other actors dealing with educational fairs or the transition of young people to study or training. The interviews for knife organisers and institutional representatives are in the parts: General data, preparation, exhibition execution and follow-up of trade fairs. The Messer Preparation section focuses on the selection of the location as well as the promotion of the trade fair as well as the financing. The exhibition implementation section focuses on the effects of educational fairs for exhibitors and visitors, as well as the further development of this marketing instrument and the effects on the location. The trade fair follow-up is the fourth part of the interview with the focus on evaluation and cost-effectiveness analyses as well as the wishes regarding the linking of the trade fair location, city/region and the education fair. [10-12] The expert questionnaire for other stakeholders is structured in a similar way, but is not broken down into the above sections. Relevant questions were taken from each section on the organisation of educational fairs, future development and the interaction between the educational fair and the venue.

According to the respondents (classification scale 1 most important, 10 least important), visitors use trade fairs mainly because there are direct and personal discussions on site and many suppliers are in one place. Apart from that, networking (making contacts) and experiencing/trying out professions followed by checking out the study and career aspirations are particularly emphasised.

It is interesting to note that companies feel that trying things out at the stand is not very important, while trade fair organisers feel that making contacts among the exhibitors is particularly important.

Fig. 1. Why do visitors use educational fairs? [Own presentation]

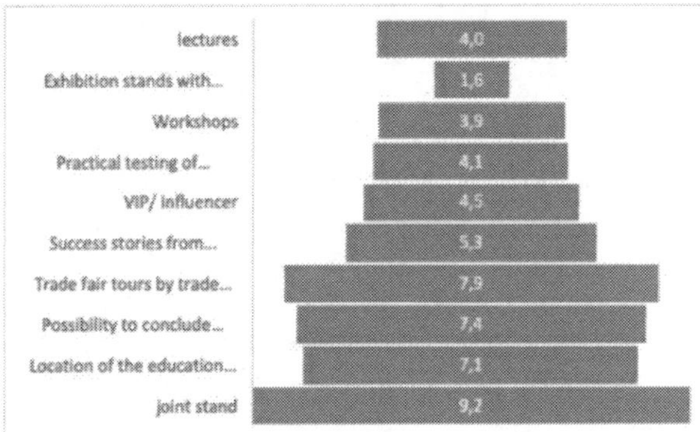

Fig. 2. What does trade fair visitors speak to? [Own presentation]

Respondents should assess (most importantly) on a scale of 1 (most important) to 10 (most unimportant) what particularly appeals to trade fair visitors. The trade fair stand was clearly underlined with personal advice, workshops, lectures and practical testing of professions and courses of study.

Education fairs are therefore an important tool for exhibitors to attract young talent and an important place for network maintenance. In particular, however, the contribution to the more targeted career or study choice is confirmed.

Exhibitors and visitors use the educational fairs for different and identical motives. In the following overview, the results of the expert interviews are highlighted in order to find support in the decision for or against an educational fair.

Table 1. Expectations Exhibitors and visitors at educational fairs.

Expectation	Exhibitor	Visitors
Particularly important	**Direct and personal target group approach Getting** to know the target group/ "feeling"	many offers and exhibitors in one place/overview provide active **/personal conversation direct information**
Important	increase recognition company Network Making Advertising for the company feedback by Visitor Regional Orientation/Connection	parents advise participation Prescription/vocation or study request check career/studies choice obligatory event school
Less important	competititors are also there/show presence training must be made/ secure knowledge concrete job interviews arrange company proximity to the trade fair location	Making contacts experience and trying out professions

[Own presentation]

The most important match of the expectations of exhibitors and visitors is to obtain direct/personal contact as well as direct information. The first motive in the decision examination of the use of an educational fair as a marketing tool for exhibitors is the possibility of direct contact with potential junior staff as well as direct contact with target groups.

With the help of the PESTEL analysis, decision criteria for participation as an exhibitor at an educational fair resulting from the expert interviews and the online survey can be summarized.

Fig. 3. Pestel analysis Decision criteria Participation Exhibitors [Own presentation]

A number of factors must or should be included as an exhibitor before taking part in an educational fair. Gartner's CMO Spend Survey 2018-2019 surveyed over 600 marketing managers in the UK and North America for their assessment of the development of marketing activities, spending, or innovation. About 5 percent of the marketing budget is invested in events (including trade fairs). [13] Marketing managers are asked to constantly review their used instruments and to decide on their use. [13] It is described as essential here the examination of known instruments ("we have always done this") and the linking of indicators such as turnover and return in investment (ROI) as far as this is possible and measurable in marketing. [13]

Also educational fairs as part of the toolbox marketing should be planned, prepared and evaluated. The following factors shall be taken into account.

1. Objectives of participation in an educational fair
   - What do I want to achieve with my trade fair presence?
   - Which target groups do I address?
   - What alternative marketing tools can I use?
2. Market observation
   - Data and facts of comparable and previous educational fairs
   - Target Groups and Concept of the Education Fair – Does my target group use the trade fair?
   - Current research results on trade fairs in the field of education/trends
   - Which exhibitors will be on site? (There may be a "compulsion of participation")
3. Budget planning
   - Existing budget/share of total budget
   - Stand fees and related costs (Internet connection, electricity, furniture, entrance fees, parking tickets, supply, garbage/cleaning, etc.)
   - Exhibition equipment (stand construction, exhibition stand, equipment with company materials)
   - Trends such as augmented reality, practical offers at the exhibition stand
   - Trade fair personnel
   - Logistics: Delivery to the fair and associated costs
   - Costs of own application of participation in the trade fair
4. Examination of contractual conditions and trade fair booking
5. Implementation of the fair
   - Assembling materials, planning assembling and dismantling
   - Survey Number of trade fair conversations and possibly conversational data (if possible under data protection law), duration of talks, type of conversation, abnormalities
   - Feedback on the trade fair from stand personnel or survey participants
   - Testing of other measurement methods: GPS or Wifi tracking (visitor streams)
6. Evaluation (evaluation of the trade fair afterwards)
   - Have I achieved my goals?
   - How did the trade fair go compared to previous years? (Number and quality conversations, development costs etc.)
   - Which trade fair visitors subsequently applied? Has the number of applications increased? Have I enrolled more students than last year?

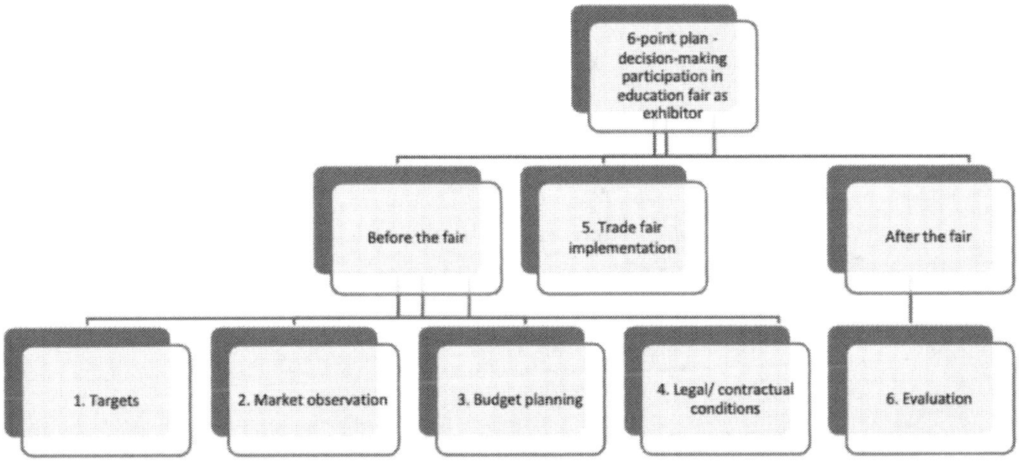

Fig. 4. 6-point plan - decision-making participation education fair exhibitors [Own presentation]

The above-mentioned 6-point plan illustrates that some points must be considered before taking part in an education fair and that educational fairs can also be evaluated. However, the evaluation is mainly qualitatively and less quantitatively possible.

Finally, participation in educational fairs as an exhibitor is merely an instrument of the marketing toolbox and often a detailed evaluation or quantitative impact, for example, on applicant numbers can only be traced back to the fair to a limited extent. However, through direct target group approach and direct feedback/impression of the target group, trade fairs are an essential part of marketing.

Philip Harting (Chairman of the AUMA Trade Fair Association) also confirms theimpact of personal contact on presence fairs. Offers can also be presented virtually and the visitor gets an overview, but cannot "test" or touch the product. In particular, the personal persuasive power of the trade fair exhibitor is lost. [14] 35 percent of respondents declared interest or already the use of online trade fairs. This has changed or strengthened due to the Corona pandemic around Covid-19, which started shortly after the survey. The survey was completed in March 2020, for example virtual Vocatium fairs or the study fair ERZ (Annaberg-Buchholz) took place in July 2020. Harting (see above) confirms this: 165 trade fairs could not be held in 2020 (current June 2020), 110 even failed. [14]

The trade fair location should above all have sufficient space for the exhibitors and good connection to the public after transport. In particular, the municipal role was seen as important or expandable, with regard to: Financing (trade fair should also remain business model with profit intention), own stand presentation, job and training/study offers on site, promotion of the trade fair, link between company and trade fair and advantage that visitors spend locally. In addition, according to the respondents, the trade fair contributes to the local acquisition of skilled workers, less to the company's location. The regional catchment area was underlined. The companies themselves consider themselves responsible for the acquisition of skilled workers.

It is undisputed that education fairs play an important role in helping young people's careers and their important personal contact.

5 Discussion and summary

In spite of the intensive development of digital formats, educational fairs are still of high relevance as a presence offer. Talking personally with a consultant at the trade fair or anonymously gaining an overview of the range of services is also an important marketing tool in the 21st century. Visitors and exhibitors also receive direct feedback from their counterpart and exhibitors learn a lot about the target group. In the opinion of the experts surveyed, this will continue to exist in the future. [15, 16] It is expected to develop more qualitatively with additional offers rather than quantitatively.

Further research should strategically assess the long-term impact of the Corona pandemic on education fairs. Will more hybrid trade fair offers persist or will there be a return to the previous situation with almost exclusive presence fairs?

Exhibitors should combine concrete goals with the participation of the trade fair and beforehand also deal with the measurability of these, in order to evaluate the trade fair later. Furthermore, it is necessary to examine how the trade fair has developed in previous years and how competitors use this trade fair.

Acknowledgement

The article is an output of research project VEGA (1/0240/20) "Financial Aspects of Sustainable Business – Enterprise Succession Solution for Small and Medium-sized Enterprises".

References

1. AUMA Exhibition and Exhibition Committee of Deutsche Wirtschaft e.V. (eds.), 2018. AUMA Edition, *The Economic Importance of Trade Fairs and Exhibitions in Germany*. Investigation of the ifo Institute, Munich
2. A. Rubens et al., Using Industry Analysis and Manager Interviews for Career Planning: A Classroom Exercise. *E-Journal of Business Education & Scholarship of Teaching* **1**, 14 (2020)
3. M. Banerjee et al., Does Career Planning Drive Agri-entrepreneurship Intention Among University Students? *Global Business Review* 10.1177/0972150920961266 (2020)
4. European INSTITUT für TagungsWirtschaft GmbH (EITW) an der Hochschule Harz, 2018, Meeting- und Event-Barometer Deutschland 2017/2018, Die Deutschland-Studie des Kongress- und Veranstaltungsmarktes
5. J. Giesinger, Education, Fair Competition, and Concern for the Worst Off. *Educational Theory* **1**, 61 (2011)
6. https://www.expodatabase.de/messen-deutschland/services-handel-beruf/erziehung-bildung-beruf/, retrieved 08.03.2019
7. https://www.messeninfo.de/Ausbildungsmessen-Y25-S1.html, retrieved 10.04.2019
8. Y. Liu, Research of China's Rural Education Fair Problem. *International Conference on Education Reform and Management Innovation* **96** (2017)
9. YM. Liu, JB. Gao, Under the new normal vocational education fair path selection. Proceedings of the 2016 *International Conference on Education, Management and Computing Technology (ICEMCT-16)* **59** (2016)
10. A. Kiani et al., Impact of Future Time Perspective on Entrepreneurial Career Intention for Individual Sustainable Career Development: The Roles of Learning Orientation and Entrepreneurial Passion. *Sustainability* **12**, 3864 (2020)
11. S. Schelfhout et al., From Interest Assessment to Study Orientation: An Empirical Advice Set Engine. *Journal of Experimental Education* **1**, 89 (2021)

12. NT. Hung, Higher education in business: a model for international students' choice. *Business Process Management Journal* 10.1108/BPMJ-11-2020-0516 (2021)
13. E. McIntyre, A. M. Virzi, CMO Spend Survey 2018-2019. http://cmo-spend-survey-free-research-2018-2019.pdf, retrieved 31.03.2021
14. K. Terpitz, Real trade fairs have a different force. *Handelsblatt* **111** (2020)
15. Y. Park, DH Lim, J. Lee, Moderating effect of career planning on job support and motivational process of training transfer. *European Journal of Training and Development*, 10.1108/EJTD-11-2020-0159 (2021)
16. AMM. Salleh et al., The Impact of Career Planning and Career Satisfaction on Employees' Turnover Intention. *Entrepreneurship and Sustainability Issues* **1**, 8 (2020)

Assessment of financial health of Slovak family businesses using models predicting financial distress

Mária Trúchliková[1,*]

[1]University of Economics in Bratislava, Faculty of Business Management, Department of Business Economy, Dolnozemská cesta1, 852 35 Bratislava, Slovak republic

Abstract. The financial health of a company can be seen as the ability to maintain a balance against changing conditions in the environment and companies should pay more attention to the financial viability and risk management. There many models for predicting of financial problems of the companies, especially Altman, Ohlson or Zmijewski are the most cited ones. The main objective of the article is the review and assessment of the level of financial health of Slovak family business in selected industries. The data was obtained from Finstat database and financial statements from 2017, 2018 and 2019 were analysed. For assessing the financial health of selected family businesses 3 models predicting financial distress were used: Kralicek Quick Test, Taffler model and Virág-Hajdu model. The results show how many family businesses are facing to the financial problems using different types of predicting models.

1 Introduction

Financial viability is the one of the key issues for every enterprise, because current economic environment shaped by globalization, quick economic changes and fierce competition impose challenging conditions for businesses and their prosperity. That's the reason, why predicting corporate bankruptcy is a crucial task for modern risk management. Different models and approaches predicting corporate financial distress or viability can be classified as a specific tool of financial analysis. The paper focuses on the analysis of financial health in family businesses using selected prediction models.

2 Comparison of Prediction Models

The financial distress of business entities is closely connected with unpleasant consequences, and these are the main motivation factors for managers or financial analysts to search for the methods that can predict possible financial problems or bankruptcy in advance [1]. A number of models to detect financial distress have been developed in the past. In the literature of bankruptcy prediction, the models of Altman [2], Ohlson [3] and

* Corresponding author: maria.truchlikova@euba.sk

Zmijewski [4] are the most cited ones that are based on accounting variables [5]. The pioneers of the empirical approach are Beaver [6] and Altman [2]. Beaver was one of the first researchers to study the prediction of bankruptcy using financial statement data. Beaver [6,7] found that a number of indicators could discriminate between matched samples of failed and non-failed firms for as long as five years prior to failure. The single dimensional model is different from reality, when the financial situation of company is seen as one unit, which cannot be evaluated only according one rate index [8]. Tamari [9] created so-called risk index model that presents very simple point system, including various rate indexes, which are generally accepted as indexes of the financial health of the company [8]. The breakthrough success was achieved by the world-famous multivariate DA model of Altman [2]. He considered simultaneous impact of several indicators on the financial condition of the company by combining them into a single measure (Z-score) [10]. Altman develop a five-factor model applying the linear discriminant analysis. The model was originally developed for publicly traded manufacturing enterprises and later was extended to unlisted manufacturing companies and to other models including non-manufacturing companies [11]. Ohlson [3] proposed a new model based on logit analysis with a set of nine accounting ratios [12] and was the first who used the method of logistic regression for creating the model to predict the probability of a company failure [7]. In 1978 was developed Springate model, that uses the ratio of multiple discriminant analysis. To determine the ratios anywhere that can detect the possibility of bankruptcy, Springate use multiple discriminant analysis to choose 4 ratios of 19 financial ratios were popular in the literature, which was able to distinguish between sound business insolvent and bankrupt [13]. Zmijewski [4] employed probit analysis and developed a three-variable distress prediction model. He used financial ratio analysis that measures the performance of debt or leverage and liquidity of a company. In his analysis, Zmijewski used probit analysis as applied to 40 companies in a state of bankruptcy and 800 companies that are still operating at the time, and then develop a model by using ROA, leverage, and liquidity ratios [14].

Table 1. Prediction accuracy of models

Model	Analysis techniques	Description
Altman model (Z-score)	Multiple Discriminant Analysis	Working capital/Total assets Retained earnings/Total assets Earnings before interest & taxes/Total assets Market value of equity/Book value of total liabilities Sales/Total assets
Taffler model	Multiple Discriminant Analysis	Earnings before taxes/short-term liabilities Current assets/total liabilities Short-term liabilities/total assets Asset turnover (sales over assets).
Spingate model	Multiple Discriminant Analysis	Working capital/total assets Earnings before interest and taxes/total assets Profit before tax/current liabilities Sales/total assets
Virág-Hajdu	Discriminant Analysis, Logit	Quick liquidity ratio Return on sales Cash flow /Total debts Current assets /Total assets Acc. Receivable/Acc. payable
Ohlson model (O-Score)	Logit	Log (Total assets/GNP price-level index) Total liabilities/Total assets. Working capital/Total assets Current liabilities/Current assets.

<table>
<tr><td></td><td></td><td>One if total liabilities exceed total assets, zero otherwise
Net income/Total assets
Funds provided by operations/Total liabilities
One if net income was negative for the last two years, zero otherwise
NIt−NIt_1/|NIt|+|NIt_1| where NIt and NIt_1 is the net income for the most recent and the preceding year respectively. The variable measures the change in net income.</td></tr>
<tr><td>Zmijewski
(Probit model)</td><td>Probit</td><td>Net income/Total assets
Total liabilities/Total assets
Current assets/Current liabilities</td></tr>
</table>

Source: own processing

There are many other models used to detect financial distress. In Czech Republic the first attempt at developing a national bankruptcy risk assessment model was made by Neumaiers in 1995 (so-called IN95 model) [15]. In the following years, the same authors, having a larger sample of companies, built other national models—IN99, IN01 and IN05 [9]. In Hungary Virág and Hajdu [16] created an early warning bankruptcy model family in 1996 indicating bankruptcy dangers for different sectors and branches of the economy using DA, based on the financial data of 10,000 economic units [17]. The most commonly model used in Germany, Switzerland and Austria is Kralicek Quick test. Peter Kralicek developed a model on a sample of companies from German-speaking countries. The Kralicek Quick Test includes both dynamic and static indicators [18]. In Slovakia, there are also several representatives focusing on prediction models. Chrastinová (CH-index) and Gurčík (G-index) were the first who applied the methodology of financial health prediction in agricultural companies [10,19,20].

2.1 Family firms and financial distress

Family businesses are described as firms that are managed or controlled by founding families. The topic that often discussed in the literature is whether a family business has a better performance than a non-family business, so that they do not experience financial distress. Several studies addressing the issues of the determinants of a firm's financial distress, or family firms' performance refer to Agency Theory, which asserts that the separation of ownership and control results in conflicts of interest and asymmetric information between owners and managers [21]. The other problem is that family businesses intend to keep the control of the firm within their family. Some studies on family firms' capital structure point out that family firms are risk averse, pursue stability and avoid risky investment strategies. As a matter of fact, families are "risk willing", in terms of performance hazard, in order to maintain a firm's control and preserve the non-financial returns they derive from the business, but, at the same time, they are averse to entrepreneurial risk [21,22]. Characteristics of family owned companies could be a possible reason for family business risk aversion and the choice of capital structure [23].

3 Methodology and data

The data was obtained from Finstat database. This database contains financial data about all subjects with a registered ID in the Slovak Republic. We analysed financial situation of family businesses in manufacturing (SK NACE 20-Manufacture of chemicals and chemical products, 21-Manufacture of basic pharmaceutical products and pharmaceutical preparations, 22- Manufacture of rubber and plastic products, 23-Manufacture of other non-metallic mineral products, 24- Manufacture of basic metals, 25- Manufacture of fabricated metal products, except machinery and equipment, 26-Manufacture of computer, electronic and optical products, 27- Manufacture of electrical equipment, 28-Manufacture of machinery and equipment n.e.c., 29-Manufacture of motor vehicles, trailers and semi-trailers). We analysed a sample of 50 family businesses in Slovakia. Financial statements from 2017, 2018 and 2019 were analysed. The analysis carried out was based on models predicting financial distress. Methods such as linear discriminant analysis and logistic regression belong to classical statistical methods applied in prediction of corporate default risk. For assessing the financial health of selected family businesses 3 models predicting financial distress were used: Kralicek Quick Test, Taffler model and Virág-Hajdu model.

Kralicek Quick Test evaluates the company's financial and revenue position. The following aspects of a company's position are then evaluated:

- Financial stability: $(X_1 + X_2) / 2$
- Revenue position: $(X_3 + X_4) / 2$
- Overall position: (Financial stability + Revenue position) / 2

The company can be assigned with points ranging from one to five. The less points the company is assigned, the better the financial prognosis of the company.

Table 2. Kralicek Quick Test indicators

Indicator	1 Very good	2 Good	3 Average	4 Weak	5 Very weak
Equity / Assets (X_1)	>30%	>20%	>10%	>0%	Negat.
(Liabilities + loans) / Operating cash-flow (X_2)	<3	<5	<12	>12	>30
Financial stability					
EBIT/Assets (X_3)	>15%	>12%	>8%	>0%	Negat.
Operating cash-flow / Sales (X_4)	>10%	>8%	>5%	>0%	Negat.
Revenue position					

Source: own processing

The Taffler model is based on calculating the following score: $Z = 0.53\ X_1 + 0.13\ X_2 + 0.18\ X_3 + 0.16\ X_4$

X_1: earnings before taxes (EBT) over short-term liabilities,
X_2: denotes current assets over total liabilities,
X_3: denotes short-term liabilities over total assets
X_4: asset turnover (sales over assets).

$Z > 0.3$ Companies with a lower probability of bankruptcy
$Z < 0.2$ Companies with a higher probability of bankruptcy

The Virág-Hajdu model is based on logistic regression.
X_1: Quick ratio
X_2: Return on sales
X_3: Cash-flow/ Total liabilities
X_4: Current assets ratio
X_5: Acc. receivable/ Acc. payable

In the logistic regression model, the starting point should be the so-called "odds" ratio which is the ratio of success and failure probabilities. Taking the logarithm of odds, this measure is called logit. The equation of the logistic regression model:

$$\ln(odds) = \ln\left(\frac{\widehat{p}_i}{1-\widehat{p}_i}\right) = \beta_0 + \beta_1 X_1 + \beta_2 X_2 + \ldots + \beta_n X_n \tag{1}$$

The model works with regression function:

$$\widehat{p} = \frac{e^z}{1+e^z} = \frac{e^{\beta_0 + \sum_{i=0}^{n}(\beta_i X_i)}}{1+e^{\beta_0 + \sum_{i=0}^{n}(\beta_i X_i)}} \tag{2}$$

Estimated parameter values of the logistic regression function are as follows: b_0 = 3.432, b_1, = -10.32, b_2 = 0.01439, b_3 = -4.438, b_4 = -0.02992, b_5 = 8.17.

The measurement of predictive performance is often based on the analysis of data in a confusion matrix. This matrix reports how the model classified the different fault categories compared to their actual classification (i.e., predicted versus observed). This is represented by four pieces of data:

- True Positive (TP): An item is predicted as faulty, and it is faulty
- False Positive (FP): An item is predicted as faulty, and it is not faulty
- True Negative (TN): An item is predicted as not faulty, and it is not faulty
- False Negative (FN): An item is predicted as not faulty, and it is faulty

If company has more than one creditor and the value of its liabilities exceeds the value of its assets (i.e., has negative equity) or is unable to pay at least two financial obligations to more than one creditor for 30 days after the due date. We approximated this fact by the variable Current Assets/Current Liabilities, determining the breakpoint of this indicator at level 1. Companies are at risk of bankruptcy if the solvency ratio is less than 0.06 in 2017, or less than 0.08 in 2018 (Act no. 513/1991 Coll.). We identified non-prosperous companies according to criteria: equity ratio (less than 0.06 and 0.08), liquidity ratio (less than 1) and loss of company.

We also calculated:

- Sensitivity, what is the proportion of actual positives which are correctly identified as such: TP / (TP + FN)
- Specificity, what is the proportion of negatives which are correctly identified as such: TN /(FP + TN)
- Accuracy: (TP+TN)/(TP+FP+FN+TN)
- Empirical Error I evaluatse the number of true positives which were classified as true negatives: FN/TP + FN
- Empirical Error II shows how many of true negatives were classified false positives: FP/FP +TN

Table 3. Confusion matrix

		Current category	
Predicted category		T	F
	T	True positives (TP)	False positives (FP)
	F	False negatives (FN)	True negatives (TN)

Source: own processing

4 Results

The models applied to assess the financial performance of family businesses in selected industries and reveal the risk of bankruptcy use different financial ratios as predictors. The models use together 13 financial-economic ratios. Table 3 presents their descriptive statistics—mean, median and standard deviation.

Table 4. Descriptive statistics of selected ratios

Financial–Economic Ratios	*Mean*	*Median*	*St. Deviation*
Assets / Equity	0,43	0,41	0,26
(Liabilities + loans) / Operating cash	104,08	-0,20	1017,67
EBIT/Assets	0,10	0,07	0,13
Operating cash-flow / Sales	-0,28	0,09	4,76
Earnings before taxes (EBT) over short-term liabilities	0,53	0,17	1,59
Current assets over total liabilities	3,43	1,04	14,45
Short-term liabilities over total assets	0,42	0,40	0,25
Asset turnover	1,62	1,33	1,27
Quick ratio	3,66	0,90	17,50
Return on sales	-35,20	3,71	499,51
Current assets ratio	56,84	57,57	23,35
Acc. receivable/ Acc. payable	2,57	1,05	5,86
Cash-flow/ Total liabilities	0,52	0,22	1,18

Source: own processing

Table 5. Descriptive statistics for Taffler model

Taffler model								
2019			**2018**			**2017**		
Mean	*Median*	*St. Dev.*	*Mean*	*Median*	*St. Dev.*	*Mean*	*Median*	*St. Dev.*
0.76	0.59	1.10	1.52	0.63	4.35	0.91	0.63	0.97
Z score								
Z < 0.2	*(0,2;0,3)*	*Z > 0.3*	*Z < 0.2*	*(0,2;0,3)*	*Z > 0.3*	*Z < 0.2*	*(0,2;0,3)*	*Z > 0.3*
6	3	41	2	1	47	0	2	48
higher prob. of bankr.	*Grey zone*	*lower prob. of bank.*	*higher prob. of bank.*	*Grey zone*	*lower prob. of bank.*	*higher prob. of bank.*	*Grey zone*	*lower prob. of bank.*

Source: own processing

Statistical characteristics, such as the mean, median, and stabdard deviation, were for each year. Table 5 shows an example of the results when applying the Taffler model for the companies with lower and higher probability of bankruptcy. This model and results will be compared with Kralicek Quick test and Virág-Hajdu model.

Table 6. Descriptive statistics for Kralicek Quick Test

Kralicek Quick Test								
2019			**2018**			**2017**		
Mean	*Median*	*St. Dev.*	*Mean*	*Median*	*St. Dev.*	*Mean*	*Median*	*St. Dev.*
2.45	2.38	0.96	2.31	2.25	0.89	2.30	2.25	0.76
Quick Test (points)								

1-4	*4-5*	*5*	*1-4*	*4-5*	*5*	*1-4*	*4-5*	*5*
48	2	0	48	2	0	49	1	0
Very good-weak	*Weak - very weak*	*Very weak*	*Very good-weak*	*Weak - very weak*	*Very weak*	*Very good-weak*	*Weak - very weak*	*Very weak*

Source: own processing

As it is clearly seen from the table 6 that represents the results of Kralicek Quick Test for years 2017-2019, only 2 (1 in 2017) companies are threatened by financial distress, because higher score implies the thread for the company, and lower score represents the financial stability of the company.

Table 7. Descriptive statistics for Virág – Hajdu model

Virág – Hajdu model								
2019			**2018**			**2017**		
Mean	*Median*	*St. Dev.*	*Mean*	*Median*	*St. Dev.*	*Mean*	*Median*	*St. Dev.*
0.58	0.74	0.45	0.59	0.87	0.45	0.55	0.80	0.45
$\hat{p}$ - value								
$\hat{p}_{<0.525}$	$\hat{p}_{>0.525}$	*Immeras.*	$\hat{p}_{<0.525}$	$\hat{p}_{>0.525}$	*Immeras.*	$\hat{p}_{<0.525}$	$\hat{p}_{>0.525}$	*Immeras.*
20	29	1	21	28	1	23	26	1
Prosper	*Non-pr.*	*X*	*Prosper*	*Non-pr.*	*X*	*Prosper*	*Non-pr.*	*X*

Source: own processing

The results of Virág-Hajdu model are shown in Table 7. the function received as an outcome of the logistic regression resulted in a higher classification accuracy in the case of both groups. Virág - Hajdu discriminant analysis worked with the whole error of 22.10%, their logistic regression function, on the other hand, affects the lower overall error only 18.20%.

Table 8. Prosperous and non - prosperous companies

Taffler model			**Virág – Hajdu model**			**Kralicek Quick Test**		
	Non-prosp.	Prosp.		Non-prosp.	Prosp.		Non-prosp.	Prosp.
Non-prosper. **2019** **2018** **2017**	6 2 0	17 17 22	Non-prosper. **2019** **2018** **2017**	23 19 22	0 2 0	Non-prosper. **2019** **2018** **2017**	2 2 1	21 17 21
Prosper. **2019** **2018** **2017**	0 0 0	27 31 28	Prosper. **2019** **2018** **2017**	7 10 5	20 19 23	Prosper. **2019** **2018** **2017**	0 0 0	27 31 28

Source: own processing

According to Taffler model more than 54% (in 2019) and 62% (in 2018) and 56% (in 2017) of companies are prosperous and the rest of them are in financial troubles. According to Virág-Hajdu model more than 40% (in 2019) and 38% (in 2018) and 46% (in 2017). According to Kralicek Quick test more than 54% (in 2019) and 62% (in 2018) and 56% (in 2017).

Table 9. Model´s Comparison

Model	2019			2018			2017		
	TM	VHM	KQT	TM	VHM	KQT	TM	VHM	KQT
Sensit.	100%	74,1%	100%	100%	65,5%	100%	100%	82,1%	100%
Specif.	26,1%	100%	8,7%	10,5%	90,5%	10,5%	0%	100%	4,5%
Accur.	66%	86%	58%	68%	76%	66%	56%	90%	58%
Type I Err.	0%	25,9%	0%	0%	34,5%	0%	0%	17,9%	0%
Type II Err.	73,9%	0%	91,3%	89,5%	9,5%	89,5%	100%	0%	95,5%

TM- Taffler model, VHM – Virág-Hajdu model, KQT- Králicek Quick Test, Sensit.- sensitivity, Specif.-specificity, Accur. – accuracy, Type I Err.- Type I Error, Type II Err. – Type II Error.

Source: own processing

The comparison is based on working with terms as Type I Error and Type II Error. Results of models' comparison are including in table 9. The value of Type I Error is 0% in the case of Taffler model and Kralicek Quick Test, 2,59 % in the case of logistic regression – Virág Hajdu model. The value of Type II Error is higher than the value of Type I Error for all models except Virág Hajdu model. It means that models predicting financial distress have difficulties to classify safe business entities correctly. The models with high prediction ability are generally connected with the statement that the value of errors cannot exceed 20-25%.

5 Discussion

Our research showed that it is possible to predict financial distress in the family businesses in selected industry. For assessing the financial health of selected family businesses 3 models predicting financial distress were used: Kralicek Quick Test, Taffler model and Virág-Hajdu model. The Taffler model is based on calculating Z score and it is one the most cited model, Kralicek Quick Test which evaluates the company's financial and revenue position and company can be assigned with points ranging from one to five. The last one was Virág-Hajdu model based on logistic regression. Financial measures represented by Virág-Hajdu model have a good explanatory power (according to Type II Error). Prediction models contain a combination of indicators, including their weights suitable for evaluating the performance and credibility of the company. Determining the prosperity od the company and diagnose the financial health of the company should be one of the key issues of the company.

It is necessary to emphasize that the models predicting corporate financial distress should provide quick and inexpensive recommendations. On the one hand they can help in decision making and strategy creation but on the other hand they should not be used as the only tool and they work with limitations.

This paper is the output of the internal grant project of the University of Economics in Bratislava (Faculty of Business Management) no. I-21-109-11-00 called "Research of Family Businesses in Context of Sustainable Development" (100 %).

References

1. E. Gregova, K. Valaskova, P. Adamako, M. Tumpach, J. Jaros, Predicting Financial Distress of Slovak Enterprises: Comparison of Selected Traditional and Learning Algorithms Methods. *Sustainability*, **12,** (2020)
2. E. I. Altman, Edward, Financial Ratios, Discriminant Analysis and the Prediction of Corporate Bankruptcy. *Journal of Finance*, **23**, 589–609 (1968)
3. J.A. Ohlson, Financial Ratios and the Probabilistic Prediction of Bankruptcy. *Journal of Accounting Research,* **18**, 109–31 (1980)
4. M. E. Zmijewski, M.E., Methodological Issue Related to The Estimation of Financial Distress Prediction Models. *Journal of Accounting Research*, **22**, 59-82 (1983)
5. N. Bărbuță-Mișu, M. Madaleno, Assessment of Bankruptcy Risk of Large Companies: European Countries Evolution Analysis. *Journal of Risk and Financial Management*, **13**, (2020).
6. W. H. Beaver, Financial Ratios as Predictors of Failure. *Journal of Accounting Research,* **4**, 71–111 (1966)
7. L. Svabova, L. Michalkova, M. Durica, E. Nica, Business Failure Prediction for Slovak Small and Medium-Sized Companies. *Sustainability*, **12** (2020)
8. A. Csikosova, M. Janoskova, K. Culkova, Limitation of Financial Health Prediction in Companies from Post-Communist Countries. *Journal of Risk and Financial Management*, **12**, (2019).
9. M. Tamari, Financial ratios as a means of forecasting bankruptcy. *Management International Review,* **6**, 15–21 (1966)
10. B. Prusak, Review of research into enterprise bankruptcy prediction in selected central and eastern European countries. *International Journal of Financial Studies*, **6**, 60-88 (2018)
11. Ľ. Lesáková, P. Gundová, M. Vinczeová, The practice of use of models predicting financial distress in Slovak companies. *Journal of Eastern European and Central Asian Research*, **7**, 123-136 (2020)
12. S. Ashraf, E.G. S. Félix, Z. Serrasqueiro, Do Traditional Financial Distress Prediction Models Predict the Early Warning Signs of Financial Distress? *Journal of Risk and Financial Management*, **12**, (2019).
13. Sinarti, T.M. Sembiring, Bankruptcy Prediction Analysis of Manufacturing Companies Listed in Indonesia Stock Exchange. *International Journal of Economics and Financial Issues*, **5**, 354-359 (2015)
14. M. Fakhri Husein, Galuh Tri Pambekti, Precision of the models of Altman, Springate, Zmijewski, and Grover for predicting the financial distress. *Journal of Economics, Business, and Accountancy Ventura*, **17**, 405 – 416 (2014)
15. O. Machek, Long-term Predictive Ability of Bankruptcy Models in the Czech Republic: Evidence from 2007–2012. *Central European Business Review,* **3**, 14–17 (2014)
16. O. Hajdu, Ottó, M. Virág, Hungarian Model for Predicting Financial Bankruptcy. *Society and Economy in Central and Eastern Europe*, **23**, 28–46 (2001)
17. T. Kristóf, M. Virág, A Comprehensive Review of Corporate Bankruptcy Prediction in Hungary. *Journal of Risk and Financial Management*, **13**, (2020).
18. M. Situm, Analysis of correlation behavior of solvent and inslovent firms based on accounting ratios. *Journal of Modern Accounting and Auditing,* **11**, 233–259 (2015)

19. M. Boďa, V. Úradníček, Predicting Financial Distress of Slovak Agricultural Enterprises. *Ekonomický časopis*, **67**, 426-452 (2019).
20. K. Valaskova, P. Durana, P. Adamko, J. Jaros, Financial Compass for Slovak Enterprises: Modeling Economic Stability of Agricultural Entities. *Sustainability*, **13,** (2020)
21. P. Gottardo, A.M. Moisello, Family forms, Risk-taking and financial distress. *Problems and Perspectives in Management*, **15**, 168-177 (2017)
22. L. R. Gómez-Mejía, K.T .Haynes, M. Núñez-Nickel, K. J. Jacobson, J. Moyano-Fuentes, Socioemotional wealth and business risks in family-controlled firms: Evidence from Spanish olive oil mills. *Administrative science quarterly*, **52**, 106-137 (2007)
23. L. A. Tabot Ntoung, H. M. Santos de Oliveira, B. M. Ferreira de Sousa, L. Marques Pimentel, S. A. Moreira Carvalho Bastos, Are Family Firms Financially Healthier Than Non-Family Firm? *Journal of Risk and Financial Management*, **13**, (2020).

Reliability of the Slovak Legislative Categorization in Comparison with the Selected Prediction Models - Application in Companies from the Creative Industry – Architecture in Slovakia

Miroslav Uhliar[1,*], *Andrej* Kovalev[1]

[1]University of Economics in Bratislava, Faculty of Business Management, Department of Business Economy, Dolnozemská cesta 1/b, 852 35 Bratislava, Slovak republic

Abstract. One of the important changes in the world economies is the increasing significance of the human capital and high emphasis on the Creative Industry. Many companies have been founded within this sector, and like the classic industrial companies, these also need to use models to analyze their financial health and predict their future development. Thus, the Authors decided to verify whether the categorization of the financial situation of the companies based on the Slovak legislative corelate with renowned models like the Altman Z-Score, the IN05 and Creditworthiness Indexes, the Quick Test and the Taffler model. Within the Creative Industry, the Authors targeted the sector of Slovak Architecture and focused on the legislative categorization based on the Acts No. 513/1991 Coll. and 7/2005 of the Commercial Code, as amended. According to the statistical testing results, it was the Quick Test which showed the highest rate of correlation to the legislative Acts. However, the current design of the Quick Test is not sufficient enough to assess the companies in architecture which, based on their results, fall into the so-called grey zone. Therefore, the Authors declare that it might be necessary to form a new predictive model for the assessment of a company's financial health for the Creative Industry.

1 Introduction

The development of the world, society and economies keeps bringing new challenges and pressure to increase the competitiveness of the companies within their environment. New trends and approaches appear, the processes get faster and more effective, more emphasis is put on the human capital and creativity. The importance of the human capital keeps growing, as it becomes one of the key factors for the future of the companies. New investment opportunities arise. The human capital and its quality are the decisive factors of the business activities in the companies which belong to a relatively new sector known as

* Corresponding author: miroslav.uhliar@euba.sk

creative industry. The companies of the Slovak Architecture sector also belong to this industry.

2 Literature Review

The creative industry includes all those companies which base their activities on the individual creativity, skills and talents of a person (employee). Goods and services, which directly create and form the environment that we live in (architecture, design, mass media, fashion, …) are the results of the activities done by the creative industry companies. Their economic potential mostly comes from the ability to use an individual talent or realize an idea thus creating goods and services with above-average added value. Thanks to this, they can become and interesting investment opportunity or a subject to financial support. For these reasons it is necessary to assess their financial health and to be able to predict their future in order to minimize the business or investment risks. However, there is no predictive model which would target the creative industry companies. There is also a question if and how the current classification of companies (according to the Slovak legislative parameters given by the Act no. 513/1991 Coll., Commercial code, as amended, and the Act no. 7/2005 Coll. on Bankruptcy and Restructuring, as amended, and other acts) reflects the categorization of the creative industry companies in line with the generally most used predictive models.

For the first time, the content of the term creative industry can be identified in the 90s of the 20th century. In this time there is a development in many economics of processes based on the use of knowledge and ideas.[1] It is a trend of the so-called knowledge orientation of the economy with a stronger focus on production with higher added value. Simultaneously, the concept of the creative industry, resp. creative economy also includes companies or organizations focused on the provision and production of cultural activities and services.

O´Connor is of the opinion that the broadest concept is the creative economy, that includes the cultural and creative industries sectors.[2] Florida considers the creative industry as economic sectors based on resources such as people, their individual creativity, talents and abilities.[3] According to Howkins, the creative economy includes products and services, that originate in creativity and have economic value, such as: copyrights, patents, trademarks and design. [4] Caves says, that an essential characteristic of the creative industry companies is the fact that their method of production and consumption differs from other companies, mainly in the form and complexity of commercial contracts.[5] According to Šagátová, companies from the creative industry focus on the production of goods and services that are somehow connected with cultural, artistic or entertaining value.[6] Kloudová states that the creative industry is a segment of business in which outputs are strongly dependent on ideas, innovation, artistic rendering and research.[7]

According to the Profile of the Slovak Culture, the business activities of the creative industry include: advertising, architecture, art, antiques, computer and video games, crafts, design, design fashion, film and video, music, performing arts, publishing, software, television and radio.[8] The common point is that these are business activities carried out by companies that have the potential to create jobs and capital through the use of intellectual property.[9]

The scale of dynamics of the global creative sector is rich of growth potential and the creative companies are less sensitive to financial and economic crises then other traditional sectors companies. [10]. This hypothesis was confirmed by the analysis carried out on the sample of Portuguese companies and their data from 2004 to 2011, thus including the period of the worldwide economic crisis. The results of this analysis showed that the cultural and creative companies were only temporarily affected by the economic crisis. We can say that these companies are more dynamic then the companies of the more traditional sectors.[11]

Kmety Barteková states that in 2015, within the V4 countries, the highest number of the companies in the creative industry was identified in Poland (46 %) and the lowest in Slovakia (8 %).[12] Since 2015, according to the EUROSTAT data, the number of enterprises has increased in Slovakia from 11690 in 2015 to 13615 in 2017.[13]

The number of subjects in the creative sector of Architecture in Slovakia has been increasing constantly since 2013. The sector significantly favors natural persons but also the number of corporations from this sector grows. Based on the data from 2019, there were almost 1500 of them and the trend is growing.[14]

3 Research Goal and Methodology

The research design of the paper is based on three main quality criteria of the empirical process: objectivity, reliability and validity of the results.

In 2015, Rezňáková and Karas [15] presented the results of their research of the discriminative ability of the Altman Model. They examined the possibilities of the portability of models which originated in a different environment or in a different time period. The tests of the Altman Model's accuracy on a sample of data from the V4 industrial enterprises showed that the original Model works with lower statistical accuracy than stated by the Author of the Model, and with a high proportion of unrated companies. The researchers also found out that by re-evaluating the weight of the individual coefficients of the Model and the grey zone boundaries, based on conditions specific to each V4 country and by maintaining the Model variables, it is possible to increase the resolution of the Model, meaning the ability to correctly distinguish failing businesses from prosperous. The examined sample of data showed that the variables contained in the Altman Model have a very different discriminatory capability in different countries, therefore this Model is not transferable between environments. According to the authors, this points to the fact that, in a specific environment, it is necessary to find your own optimal combination of indicators and create your own original models.

We decided to test these findings on the sector of architecture, which falls under the creative industry in the Slovak Republic. We asked a research question – whether the categorization of the companies based on the Slovak legislation is significant with the results of the most used mathematical models, which are available at the web portal https://finstat.sk/. These mathematical models use mainly those rations for the clarification of a company's financial health, which categorize the company as prosperous, non-prosperous or a company, for which it is not possible to determine the affiliation to one of the mentioned groups.

3.1 Research Goal

The goal is to compare the reliability of prognosing the categorization of the companies in the sector of architecture in the Slovak Republic according to the Slovak legislation, while compared to the resolutory capabilities of the most renowned selected predictive models, which have their structure based on the linear discriminative analysis or the logistic regression. This paper is linked to a previous similar analysis of the Authors, which targeted a different segment of the creative industry: Media, Publishing and Culture.[16] The legislative categorization is based on the criteria from the Commercial Code and the Act on Bankruptcy and Restructuring. Based on this, the analyzed subjects are classified within three groups of companies: Non-Prosperous, Prosperous, Average.

3.2 Methodology and Research Object Definition

Companies, which belong to the creative industry sector – Architecture, code SK NACE [71110], are the objects of this research. This sector of economy was chosen with regards to the participation of the Authors of this Paper in the VEGA 1/0340/19 research project.

The Slovak legislation defines the criteria which allow us to define the companies as non-prosperous for the purpose of our analysis. According to the provisions of the Act no. 7/2005 Coll. on Bankruptcy and Restructuring and on amendments to certain acts, a debtor is in the danger of bankruptcy, especially if they are in crisis. The Act no. 513/1991 Coll., Commercial Code, § 67, states that a company is in crisis when:

- it is bankrupt or under the threat of bankruptcy;
- it is under the threat of bankruptcy → equity : liabilities < 8 : 100.

The Act on Bankruptcy and Restructuring states that a debtor is bankrupt when they are insolvent or over-indebted.

• A legal entity is considered insolvent when they are not able to clear at least 2 financial liabilities, which are 30 days overdue, to more than one creditor. When judging the debtor's ability to pay, all the receivables, which belonged to a single creditor 90 days prior to the file for bankruptcy, are considered a single receivable.

• Over-indebted are those who are required to manage their accounting according to special regulations, they have more than one creditor and the liabilities surpass the assets. The value of liabilities and assets is based on the accounting data or the value given in an expert report.

Prosperous companies are not legislatively defined in Slovakia. According to the above-mentioned definition of bankrupt companies (non-prosperous companies thereinafter), a company is considered prosperous when it does not fulfill the criteria given for the non-prosperous companies. Therefore, in terms of the afore-mentioned criteria, prosperous companies are those which are able to pay their dues, the value of assets > liabilities and the ratio of their equity and liabilities is > 8 : 100.

This classification of prosperous and non-prosperous companies is not sufficient for our needs. The above-mentioned definition of a prosperous company is too wide for our research. The predictive models used in the analysis (Altman Z-score, INDEX 05, Creditworthiness Index, Quick Test, Taffler Model), according to their methodology for the evaluation of a company's financial health, divide the companies in three groups: prosperous, non-prosperous and average.

The companies, which will not fall into the non-prosperous category, will go through another selection process in order to create a sample of companies in the average and prosperous categories for our research. The definition of companies as average or prosperous is based on how they reach their revenue profitability – whether it is above the level of the median for the analyzed creative sector, or they have a positive revenues profitability, however, it is lower than the calculated median. The complex systemic criteria, which we defined or order to classify a company as average or prosperous, are listed in the following table.

Table 1. Criteria for the division of companies in research groups

Non-prosperous businesses	Average businesses	Prosperous businesses
Liabilities > Assets Equity / Liabilities < 8:100 Equity < 0 Profitability of Sales < 0	Liabilities < Assets Equity / Liabilities > 8:100 Equity > 0 Profitability of Sales < Med	Liabilities < Assets Equity / Liabilities > 8:100 Equity > 0 Profitability of Sales > Med

Note: Med – value of the median of the revenues profitability for the segments of the creative industry in the Slovak Republic - Architecture

Such parameter settings of the research objects allowed the resulting categorization of the researched object to reflect the parameters of the evaluative zones of the analyzed predictive models.

In the first step of the research, we defined the basic file of the company test sample according to strictly set criteria:

- companies from the sector of Architecture, code SK NACE [71110];
- small and middle-sized companies defined according to the business subject categorization with reference to the European Commission's Recommendation No. 2003/361/EC;
- companies managing their accounting through double-entry bookkeeping according to the Act no. 431/2002 Coll. on Accounting, and the Order of the Ministry of Finance of the Slovak Republic no. 23054/2002-92;
- companies established before 2017;
- companies with available data from financial statements for the period of 2019;
- companies which are still in the business in 2021.

The basic file of the test sample contained 300 companies. We applied filters to select those companies which have a history of business activities longer than 5 years, then we removed those companies which had not provided basic data necessary to apply the parameters for the company categorization and the final number of the companies in the research sample was set to 135.

3.3 Used Methods

The sample of 135 studied companies was analyzed through methods, which helped reflect the objective characteristics of the studied object and maximally draw near to the equivalent relations in the research object.

The structure of the research design was conditioned to the following sequence of methods:

- one-dimensional empirical division of values;
- statistical measures of variability;
- bivariate division of the sample;
- ordinal scale of variables;
- statistical measures for two-dimensional division of the sample;
- contingency links based on forecasts;
- statistical probability of a random event.

We used MS Excel and mainly IBM SPSS Statistics ver. 25 to process the data and to evaluate them statistically.

4 Results of the Study

According to the methods defined in the previous methodologic part of the Paper, we categorized the selected file of 135 companies in line with the chosen criteria of the indicators: *liability/assets ratio, equity/assets ratio, value of equity, profitability of revenues*. Afterwards, the obtained legislative categorization of the companies went through the statistical testing for two-dimensional division of the sample with the Pearson correlation coefficient.

Tab. 2. Correlation Coefficients of the Schematic Structure of the Selection

		Altman Z-score	INDEX 05	Creditworthiness Index	Quick Test	Taffler Model
Legislative Categorization of the Selection	Pearson Correlation	,696*	,748*	,850*	,862*	,323*
	Sig. (2-tailed)	,000	,000	,000	,000	,000
	N	136	136	136	136	136

source: own processing
Note: * Correlation is significant at the 0.01 level (2-tailed).

Thanks to this analysis we found out that the legislative categorization of the tested sample of companies has a positive, strictly linear correlation with the predictive Quick Test model (,862) and with the Creditworthiness Index model (,850) as well. Weaker correlation was proved with the Index 05, Altman Z-score and only a small correlation with the Taffler model. The next steps of the research were focused on the reliability prognosis for the legislative categorization of the creative industry companies in comparison to the design of the predictive model for a company's financial health made through the Quick Test, where the correlation rate had been the strongest.

In the next step we tested the connections between the nominally scaled variables of the company legislative categorization and the Quick Test predictive model. The results of this test are presented in the following part.

Table 3. Matrix of the connections between the legislative categorization of the selection and the Quick Test

			non-prosperous	average	prosperous	Total
Legislative categorization of selection	**non-prosperous**	Count	14	3	0	17
		% within LKV	82.4%	17.6%	0.0%	100.0%
		% within Quick	100.0%	10.7%	0.0%	12.3%
		% of Total	10.1%	2.2%	0.0%	12.3%
	average	Count	0	17	7	24
		% within LKV	0.0%	70.8%	29.2%	100.0%
		% within Quick	0.0%	60.7%	7.3%	17.4%
		% of Total	0.0%	12.3%	5.1%	17.4%
	prosperous	Count	0	8	89	97
		% within LKV	0.0%	8.2%	91.8%	100.0%
		% within Quick	0.0%	28.6%	92.7%	70.3%
		% of Total	0.0%	5.8%	64.5%	70.3%
Total		Count	14	28	96	138
		% within LKV	10.1%	20.3%	69.6%	100.0%
		% within Quick	100.0%	100.0%	100.0%	100.0%
		% of Total	10.1%	20.3%	69.6%	100.0%

source: own processing

The statistical analysis showed that the legislative categorization of the companies could correctly determine 100 % of non-prosperous companies, 60,7 % of average and 92,7 % of prosperous companies. The legislative categorization of the companies reached its limit value for the assessment of the classificatory ability and the predictive power of the Quick Test model in relation to the non-prosperous companies.

In the final step of our statistical testing, we form the prognosis model, which is based on the data from the table with the two-dimensional division of the sample of the selection's legislative categorization and the Quick Test predictive model.

Table 4. Prognosis Model Based on the Two-Dimensional Division of the Sample

			non-prosperous	average	prosperous	Total
Legislative categorization of selection	non-prosperous	Count	14	3	0	17
		% within Quick	100.0%	10.7%	0.0%	12.3%
	average	Count	0	17	7	24
		% within Quick	0.0%	60.7%	7.3%	17.4%
	prosperous	Count	0	8	89	97
		% within Quick	0.0%	28.6%	92.7%	70.3%

source: own processing

According to the prognosis model, we found out that the total reliability of the legislative categorization for the sample of companies from the creative sector of Architecture reaches the quality of the prognosis for a company's financial health in agreement with the Quick Test predictive model at the level of 79,13 %, i.e. (14 + 10,319 + 82,503) / 135.

5 Discussion

The results made by the research of companies from the creative sector of Architecture proved the significance and reliability of the categorization according to the legislative provisions of the Act no. 7/2005 Coll. on Bankruptcy and Restructuring and on amendments to certain acts, and the Act no. 513/1991 Coll., Commercial Code, as amended, namely in the section related to non-prosperous companies. The success rate of recognizing a non-prosperous company was 100% when compared to the Quick Test. The success rate of recognizing a prosperous company was 84.3% when compared to the Quick Test. The limiting aspect for a complex acceptance of the division according to the above-mentioned legislative acts is given by insufficient reliability for the companies categorized in the section of neutral results (i.e. average company or the so-called grey zone) and partly in the category of prosperous companies as well. In order to increase the total reliability of the correct quantification of companies into different categories, it will be necessary to assess companies according to multiple, perhaps even alternative parameters different from the current design of the Quick Test predictive model.

Another possibility to increase the reliability for the division of companies based on their economic results is to create a new predictive model, which will respect the specifics of the assessed sector of the economy, or the specifics of how the Slovak business environment is arranged, with all its main characteristics and features. The Authors' goal is to create a new predictive model in the future which could be applied to the area of the culture and creative industry companies in such a way that it would be able to divide companies into different categories (i.e. prosperous, non-prosperous and average) with a high rate of reliability. The group of average companies represents companies with neutral results, which can fall into one of the two boundary categories in the near unspecified future. Therefore, in terms of the predictive model's practical use, it is important to identify these companies correctly.

6 Conclusion

The creative industry companies are oriented on the production of goods and services with high added value. For this reason, they can be seen as new investment opportunities for the investors. This is valid for the architecture companies as well, since they belong to the sector of the creative industry. Currently there is no rating predictive model reliable enough to assess and categorize companies of the creative industry based on their historical development with the purpose of minimizing the failures related to the investments in their business activities. At the same time, it has become clear that the most renowned models, which were created in different environments, other sectors, different types of companies, etc., may not be precise enough when assessing and predicting the financial situation and the future of the companies from the creative industry.

In their scientific study the Authors evaluated the legislative definition of a non-prosperous company in accordance with the Act no. 513/1991 Coll., Commercial Code, as amended, and the Act no. 7/2005 Coll. on Bankruptcy and Restructuring and other acts as amended, in comparison to the design of selected predictive models. The following models were assessed: Altman Z-score, INDEX 05, Creditworthiness Index, Quick Test and Taffler Model. The results of the realized analyses showed that the Quick Test predictive model had the highest rate of correlation to the legislative categorization of a non-prosperous company in Slovakia. The Creditworthiness Index model had a somewhat lower rate of correlation to the legislative categorization of a non-prosperous company.

The results of the statistical testing showed that the legislative categorization of the companies, when compared to the Quick Test, correctly identified 100 % of non-prosperous companies, 64,3 % of average companies and 75,8 % of prosperous companies. The legislative categorization of the companies reached its limit value for the assessment of the classificatory ability and predictive power of the Quick Test model in relation to the non-prosperous companies. The total reliability of the legislative categorization for the sample of companies from the creative sector of Architecture was, according to the Quick Test predictive model, at the level of 79,13 %, i.e. (14 + 10,319 + 82,503)/135.

The results of these research activities proved the significance and reliability of the company categorization based on the legislative provisions defined above. The limiting aspect for a complex acceptance of the division according to the above-mentioned legislative acts is given by insufficient reliability for the companies categorized in the section of neutral results (i.e. average company or the so-called grey zone) and partly in the category of prosperous companies as well. In order to increase the total reliability of the correct quantification of companies into different categories, it will be necessary to assess companies according to multiple, perhaps even alternative parameters different from the current design of the Quick Test predictive model.

One possibility to increase the reliability for the division of companies based on their economic results is to create a new predictive model, which will respect the specifics of the assessed sector of the economy, or the specifics of how the Slovak business environment is arranged. The Authors' goal is to create a new specialized predictive model in the future which could be applied to the area of the culture and creative industry companies in such a way that it would be able to divide companies into different categories (i.e. prosperous, non-prosperous and average) with a high rate of reliability.

This article is a partial output of the VEGA MŠ SR 1/0340/19 project – "Podnikateľský rozmer subjektov kreatívneho priemyslu v kontexte inovácií a inteligentného rastu" to the extent of 100%.

References

1. B. A. Matheson. Culture of creativity: Design education and the creative industries. In: *Journal of Management Development*, **25**, pp. 55-64. (2006)

2. J. O´Connor. The definition of the „Cultural Industries". In: *The European Journal of Arts Education*, Vol. **2** No. 3, February 2000. pp. 15-27, (2000)

3. R. Florida. The Rise of the Creative Class: And How It's Transforming Work, Leisure and Everyday Life. New York: Basic Books. (2002)

4. J. Howkins. The Creative Economy: How people Make Money from Ideas. London: Penguin Books. ISBN 9780141977034. (2001)

5. E. R. Caves. Contracts Between Art and Commerce. In: *Journal of Economic Perspectives*, Vol. **17**, No. 2, pp. 73-78. (2003)

6. S. Šagátová. Vývoj vybraných sektorov kreatívneho priemyslu v európskych krajinách: Development of Selected Sector of the Creative Industry in European Countries. *Ekonomika, financie a manažment podniku 2019: zborník vedeckých prác = Proceedings of Scientific Papers*, pp. 539-544. (2019)

7. J. Kloudová. Kreativní ekonomika: Trendy, výzvy, příležitosti. Praha: Grada Publishing a.s., ISBN 978-80-247-3608-2. (2010)

8. Profil Slovenskej kultúry 2019. *Definícia kreatívneho priemyslu*. https://profil.kultury.sk/sk/kreativny-priemysel/, [accessed 15.04.2021]

9. D. Rybárová. Creative Industry as a Key Creative Component of the Slovak Economy. *Globalization and its Socio-Economic Consequences 2019: Sustainability in the Global-Knowledge Economy*, pp. 1-8. (2019)

10. I. Dronyuk, I. Moiseienko, J. Greguš. Analysis of Creative Industries Activities in European Union Countries. In: *Procedia Computer Science,* Vol. **160**, November 04-07, 2019. pp. 479-484. (2019)

11. E. Fontainha, E. Lazzano. Cultural and Creative Enterpreneurs in Financial Crises: Sailing Against the Tide? In: *Scientific Annals of Economics and Business*, 66 (Special issue), pp. 73-89. (2019)

12. M. Kmety Barteková, H. Majdúchová. Evaluation of the Policies Supporting Cultural and Creative Industries in the Slovak Republic. In: *European Financial Systems 2019, Proceedings of the 16th International Scientific Conference, June 24-25, 2019 Brno, Czech Republic*, pp. 30-39. (2019)

13. Eurostat. (2019). https://appsso.eurostat.ec.europa.eu/nui/submitViewTableAction.do, [accessed 15.04.2021]

14. M. Uhliar. Specifics of Creative Business in Architecture. *In: Current Problems of the Corporate Sector 2020: 17th International Scientific Conference,* pp. 1-12. (2020)

15. M. Rezňáková, M. Karas. The Prediction Capabilities of Bankruptcy Models in a Different Environment: An example of the Altman Model under the Conditions in the Visegrad Group Countries. In *Ekonomický časopis/Journal of Economics*, vol. **63**, no.6, pp. 617-633. (2015)

16. A. Kovalev, J. Kintler, M. Uhliar. Reliability Prognosis of the Legislative Cathegorization of Enterprises in Creative Industries in Comparison with the Design of Mathematical Prediction Methods. *Ekonomika a manažment: vedecký časopis Fakulty podnikového manažmentu Ekonomickej univerzity v Bratislave*, **16**, pp. 9-26. (2019)

Environment Charge and Covid

Katarina Vavrová[1,*] *Miloš* Bikár[1]

[1]University of Economics in Bratislava, Faculty of Business Management, Dolnozemská 1, 85235 Bratislava, Slovak Republic

Abstract. The primary goal of the paper is to examine the trends of global development and also global changes in the context of selective excise taxes, which are part of the Slovak tax system. First, it is necessary to describe the theoretical research, for clarity in this topic. To this end, we will provide an overview of developments in excise duties and environmental taxes. We will also focus on key events in the global economy that have had an impact on the formation of the Slovak tax system and also on legislative changes in the field of selective excise taxes. Next, we focus on the specification of the main goal of the paper. We will further specify the mathematical and statistical methods used. The result of the paper represents the latest legislative changes that are relevant from our point of view. In the next part of the article, we are going to examine the financial results of the company in connection with the applied environmental policy. In the last part of the article, we summarize the most important findings of our analysis and also point out the impacts of the environmental policy applied by the state on the indicators of green growth in the time of Covid.

1 INTRODUCTION

Manufacturing companies strive to make the highest possible profits and do not think about the future in relation to the environment. Countries are also trying to solve this unfavorable situation through taxes. Environmental taxes are key tools for achieving sustainability in the economy by increasing the prices of environmentally harmful goods or the cost of production inputs. Developed countries, such as the European Union, have recently been drawing more and more attention to this area of the economy.

Growing concerns about climate change are putting environmental issues at the forefront of the economic agenda in many European countries. Global warming is one of the most important challenges facing the world [1]. Taxes, fees, tradable permits and other economic instruments play an important role in achieving cost-effective control of greenhouse gas emissions. Their potential scope and revenues increase the much broader implications of economic and fiscal policy. European countries introduced carbon taxes in the 1990s, although the draft European energy tax was ultimately unsuccessful. Recently, the focus has been on emissions trading. Several tax measures have been introduced in the EU, in particular with environmental objectives. They included national environmental

* Corresponding author: katarina.vavrova@euba.sk

taxes along with the existence of environmental tax measures. The increasing use of environmental taxes, emissions trading and other economic instruments was partly conditioned by the recognition of the limitations of conventional environmental regulation. They can initiate the necessary changes in the economy, resource use, behavior and general approach to nature [2]. The relationship between environmental knowledge, attitudes and behavior is complex. In addition, electricity consumption in the six Gulf Cooperation Council countries has increased rapidly in recent decades due to rapid population growth and relatively rapid economic growth [3]. Energy is an integral part of our daily lives in a market economy society. We need it for heating, cooling, lighting and movement; it is important for the functioning of our offices, workplaces and the whole economy [4]. This is one of the reasons why the EU Commission has proposed its Energy Union Strategy.

2 Data and methods

The main object of our analysis is the company Mondi SCP, a. s., operating in the territory of the Slovak Republic. As the subject of the company's activity is the production of paper and pulp, it is a suitable object for the implementation of our research, which we want to primarily determine the degree of involvement of the company in environmental protection and the consequences of this activity on the achieved financial results. Another subject of the survey can be considered the Slovak Republic, specifically in the field of taxation of its entities and also the European Union, which consisted of 28 member states, as in the analyzed period was also part of the United Kingdom. It is also a comprehensive source of information on the state and development of the environment and related aspects for the general public [5]. The Slovak Environmental Agency processes and regularly evaluates various sets of indicators [6]:

- **Sustainable Development (TUR) indicators.**
- **Green growth indicators, environmental indicators.**
- **Resource efficiency indicators.**

To obtain current data, we also drew from the OECD library, where several publications focused on the topic of environmental policy of states, tax policy in general, as well as data on the effects of the coronavirus pandemic on taxation were published. We also drew information from articles in professional journals and from scientific papers. Data related to changes in legislation were available on the MFSR website. We had quantitative data on the development of selective excise duties within the Slovak Republic and the European Union at our disposal thanks to the EUROSTAT database, which falls under EU control. Financial data of Mondi SCP, a. with. we drew from the published financial statements. Information on implemented projects was specified on the company's website and in published annual reports. We obtained data for the analysis of green growth indicators of the country from the DATACUBE database published by the Statistical Office of the Slovak Republic. The procedure for the calculation of indicators is published by the Ministry of the Environment of the Slovak Republic. From the Ministry of Economy of the Slovak Republic, we obtained data for quantifying the efficiency of energy consumed by Mondi SCP. Based on the analysis of indicators and regular evaluation of the European Environment Agency (EEA) [7], the evaluation options of the Organization for Economic Development and Economic Cooperation (OECD) and of the European Union Office of Indicators (Eurostat) were evaluated. Flexible and liquid markets can enable supply and demand to be more efficiently adjusted, reducing production costs and therefore prices. Such situations should also trigger bilateral contract prices in the most developed markets [8]. To interpret the obtained data, we mainly used descriptive statistics, through which we divided the quantum of data into clear tables and graphs. We mainly used quantitative indicators in absolute terms. For a clearer illustration of the relationships between quantities, in some cases we considered it appropriate to use the share of individual parts in the whole. We also used the method of comparison, specifically we compared data for different time periods. Within the analyzed company, we looked for an analogy between the procurement of equipment relevant from an ecological point of view and the achieved results of the company in the area of costs, revenues and emissions of harmful substances into the air. Looking at selective excise taxes as part of the Slovak tax system, we looked for an analogy between the development of this type of taxes and global changes. To express the company's performance, we used relative profitability indicators, which are part of the ex-post analysis. To express the development of CO2 emissions produced into the air in the company after the introduction of more environmentally friendly technology, we expressed by means of a moving average. Its positive is mainly the exclusion of random influences from time series. Only after this adjustment is it possible to objectively assess the data. The charts also includes a forecast based on historical data. We also used time series for the chronological arrangement of the sequence of comparable values, while applying the long-term periodicity of their monitoring. The analysis allowed us to select the most important ones from the amount of available data. Through synthesis, on the other hand, we assessed their interrelationships.

2.1 Environmental charge, indicators

In this part of the paper we will analyze the individual environmental charges from a theoretical point of view. Indicators are measurable quantities that provide information on the development and trends of phenomena and processes from a quantitative and qualitative point of view. In terms of indicators, the basic aim of research is to identify key environmental indicators that reliably reflect environmental policy. The basic starting point for their identification was environmental indicators obtained from data of individual sectors (selective excise tax trends). Environmental taxes are one of the most important instruments of the European Union's environmental policy.

- **Sustainable Development (TUR) indicators.**

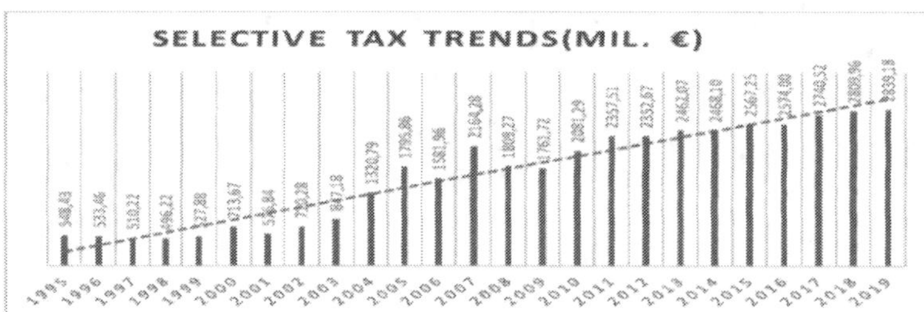

Fig.1. Selective tax trends in Slovac republic, 1995 -2019. Source: Eurostat

Their values ranged from 548.43 mil. EUR (in 1995) and 2,839.18 mil. € (in 2019). This significant difference is also due to the fact that energy taxes were at the level of 0% until 2007, and starting this year, the total value of revenues began to exceed the level of 2,000 mil. EUR. In addition, a green energy tax was introduced in 2010.

- **Green growth indicators, environmental indicators.**

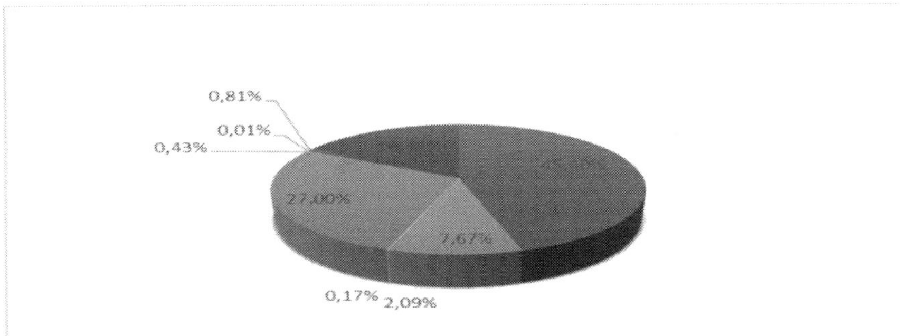

Fig. 2. Structure of revenues from selective excise taxes in the Slovak Republic (2019).
Source: Eurostat

The chart shows that almost half, namely 45.40% of total revenues are accounted for by mineral oil tax revenues, which also dominated in the previous monitored periods. The second highest share falls on taxes levied on tobacco and tobacco products at 27%. This item can also be considered important as its share in total revenues has consistently exceeded the level of 20% since 1995 and its value has been constantly growing from 1995 to 2004, this growth was accompanied by fluctuations in subsequent periods. The last item, the value of which was above the level of 10% (16.41%) in 2019, was the tax on green energy. As we have already mentioned, this was the last tax introduced from the category of excise duties, although, as we can see, it represents a significant part of revenues. Its share of excise revenue has doubled in the last 10 years. The opposite trend can be observed for the alcohol tax, which accounted for 16.25% in 1995 and is currently at around 7.67%. The percentages of other types of excise duties are below 3%, with the lowest share of coal tax at 0.01%.

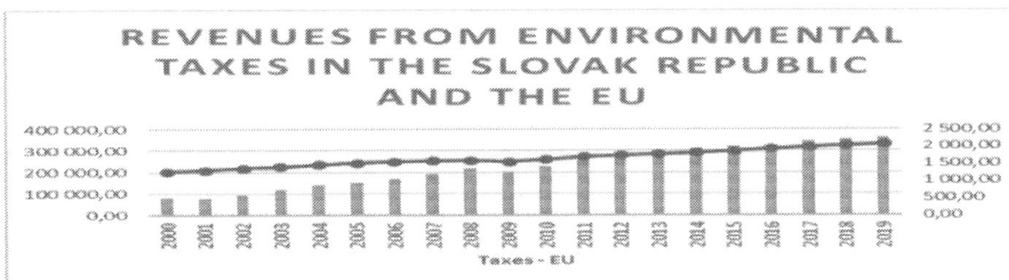

Fig. 3. Trends in European Union, electricity prices. Source: Eurostat

In the European Union, the value of these revenues in 2019 reached the level of 389,406.79 mil. EUR, while in the Slovak Republic revenues amounted to 2,245.98 mil. EUR. In the European Union, the value of these revenues in 2019 reached the level of 389,406.79 mil. EUR, while in the Slovak Republic revenues reached 2,245.98 mil. EUR. Current results cannot be analyzed due to a pandemic Covid. As economic growth has slowed significantly and even stopped production altogether for some time, we expect its effects to be significant.

- **Resource efficiency indicators, implementation of environmental measures.**

The basic starting point for their identification was environmental indicators obtained from data of individual sectors (selective excise tax trends). The picture shows the development of total revenues from this tax category. Over the last ten years, total EU environmental tax revenues have increased between 2008 and 2019. Data show that total EU environmental tax revenues are increasing [9]. The justification for revenue growth stems from rising

environmental tax rates. Exceptions to this rising trend are 2007 and 2009, during which Europe went through a global financial crisis [10]. Environmental taxes are closely linked to the effective tax rate.
The chart also shows the effective tax rate, which is calculated as the ratio between the income tax and the company's profit, which we found before tax. The difference between the income tax and the actual tax rate arises when items such as permanent, temporal or other differences and the tax credit are taken into account. It may be affected by changes in statutory income tax rates.

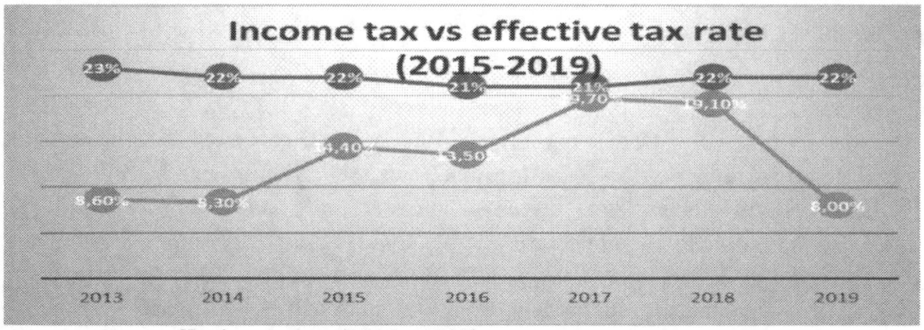

Fig. 4. Income tax vs effective tax rate (2016 - 2019).
Source: own processing based on data from: www.registeruz.sk
The OECD uses income data provided annually by correspondents from national ministries of finance, tax administrations or national statistical offices [11]. Although preliminary data are available for most countries with a lag of about six months, finished data are available with a lag of about one and a half years. Final data on income for 2017 were received in the period May - August 2019. In OECD countries, the reported year coincides with the calendar year. This will be achieved through flexible recruitment procedures designed to attract quality candidates, a focus on training and the development and development of systems that capture and enable best practice to be repeated. As they say, globalization is also one of the most important facts [12]. The modern trend in tax agencies is for more highly skilled workers who have merit based on merit and performance management and development systems [13]. In the context of energy consumption, the issue of greenhouse gas emissions needs to be addressed. For the development of emissions, we have compiled a graph that interprets the results of the ***moving average of emissions for the period from 2013 to 2019.***

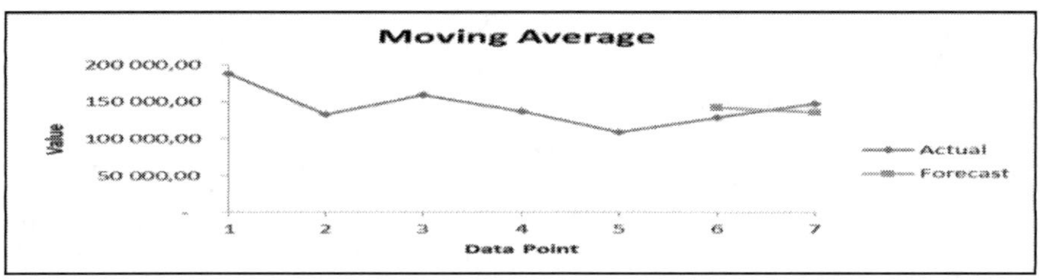

Fig. 5. Moving average of CO2 emissions in Mondi SCP, a. with. (2013 - 2019)
Source: own processing based on data from: https://www.mondigroup.com/en/sustainability/

Mondi SCP, a. with. after its latest investment in a renewable boiler, it is 100% a self-sufficient company in the field of energy production, with up to 94% of it coming from renewable sources. It is essential for society to pay attention to climate change. The climate crisis is affecting the company's operations and supply chain, through effects on water, weather, carbon regulation and taxation, energy availability and price. Increasing self-sufficiency in energy production contributes to improving energy profitability and security. The production of pulp, paper and packaging is energy-intensive, with energy generation being one of the main sources of greenhouse gases. The highest decrease compared to 2013 was recorded in 2017, when the value of emissions was up to 42% lower than before the introduction of the recovery boiler.

3 Results and Discussions

Environmental charges are used to influence the behavior of economic operators, whether producers or consumers. The EU favors these instruments because they provide flexible and cost-effective means. The use of economic instruments for the environment is promoted in the EU Environment Action Program 2020, the EU's Sustainable Development Goals and the Europe 2020 Strategy. Economic instruments for pollution control and natural resource management are an important part of environmental policy in EU Member States [14]. Environmental indicators, together with those supporting the Europe 2020 strategy, continue a series of important Eurostat publications in support of the Europe 2020 strategy by monitoring progress towards the objectives set under the three mutually reinforcing priorities of smart, sustainable and inclusive growth. The analysis is based on the Europe 2020 headline indicators selected to monitor progress towards the strategy's objectives. Other breakdowns focusing on specific subgroups of society or the economy are also used to deepen the analysis and provide more detailed information. The data come mainly from official statistics compiled by the European Statistical System and published by Eurostat. To support this statement, we present graph no. 7, which shows data on year-on-year growth, resp. decline in GDP at constant prices, CO2 and CO2 productivity, which we have quantified as the ratio between GDP and CO2 emissions. It is evident that in 2009 there was a year-on-year decrease in both GDP and CO2 emissions, which resulted in an increase in the value of the CO2 productivity indicator. However, in the following year 2010 we can already see an increase in the volume of production. The increased activity of economic entities was also reflected in the increased volume of produced harmful emissions. In the following years, fluctuations can be observed, from 2014 to 2017, even a decline in greenhouse gas productivity, which evokes that the state's environmental policy would probably need to be reconsidered. However, in 2018, for which we had the latest data, we can observe an increase in the value of the indicator. The positive impact was mainly due to the higher volume of gross domestic product in absolute terms than the volume of CO2 emissions, while CO2 emissions recorded a significant year-on-year decrease.

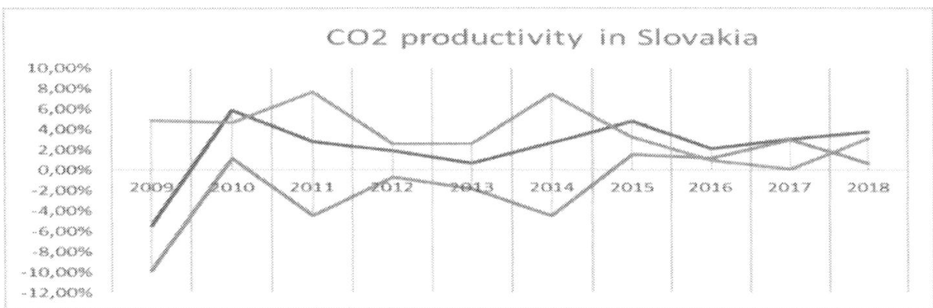

Fig. 6. Total environmental taxes in countries European Union (mill. Eura). Source: Eurostat

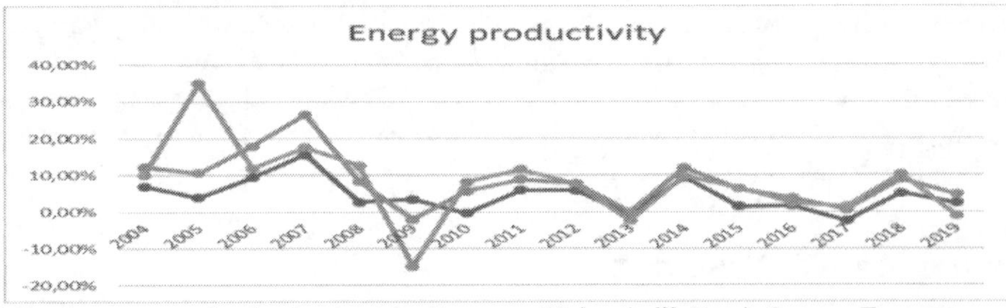

Fig. 7. Total energy production in countries European Union (mill. Eura). Source: Eurostat

A similar indicator is energy efficiency, which represents the share of energy consumption in gross domestic product. We can observe a similar development, in 2009 there was a year-on-year increase in energy efficiency. In the context of environmental indicators, needs considered in energy. Primary energy production is any extraction of energy products in a usable form from natural sources [15].

4 Conclusion

Our claims are supported by the fact that, after each global crisis, the economy has recovered, accompanied by increased production, which has led to a renewed increase in gas emissions. The proof is the following picture:

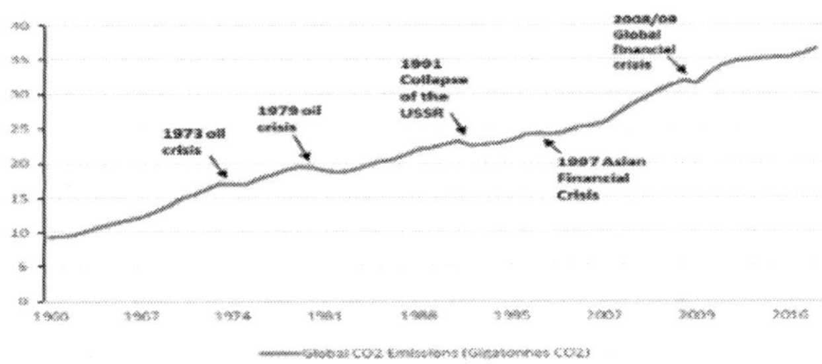

Fig. 8. Impact of economic crises on CO2 emissions. Source: Eurostat

Great competition is a reality. The market has changed, competitiveness is higher. Slovak manufacturing companies must carefully monitor changes in the market and respond to the requirements of domestic and international markets. In particular, they must consider meeting the conditions in foreign markets [17]. Businesses must also focus on the environmental side of production [16]. they must consider digitization production and logistics [18].

Acknowledgment
The presented working paper is the output of the scientific grants VEGA n. 1/0007/19 "Allocation of assets in the environment of low interest rates in financial and non-financial corporations in Slovakia" in the range of 100%.

References

1. N. Suwirman, Environmental Altruistic Behavioral. *International Journal of Environmental Research and Development* **4**, 1 (2014)
2. F. Aidyn, Secondary school students' perceptions towards global warming: a phenomenographic analysis. *American Journal of Nursing Science,* Essays **5**, 2328 (2017)

3. J.M. Hines, H.R. Hungerford, A.N. Tomera, Analysis and Synthesis of Research on Responsible Environmental Behavior: A Meta-Analysis. *The Journal of Environmental Education* **18**, 1 (2010)
4. J. Squalli, Electricity consumption and economic growth: bounds and causality analyses of OPEC members. *Energy Economics Journal* **29**, 1192 (2007)
5. C. Fischer, W.H. Parry, W.A. Pizer, Instrument choice for environmental protection when technological innovation is endogenous. *Journal of Environmental Economics and Management* **45**, 523 (2003)
6. Energy, transport and environment statistics 2019 edition, *The Statistical books, Eurostat, Printed by Imprimeries Bietlot Freres in Belgium*, European Union, (2019).
7. I. Ioannou, G. Serafeim, What drives corporate social performance, International evidence from social, environmental and governance scores. *Journal of International Business Studies* **43**, 834 (2012)
8. P. Deng, A. Delios, M.P. Pemg, A geographic relational perspective on the internationalization of emerging market firms. *Journal of International Business Studies* **51**, 50 (2020)
9. G. Tomas, M. Hult, M.A. Gonzalez-Perez, K. Lagerstrom, The theoretical evolution and use of the Uppsala Model of internationalization in the international business ecosystem. *Journal of International Business Studies* **51**, 38 (2020)
10. M.A. Marinov, S.T. Marinova, J.A. Larimo, T. Lepovsky, Characteristics and Future Trends. *Internetional Business and Emerging Economy Firms* **1**, 1 (2020)
11. J. Kwan Kim, R. Mudambi, An ecosystem-based analysis of design innovation infringements : South Korea and China in the global tire industry. *Journal of International Business policy* **3**, 38 (2020)
12. M. Minárová, D. Malá, M. Sedliačiková, Emotional Intelligence of Manageres. 4th Conference on Business. *Economics and Management Journal* (WCBEM). *Ephesus, Turkey* **124**, 1119 (2015)
13. J. H. Rogers, C. Scotti, J. H. Wright, Evaluating asset-market effects of unconventional monetary policy: a cross-country comparison. *Journal Economic Policy* **29**, 749 (2014)
14. J. Dobrovič, A. Korauš, L. Dančišinová, Sustainable economic development of Slovakia: Factors determining optimal tax collection. *Journal of Security and Sustainability Issuess, International Entrepreneurial perspectives and Innovative Outcomes* **5**, 533 (2016)
15. D. Youová, Environmental taxation. *Brussels,* [electronic source]. Available on the Internet: [cit. 2018-02-02]
16. K. Mc Cullagh, O. Tambou, S. Bourton, (Eds.), National Adaptations of the GDPR, *Collection Open Access Book, Blogdroiteuropeen*, Luxembourg, 1- 130 (2019)
17. K. Porubanová, K., P. Richnák, Usability of Intangible Assets in the Terms of Manufacturing Companies. Ad Alta. *Journal Of Interdisciplinary Research*, **7(2)**, 177-180 (2017)
18. A. Dupaľ, P. Richnák, L. Szabo, K. Porubanová, Modern Trends in Logistics of Agricultural Enterprises. *Agricultural Economics*, ***65*(8)**, 359-365 (2019)

Human Resource Management Department Development and its Organizational Structures

Lasha Abuladze[1,*] *Zuzana Skorková*[1]

[1]Department of Management, Faculty of Business Management, University of Economics in Bratislava, Slovak Republic, Dolnozemská cesta 1, 852 35 Petržalka

Abstract. The purpose of our study represents to determine how HRM's organizational structures has been changed in the past and where HRM is positioned in present. For obtain theoretical knowledge, we used methods: collecting, gathering, sorting, content analysis, synthesis, comparative and deduction of information. Main sources of obtained information, knowledge represent professional articles and research. Today's HRM, its position and roles (strategic partner, administrative expert, change agent, employee champion) has undergone five key stages of evolution, changes in organizational structures. At the early stages, HRM was positioned within different departments, as *Labor/administrative manager* and *Welfare officer* with roles: administrative expert and change agent. From later to modern stages (*Personnel management*, *HRM* and *Modern HRM*), HRM, by obtaining new roles (change agent and strategic partner), become mostly complected, complicated and managed to operate as separate department, on the same level of importance as financial, strategic, marketing and other departments within organizational structure. Based on todays trend, linked with *HR-to-employee ratio*, HR ratio per 100 employees year by year decreases, which might cause another important changes to HR department's position within the organizational structure.

1 Introduction

Organizational structure is defined as a system that shows how specific activities are designed for achieving, obtaining the goals and objectives of the organization [1]. These activities consist of specific tasks, powers, roles and responsibilities. Organization structure designates formal reporting relationships, including the number of levels in the hierarchy and the span of control of managers and supervisors. Organization structure includes the design of systems to ensure effective communication, coordination and integration of efforts across departments. Through the organizational structure, we can define the way in which information moves and disseminates between different departments within the organizational structure of the company. In a centralized structure, we can recognize that

* Corresponding author: lasha.abuladze@euba.sk

decisions move, spread from top to bottom, and vice versa, while in a decentralized structure, decision-making power is divided between different levels of the organization.

Creating and setting an organizational structure allows companies to remain efficient and focused on their objectives [2]. Organizations with variety of sizes shapes make extensive use of organizational structures. They define a specific hierarchy within the organization. A functional and successful organizational structure determines employee's or group of employees' responsibilities, duties, tasks powers and importance. Shortly, the organizational structure shows, underlines who is responsible and what is their responsibility, so the organization can obtain, reach its goals, objectives.

Organizational structures usually have some kind of tall or wide pyramid shape, where most powerful positions are placed on the top of pyramid, while those with the least amount of power are at the bottom. Organizations without formal organizational structure may face difficulties for fulfilling certain activities. For example, employees may have difficulty knowing what responsibilities, power do they have, with whom they have to communicate and report. With help of organizational structure, we can increase efficiency, provide clarity for everyone at every level and make sure, that employees across different departments can be more productive, focused and time efficient.

Human Resource Management – (HRM) represents one of the main parts in today's organizational structures. HRM, alike organizational structure helps organization to reach its goals by effectively managing, motivating, mobilizing and coordinating human resources [3].

Today's HRM, as well as its roles (strategic partner, administrative expert, change agent, employee champion) [5] has undergone number of key changes in organizational structures. HRM's organizational structure also has evolved and changed time by time [6].

In our research we would like to define, determine how HRM's organizational structures has been changed in the past and where HRM is positioned in present. It is important to discuss HRM organizational structure because it is the function's framework, aligning resources with priorities defined in the business strategy and building the organization's culture. The selected organizational structure directly impacts the company's ability to attract, hire, motivate, engage and retain top talent. The demands on human resources function have been changed dramatically in last decades. HRM department is increasingly structured, we are witnessing the emergence of new, specific positions – those are as well defined in our study.

2 Methods

For obtain theoretical knowledge, we used the method of collecting, gathering and sorting information. Subsequently, we used method of content analysis of the obtained information, based on which we have selected the most important findings for our research. By using method synthesis, we combined them into one whole and used method induction to draw conclusions. We also used a comparative and deduction methods, i.e. methods of comparing and deducting information and knowledge from several authors. Main sources of obtained information, knowledge represent professional articles and research. Based on the knowledge from the theoretical basis, we defined the current state of the issue.

3 Results

3.1 Evolution of Human Resource Management

Position of HRM in organizational structures has undergone five major evolutionary stages before its present formation [6]. As we can see on the Figure 1, these five evolutionary stages are: welfare officer, labor manager, personnel management, HRM and modern HRM.

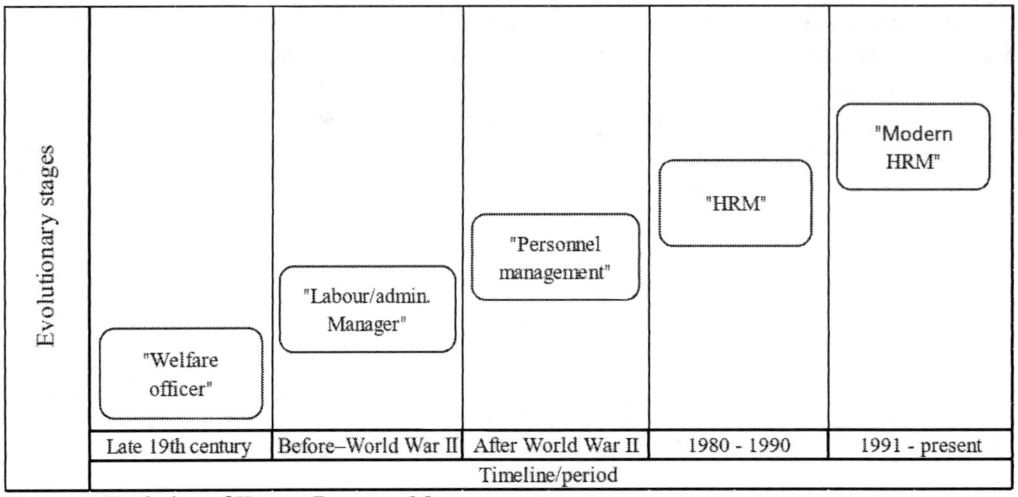

Figure 1. Evolution of Human Resource Management
Source: Own processing

3.1.1 Welfare officer and its position in the organizational structure

Between the dates 1890 – 1913, HRM first time emerged as welfare officer in organizational structure [7]. It was the time, when industrial revolution has started booming. Before that time organizations were treating their employee as machine, whose desires, feelings, needs and motives were ignored. Employees used to work 50 hours in a week with possibility to earn few cents per worked hours and without any coffee, relax brakes. Employee discrimination, firing them without any notice or even cause was regular and usual.

After the beginning of the industrial revolution, organizations realized, that the well-being and productivity of employees have a mutual relationship. They decreased weekly working hours, offer employees coffee and relax breaks and even started to pay them based on their individual performance. Shortly, organizations started to satisfy very basic and narrow needs of their employees, but they still recognize them as extend of machines.

Primary functions of welfare officer were to hire, fire, discipline, reward employees, policy development and paperwork, which made basement for one of the todays HRM's role - administrative expert with the main goal to create an effective organizational infrastructure within the organization [8]. In the role of administrative expert, welfare workers were responsible for ensuring that the proposed organizational processes were effective and that their successful implementation brings added value to the company. Of course, at that time, welfare worker as administrative expert had very simple form and pretty limited power or influence [9].

In organizational structure, welfare workers were not recognized as a separated department with the same power and importance for organization as financial, operational, strategic and other departments. Welfare workers were represented under financial, operational or administrative departments.

3.1.2 Labor/administrative manager and its position in the organizational structure

The second evolutionary stage of today's HRM is known as evolution Pre–World War II (1914 – 1939). That was the time, when HRM transformed from welfare worker into labor/administrative management. New type of HRM primarily focused on the employees' productivity and started to find proper ways, methods for efficiently using of manpower.

One of the first author who came out with new method, idea of using manpower more efficiently was Frederick W. Taylor, who introduced scientific management. This method, idea still considered employees as some part of machines, who needed scientific training and education in order to fulfil everyday tasks with higher productivity. According to Taylor, there was one best way how to maximize employee productivity and it was scientific way. His theory contributed to HRM new functions: determination of work standards, procedures, individual remuneration, division of labor, scientific selection and employee training [10].

After scientific management emerged another method, ideas how to optimize employee productivity. For example, Henry Fayol published theory of administrative management, which provided HRM new functions focused on organizing, controlling, coordinating and directing. At that time, on the market also emerged another new method called bureaucratic management defined by Max Weber. Bureaucratic management created certain rules, directions, which had to improve employee productivity. According to Max Weber, the ideal organization (bureaucracy) is based on firm standards and formal relationships. This bureaucratic style required from labor/administrative management perfectly knowing rules, directions of their organizations.

New, above mentioned methods were more concentrated on the employees' material, educational, developing and well-being needs than ever before, but despite of all, organizations still greatly were focused on productivity than employees physical or social well-being. Working conditions still continue to be harsh and unfair and governments had very narrow influences on employer/employee relationships. These unsafe, harsh and unfair working conditions forced government to create employer/employee relationship and protecting laws, also first time in history labor unions had established.

As we can see, new stage of evolution brought many changes, new functions to labor/administrative management. During times of labor/administrative management, HRM's role as administrative expert grew near to perfection, also we can see first, very basic shape of role - change agent. At that stage HRM boosted, grew very quickly from welfare worker to labor/administrative management and obtained more important position in organizational structures. Within the organizational structure, despite of its fast growing and increased importance, labor/administrative management still was not considered on the same level as financial, strategic, operational and other departments. Labor/administrative management still used to be positioned within financial, administrative and operational department or was positioned as complementary function.

3.1.3 Personnel management and its position in the organizational structure

After World War II (1945 – 1979), today's HRM moved from labor/administrative management into Personnel Management, and it was caused by new emerged behavioral science era. That was the time, when organizations, based on new behavioral theories realized, that in order to increase employees' productivity, only money and material aspects were not enough, but also social and psychological factors needed to be considered and included. Simply, at that time, for organization, employees no more were recognized as part or extend of machines, but as a manpower with their material, non-material, social and psychological needs. These new behavioral notions were proved by Hawthorne study,

Maslow's theory of needs, Herzberg's two factor theory, McGregor's theory of X and Y and so[11-14].

Post–World War II period (1945 – 1979) is not only known for emerging of behavioral theories, but also for dramatic changes in organizations' external environment. During that time, governments of different nations had started to focus even more on the employee's rights by creating, adopting new employment laws. Also, labour unions were growing and extending with speed of light.

All these new changes, challenges required from personnel management proper adaptation and reaction. During that time HRM as personnel management acquired new, third important role - employee champion, while other roles – administrative expert and change agent – become more complete. That period of time also created safe, fair and positive conditions, atmosphere at workplace. Adaptation on these changes and challenges significantly increased importance of personnel management for organizations and as the result, personnel management was placed as a separated department within the organizational structure, on the same level as financial, strategic, marketing or other departments. Also, we have to mention, that personnel management obtained its new position of organizational structure only in few big, experiment oriented organizations, but generally in most organizations concentration and position remained same until late 20th centuries.

3.1.4 HRM and its position in the organizational structure

Between the dates 1980 – 1990 personnel management transformed into human resource management by recognizing employees as most valuable asset and capital of organization. During that time, HRM started to pay even more attention to employees' need for career achievement, development and advancement. HRM also acquired and started using new technologies, EQ, IQ, hard, soft skills and so, through which they tried to concentrate on employees' productivity, well-being, motivation and engagement with the same importance.

Through new look and understanding of manpower, HRM and organization had recognized huge potential and role of human resources as main element, source for creating and obtaining competitive advantage on the market. All these supported to create new, 4th role of HRM – strategic partner [17]. Through this role, the HRM started to connect and align the human resources strategy with the company's business strategy. This means that HR managers became strategic partners by mobilizing, motivating, engaging, coordinating employees for successfully implementing and fulfilling its organizations business strategy. At that time, new HRM role – strategic partner, besides of roles administrative expert, change agent and employee champion, dramatically strengthen importance of HRM for most organizations as well as HRM's position within organizational structure became undoubtable as financial, strategic, marketing and other departments. HRM, as an separate department was positioned on the same level as mentioned ones almost in every type of organizational structures: flat, tall, centralized, decentralized, divisional, matrix and so.

3.1.5 Modern HRM and its position in the organizational structure

From 1991 till today, HRM has undergone many internal and external changes as well as acquired variety of new functions based on which modern HRM became more complex than ever. Modern HRM's importance to organization as a strategic partner went to a whole new level by understanding, adopting new trends of: *globalization* – HRM started to understand the needs, wants and motivations of a diverse and multi-generational-national workforce to attract and retain high-performers and top talents, *technological advancement and digitalization* (industry 4.0)[15,**Chyba! Nenašiel sa žiaden zdroj odkazov.**] – automate much of the work traditionally done by HR professionals. For instance, payroll and streamline the onboarding process, there are also platforms that simplify the recruiting process and talent management systems that enable companies to quickly determine whether their employees are getting the right training opportunities, e-HR; e-training, e-recruitment and etc., *personnel marketing* – retaining and attracting existing or potential employees, *organizational culture, ethics, workforce evaluation methods, Human resource planning techniques, flexible work arrangements, virtual teams, work life balance, improved networking; influence of mass media, Greene economy and many others.*

All these new changes gave HRM solid position in organizational structures. Nowadays, because of its roles – strategic partner, administrative expert, change agent and employee champion, HRM department has separate position almost in every organizational structure, on the same level as financial, operational, marketing, strategic or other departments. Also, we have to mention, that HRM's recognition and position beside another mentioned department, mostly is caused by its role of strategic partner. Otherwise, HRM might be positioned within production, administrative or financial departments. HRM's position within the organizational structures heavily depends on the following factors too: organization life cycle, business/corporate strategy, state of external environment, size of organization, technology advancement and so.

3.2 Organizational structure of Human resource department

The place of HR department in the total organizational structure of the company depends mainly on the size of the company, industry, the speed of change within the industry, and whether or not the responsibilities of HR are sourced internally or externally.

In small organizations, the focus is on company survival, HR practice may appear unstructured and unformalized, there might be no separate HR department, the human resource functions might be carried out by the office manager. According to a report produced by the Cranfield Network on International Human Resource Management [19], 70% of organizations said HR has a place on the board of directors, and 66% reported having a written HR management strategy. Responses also indicated that HR is taking sole responsibility for major policy decisions – the more it confirms its position as a strategic partner. Findings of mentioned study pointed out the creation of new positions like Chief Diversity Officer (CDO), Chief Human Resources Officer (CHRO), and changes in who the Vice President of HR or CHRO reports to [20].

For considering the position of HR in organizational structure, we should take into account HR-to employee ratio. This ratio measures the number of HR full-time equivalents (FTEs) for every 100 FTE employees in the organization [21]. Generally, as firms increase in size, the ratio of HR employees to total employees decreases. In a 2015, study of SHRM reported that in small organization (250 or fewer employees) had a significantly higher HR-to-employee ratio of 3,4; medium organizations 1,22 and large organizations had ratios of 1,03 [22]. These numbers decrease in time – last study of Workology [23] confirmed, that HR – to employee ration in small company was 3, in medium company was 1,2 and only 0,72 in large one.

Historically, several specific positions have developed at HR department. We provide an overview of them in the table 1.

Table 1. Positions of HR members by the company size - small (S), medium (M) and large (L)

Size	S	M	L
HR Ratio	3	1,2	0,72
Chief Human Resources Officer (CHRO) - Responsible for developing and executing HR strategy in support of the overall business plan and strategic direction of the organization. Position is aimed specifically on the areas of succession planning, talent management, change management, organizational and performance management, training and development, and compensation. The CHRO provides strategic leadership by articulating HR needs and plans to the executive management team, shareholders and the board of directors. 69% of CHROs report directly to CEO [20].	NO	YES	YES
Vice President of Human Resources (HR) provides executive-level leadership and guidance to the organizations HR operations. The Vice President takes responsibility for setting, enforcing, and evaluating legally compliant human resources policies, procedures, and best practices, and projecting long-range strategic talent management goals.	NO	YES	YES
Chief Diversity Officer (CDO) has responsibility for guiding efforts to conceptualize, define, assess, nurture, and cultivate diversity as an institutional and educational resource. Although duties may include affirmative action/equal employment opportunity, or the constituent needs of minorities, women, and other bounded social identity groups, chief diversity officers define their mission as providing point and coordinating leadership for diversity issues institution-wide [20].	NO	NO	YES
HR Business Partner (HRBP) – consultative position, providing resources and builds relationships while focusing on the strategy, mission, vision and main objectives of the organization.	NO	NO	YES
Recruiting Director – senior leader for hiring and recruiting. There are also other responsibilities – budgeting, allocation, and strategy planning with senior leadership on talent management. Usually reporting to Vice President of HR.	NO	YES	YES
HR Director – responsible for the budgets and decisions on financial investments related HR. It is mostly related on results driving for the company and less focused on compliance and policy.	YES	YES	YES
HR Manager - are responsible for human resources department and they are insurers of the functions and tasks being carried out by the HR team. They are the link between employees and an organization's management, as their work consists of variety of duties - interviewing, and hiring new staff providing consultation on strategic planning with top executives to recruiting.	YES	NO	YES
Compensation and Benefits Manager – plans, develops, and oversee programs to determine how much an organization pays its employees and how employees are paid. Benefits managers plan, direct, and coordinate retirement plans, health insurance, and other benefits that an organization offers its employees. They are also known as **remuneration professionals**.	NO	YES	YES
HR Specialist - *early career* human resources professionals who are specialists providing specific support function or generalists with limited experience. HR Specialists are focused on administrative tasks – e. g. payroll, interviewing, employee orientation, training. They can be specialized in talent management, employee relations, project management, etc. Large companies have several HR Specialists for these areas.	NO	YES	YES
HR Generalist - covers a wide variety of responsibilities relating HRM - recruiting and employee relations, handling benefits, compensations, forecasting and budgeting.	NO	YES	YES
Human Resource Information Specialist (HRIS) - High-tech position, which is often rather associated with the IT department than with the HR	YES	YES	YES

department, especially in small companies. HRIS supervises and maintains HR and recruitment technologies.			
Recruiter – person responsible for recruiting employees – aimed on hiring and selection for open requisitions. Large companies may have a team of recruiters, who are dedicated to specific lines of business, functionalities or regions.	NO	YES	YES
Recruiting Coordinator – person responsible for coordinating activities of other recruiters. More administrative position.	YES	YES	YES
HR Coordinator – reporting, processes payroll, events planning and organizing.	YES	YES	YES

Source: Own processing.

Organizational structure of HR department should be also considered from the centralization/decentralization point of view. Centralization is the degree to which coordination and control are managed by a core person or level of the organization, usually corporate headquarters. Centralized HR departmental structures help increase efficiency, effectiveness and consistency as well. Half of HR organizations (50%) are centralized, and very few organizations (15%) report that they are decentralized [24]. The remaining 35% use a hybrid model with some decisions and policies deployed commonly across the organization, and some locally. Decentralization is the degree to which decision-making authority and responsibility for coordination and control lies in the sub-units of the firm and individual managers, rather than corporate headquarters or one specific level of the hierarchy. Decentralized model of HR department means that administration and decisions made at the local level. There is a significant variation of policies and practices across field locations. No centralized HR support or administration activities. The last mentioned – hybrid model means, that HR department is half centralized and half decentralized.

Conclusion

The aim of our study was to determine how HRM's organizational structures has been changed in the past and where HRM is positioned in present. Organizational structure is the formal system of tasks and authority relationships that controls how people coordinate their actions and use resources to achieve certain organizational goals and strategy. The organizational structure is used to prevent chaos through an orderly set to reporting relationships and formally set up communication channel. For any organization, the best and appropriate structure is that one, that facilitates effective responses to the problems, facilitates communication and activity coordination. The particular position of HR department in the total organization setup depends on the size of the company. If we talk about small organizations, there might not be any separate HR department and all HR functions will be covered by office manager. In bigger unit, there are HR departments that are structured and aligned with proper business strategy. During the years there were several special positions created for HR department only - Chief Human Resources Officer (CHRO), Vice President of Human Resources (HR), Chief Diversity Officer (CDO), HR Business Partner (HRBP), Recruiting Director, Human Resource Information Specialist (HRIS) and many other. Majority of HR departments are centralized.

Acknowledgement

Supported by the Scientific Grant Agency of the Ministry of Education of the Slovak Republic and the Slovak Academy of Sciences VEGA. project No. 1/0412/19 Systems of Human Resources Management in the 4.0 Industry Era.

References

1. W. Mierzejewska. Directions of changes in organizational structures of SME-results of empirical research. foundations of management **6**, 83-92. DOI: 10.1515/fman-2015-0020 (2014)
2. A. Marjiani, F. T. Ardahaey. The relationship between organizational structure and organizational justice. Asian Social Science **8**, DOI:10.5539/ass.v8n4p124 (2012)
3. Y. Yue. Human resource management under enterprise strategic perspective. International Conference on Education, Management and Computing Technology (ICEMCT-14). Atlantis Press. (2014)
4. K. J. Rotich. History, evolution and development of human resource management: a contemporary perspective. Global Journal of Human Resource Management **3**, 58-73.
5. N. Sh. Saleh, H. F. Yaacob, M. Sh. Bin Rosli. Critical review: assessing and seeking the intricacy or discrepancy of Ulrich model transforming paradigm in organization human resource. Mediterranean Journal of Social Sciences **6**, 324. ISSN 2039-2117 (2015)
6. D. L. Deadrick, D. L. Stone. Human resource management: past, present, and future. Human Resource Management Review **3**, 193-195. (2014)
7. T. Ahammad. Personnel management to human resource management (HRM): how HRM functions. Journal of Modern Accounting and Auditing **13**, 412-420. DOI: 10.17265/1548-6583/2017.09.004 (2017)
8. J. Lemmergaard. From administrative expert to strategic partner. Employee Relations. DOI : 10.1108/01425450910925328 (2009)
9. R. M. Yusuf, S. Fidyawan, I. S. Wekke. Ulrich model on practices of human resource strategic roles. 2017. https://doi.org/10.31227/osf.io/ysnu8
10. T. Mulcaire. Progressive visions of war in" The Red Badge of Courage" and" The Principles of Scientific Management". American Quarterly **43**, 46-72. DOI: 10.2307/2712966 (1991)
11. G. Wickstrom, T. Bendix. The" Hawthorne effect"—what did the original Hawthorne studies actually show?. Scandinavian journal of work, environment & health, Vol. **26**, No. 4: 363-367. (2000)
12. R. J. Taormina, J. H. Gao. Maslow and the motivation hierarchy: measuring satisfaction of the needs. The American journal of psychology **126**, 155-177. DOI: 10.5406/126.2.0155 (2013)
13. L. Yaokui. Insight into Herzberg's two-factor incentive theory. 2nd International Conference on Economic Development and Education Management (ICEDEM 2018). Atlantis Press. p. 321-324. (2018)
14. E. Raush, T. L. Russ. Theory X/Y assumptions as predictors of managers' propensity for participative decision making. Management Decision. DOI : 10.1108/00251741111130887 (2011)

15. K. Fettig, et al. Impact of industry 4.0 on organizational structures. In: 2018 IEEE International Conference on Engineering, Technology and Innovation (ICE/ITMC). IEEE, p. 1-8. (2018)
16. S. Shamim, et al. Management approaches for industry 4.0: A human resource management perspective. In: 2016 IEEE Congress on Evolutionary Computation (CEC). IEEE, p. 5309-5316. (2016)
17. C. Sheehan,. A model for HRM strategic integration. Personnel review. DOI: 10.1108/00483480510579420 (2005)
18. SHRM - Society for Human Resource Management. *How Organizational Staff Size Influence HR Metrics*, online https://www.shrm.org/ResourcesAndTools/business-solutions/Documents/Organizational%20Staff%20Size.pdf (2015)
19. E. Farndale, M. Vidovic, E. Rockey, *Human resource management policies and practices*, United States, Cranet, online https://www.shrm.org/hr-today/news/hr-magazine/documents/cranet%20reporthr%20management%20policies%20and%20practices_final.pdf (2015)
20. K. Shellenback, mercer select intelligence, https://www.imercer.com/uploads/Global/hr_structures_today.pdf (2017)
21. D. Williams, K. Wade-Golden, The chief diversity officer. *CUPA Journal,* **58**, 38-48 (2007)
22. K. Brustowicz, D. Delano, D. Gabor, B. Salkin, M. Wagner, R. Watson, *Mandated Benefits*. Compliance Guide (Wolters Kluwer, New York, 2020)
23. J. Miller-Merrell, *HR Organizational Chart and Department Structure,* Workology, https://workology.com/your-guide-to-the-hr-organizational-chart-and-department-structuresstructure-functions/ (2019)
24. E. Farndale, L.H. Hoeksema, J. Paauwe, In-sourcing HR: shared service centres in the Netherlands. *International Journal of Human Resource Management* **20**, 544–561 (2009)

Sustainability Balanced Scorecard as Cost Accounting Instrument for Small and Medium Sized Companies

Peter Achenbach Dipl. Ing. [1*]
[1] DG Industries GmbH, Am Berg 12, 65779 Kelkheim, Germany

Abstract. This paper deals with the sustainability management concept of the Sustainability Balanced Scorecard (SBSC), which is a further development of the Balanced Scorecard (BSC) developed by Kaplan and Norton in the 1990s. In the SBSC the four classical perspectives of the BSC, which comprise "Finance", "Customers", "Internal Processes" and "Learning & Development", are expanded to include environmental and social aspects ("Non Market") enabling sustainability aspects to be fully integrated into the strategic management of companies. After a generic presentation of the elements of a BSC this paper shows the individual steps for the implementation of an SBSC, which as a heuristic for the creation of a target system, can contribute to the transformation of sustainability strategies into actions. The "openness" of the SBSC makes it possible that sustainability is not limited to a small number of companies that are primarily active in ecologically and socially oriented sectors, but appears practicable for all those companies that want to improve their performance in all three dimensions of sustainability through increases in ecological- and social-effectiveness and rises in ecological- and social-efficiency. By overcoming the "trade-off" between profit orientation and the focus on sustainability it is thus possible to achieve through enlarged consumer- and producer surplus an important contribution to the overall increase in the welfare economy.

1 Introduction

The importance of dealing with sustainability aspects is increasing daily for companies. This is due not least to global developments such as climate change, scarcity of resources and energy, but also to increased demand from the stakeholders associated with the respective companies which ultimately issue the "Licence to operate". The efficient management of economic, ecological and social shortages is regarded as a central task of management [1]. Corporate sustainability management deals with this challenge of overcoming scarcities and thus makes a major contribution to securing the future viability of the company and also the natural and social foundations. However, if operational sustainability management is not operated effectively and efficiently considerable deficits can occur in the company. In order to overcome these deficits the use of an SBSC, which has been derived from the BSC already used in many companies and supplemented by sustainability aspects, appears to be the concept of choice. The extent to which an SBSC can contribute to the identification, alignment, measurability and control of sustainability aspects will be examined in more detail below. The focus here will be on clarifying to what extent the classic understanding of the "trade-off" between profit and sustainability can be eliminated and instrumentalized in a complementary way to the benefit of companies and society.

2 Current state of the concepts and instruments of sustainability management

Figure 1 shows the main users who use some of the concepts and instruments of sustainability management mentioned in the selected list. In total, 9 concepts can be used and 37 instruments, of which

* Corresponding author: peterachenbach@t-online.de

three concepts and sixteen instruments were selected as examples here. It can be seen that the SBSC can be used by all main users to identify, quantify and control the dominant sustainability challenges (eco-effectiveness, socio-effectiveness, eco-efficiency, socio-efficiency). With regard to the other two selected concepts (SCM - Supply Chain Management, TQEM - Total Quality Environmental Management) it becomes apparent that they do not serve the complete value chain of supporting activities in companies which further underlines the integrative function of an SBSC.
In another study, a taxonomy of integrated reporting has been developed. There, a coding catalog with criteria and key figures and the degree of conformity of business/sustainability reports with identify report types has been developed, with the result that reports with an integrated focus show in particular higher connectivity and stronger compliance with sustainability standards/guidelines and emphasize the stakeholder perspective. [2].

The explanations of the individual concepts and instruments including their strengths and limitations can be read in detail under [3].

Figure 1
Selected Concepts and Instruments of Sustainability Management

Concept/ Instruments	**Main User**								
Concept	Controlling	Purchasing	R & D	Marketing	HR	Production	Accounting	Strategic Planning	Sales / Logistics
SBSC	●	●	●	●	●	●	●	●	●
SCM	●	●		●		●	●	●	●
TQEM	●	●		●	●	●	●	●	●
Instruments									
Audit	●	●				●	●		●
Benchmarking	●		●			●	●		●
Report	●			●	●		●		
Checklist	●	●		●		●			●
Cross - Impact- Analysis	●		●					●	
Env. Shareholder Value	●						●		
KPI	●			●		●	●	●	●
Costing	●						●		
Ecological Balance	●	●				●			
Eco - Design			●	●		●			
Eco - Efficiency Analysis	●	●				●			
Eco Rating					●		●		
Risc Analysis	●					●	●	●	
Socio Balance				●	●		●	●	
Stakeholder Value					●		●	●	
Material Flow Analysis	●		●			●			

3 Methodology and delimitations

In order to deal with the topic of the presentation of an SBSC for the purpose of the measurability of sustainability aspects the BSC will be explained. On the basis of a literature search the SBSC will be theoretically described. Finally, the ecological, social and economic challenges of the SBSC are enumerated in order to illustrate the further development potential of this concept (refer to Chapter 5).

4 Sustainability and cost accounting

In the following, the concept of sustainability and its management will be described by presenting specific criteria (chapter 4.1) and methodologies (chapter 4.2).

4.1 Sustainability

This concept of sustainability received increased public attention when it was defined as follows in the 1987 UN report "Our Common Future" ("Brundtland Report"): "Sustainable development is development that meets the needs of the present without compromising the ability of future generations to meet their own needs" [4]. Profits that can be achieved in the short term, which are demanded above all by the capital markets, often contradict the long-term safeguarding of a company's existence. Incentive systems and management instruments geared to short-term goals also lead to a subordinate importance of environmentally and socially compatible business practices for the individual employee. An increasing shareholder value from the point of view of sustainability therefore demands a long-term orientation of a company through its business strategy. The consistency between sustainability indicators at national level and by companies has been investigated in [5]. Among other findings, it can be shown that indicators for each SDG are developed at the national level, but lag behind on several topics at the corporate level. Consequently, there is potential for improvement with regard to the definitions of KPIs in order to optimally align companies with the achievement of the 17 global sustainability goals. In further work, motives of small and medium-sized enterprises for the application of sustainable social and environmental practices have been investigated. The results show that SMEs use a complex mix of symbolic and substantive CSR practices to increase the legitimacy and sustainability of their activities, with the intended result of strengthening social commitment, reputation and image [6]. Furthermore, a sustainable approach to business requires the integration of ecological and social aspects in addition to economic aspects when making decisions. Depending on which of the three possible approaches to sustainability - economic, ecological or social - is chosen, a distinction can be made between various criteria which will be presented below [7].

4.1.1 Six Criteria

Figure 2 shows the six criteria of corporate sustainability.

Figure 2
Six Criteria for Corporate Sustainability

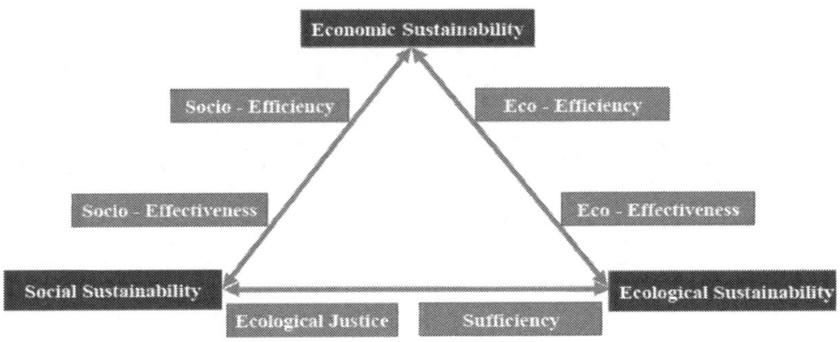

Looking at the perspective of economic sustainability the central question is how the financial success of a company can be increased through the sustainable management of natural and social capital. In essence, this is about increasing the eco- and socio-efficiency of the company. The origins of eco-efficiency can be found in the concept of efficiency according to which the aim is to achieve a maximum or a given output with a given or minimum input of resources according to the economic principle as well as in the demand for adequate consideration of ecological aspects within the system boundaries considered. The term "added value" in the form of the desired output as well as "harmful creation" are also of central importance where the (negative) environmental or social impacts of the product or process under consideration are recorded. Priority should be given to economically oriented eco-efficiency in the form of an objective based primarily on economic principles, but always with mandatory consideration of ecological aspects, which can be summarised in the concept of sustainable eco-efficiency. This manifests itself in a permanent creation of added value in the form of increased output with a simultaneous reduction of the input and reduction of the environmental impacts associated with the system. The analysis of the system-related environmental impacts can be carried out on the basis of weighted criteria specified in the form of energy consumption, material consumption, land consumption, emissions, risk potential and toxicity potential [8]. The technical term "socio-efficiency" describes the social effects of a product or process in relation to the total costs for the end consumer. This can be represented by key figures to be determined, which can find their formulation in value-added or product-related work, for example in the form of "person hours per value added" or "product life cycle costs". If both eco-efficiency and socio-efficiency analysis are available for a specific system a combination of the generated eco-efficiency and socio-efficiency portfolios can be used to create an overall graphic presentation of all evaluation results in the form of a three-dimensional illustration (abscissa: costs; ordinate: environmental impacts; z-axis: social impacts). The result is a comprehensive socio-eco-efficiency analysis of the relevant product or process which identifies, quantifies and evaluates not only the economic but also the ecological and social aspects of the analysed target. On the basis of the results obtained various product or process alternatives that generate the same benefits can then be compared in terms of costs, environmental and social aspects using relative indicators. With regard to ecological sustainability, the goal is not only a relative improvement compared to value creation but an absolute improvement of the environmental situation by optimizing the entire product life cycle with the help of an assessment of the environmental impacts according to global problematic areas. In addition to the eco-effective satisfaction of needs by the economy a change in the purchasing behaviour of consumers under the heading of sufficiency is also required in order to improve the ecological dimension of sustainability. In the context of social sustainability the socio-effectiveness of a company's activities must be designed in such a way that it can withstand assessment in the context of general societal challenges and "does the right thing". Social sustainability is completed by ecological justice describing the treatment of natural resources that includes the rights of all living beings. In addition to humans the rights of animals and plants are also included in the concept. Ecological justice in the sense of a holistic approach thus deals with social and spatial distributive and procedural justice in questions of human interaction with the biotic environment. A central task of corporate management is to manage these six criteria of operational sustainability in order to contribute to securing the future viability of the company without neglecting the natural and social foundations that form the basis of this future viability. In principle, it was primarily large multinational companies that first introduced corporate sustainability management or implemented Corporate Social Responsibility (CSR) [9]. The European Commission sees as "pioneers in the development of corporate social responsibility mainly larger companies" [10]. However, dealing with sustainability issues increasingly also applies to small and medium-sized enterprises (SMEs), which represent more than 99% of the German business community. They are the main pillar of the economy and therefore of at least equal importance for its transformation towards sustainability as large enterprises. Companies are increasingly introducing environmental and social management systems (e.g. ISO 14001, EMAS, ISO 26000, SA 8000) in order to align their sustainability management accordingly. However, these systems represent only partial solutions as they only take into account certain dimensions of sustainability with the consequence that they are isolated from general management systems. The weak points of environmental- and social- management systems therefore lie in their lack of focus on the financial success of companies. In order to make a contribution to the sustainable development of a company environmental- and social management systems must also include the economic dimension of sustainable development. A value orientation of sustainability management can guarantee this. For effective sustainability management it is also necessary to make use of strategic instruments as well as to ensure operational implementation. Furthermore, long-term risks and opportunities are often not taken into account, the management systems are not strategically oriented and often do not reach the core business. Overcoming these deficits thus becomes a decisive challenge for effective and efficient sustainability management. In the following chapter in particular one concept is presented that makes such sustainability management possible

4.2 Concepts of sustainability management

In recent years, scholarly interest in corporate social responsibility (CSR) has increased; however, research on the relationship between CSR and firm performance has revealed a complicated relationship. Part of the basis for the generally positive relationship between CSR and corporate performance comes from a reduction in agency costs (based on Behavioural Agency Theory) [11], which can be managed using appropriate sustainability concepts with suitable methods and tools. Nine concepts are listed in the literature of which the BSC and the SBSC derived from will be examined in more detail below as a sustainability management concept.

4.2.1 Balanced scorecard

The basic idea of this concept was to derive qualitative and quantitative objectives and indicators from a company's vision and strategy. In order to promote a multidimensional and balanced way of thinking in deriving individual strategic goals from the strategy, Kaplan and Norton proposed four perspectives for the BSC [12,13,14], on the basis of which managers should regard the value chain of their company: - Finance; - Customers; - Internal Processes and - Learning and Development (refer to Figure 3). The goals thus achieved are linked to each other and to the financial goals which are reflected in the financial perspective. The developed BSC focuses on achieving financial goals which is why the strategic goals of the financial perspective are at the top of the causal chain. The logic of the causal chain is that optimisation of operational performance and new products and services are directly or indirectly linked to a better financial result [15]. By creating this system of strategic goals the next step is to allocate metrics to the individual goals to create a balanced set of key figures. When developing the indicators care must be taken to ensure that, in addition to the traditional "Lagging Indicators", so-called "Leading

Indicators" are also taken into account [16]. These leading indicators, also known as "Early Indicators", measure the reasons that should lead to the desired effects in the future - measured with "Lagging Indicators". After deduction of the goals, the formation of cause-effect relationships and the generation of key figures further values are to be defined which additionally specify the corresponding strategic goals. This defines the target levels for the strategic goals which can be achieved through appropriate strategic measures. The definition of strategic measures in the form of a "Road Map" is at the same time the last step in setting up a BSC which forms a kind of „conceptual framework" [17]. The number and type of perspectives used and the content derived from the strategy must be determined individually by each company and enable the implementation of strategies in actions. When the BSC is extended from a key performance indicator system to a strategic management system the following steps can be distinguished [18]: - Translating the corporate vision into strategy: Consensus can be reached on the strategy which often exists only implicitly in the minds of managers. - Communication of the strategy: Top-down from the management to the downstream areas. Matching the objectives of the individual employees, the teams or the departments with the strategic objectives of the company or business unit strategy takes place. - Planning and targets / strategic feedback and learning: The defined strategic activities link the strategic planning with the medium-term and operative planning. A distribution of resources coordinated with the corporate strategy is achieved. Key performance indicators enable a permanent comparison between target and actual results and verification of the appropriateness of the assumptions made. This induces a learning process and enables continuous work with the BSC. - The final goal is an organization in which all employees are consistently oriented and aligned to the strategy. In line with the structure of the classic BSC the financial perspective is the primary objective responsible for "increasing shareholder value" [19], the achievement of which is measured using key performance indicators such as ROI, ROCE or EVA.

Figure 3
The Balanced Scorecard

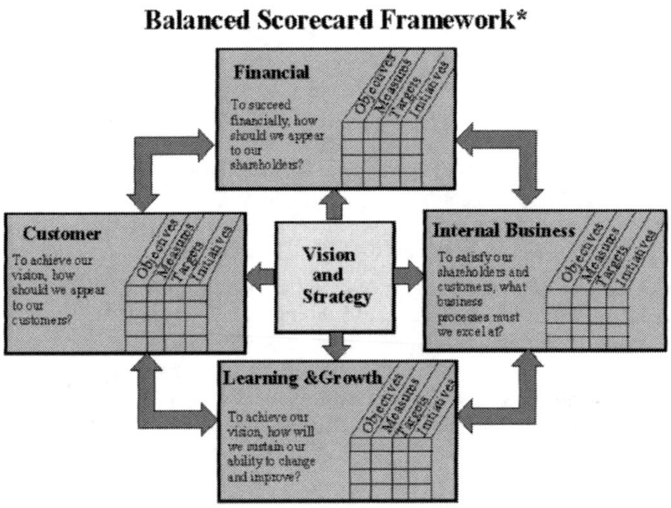

Based on Kaplan's and Norton's practical experience there are two basic strategies for achieving this goal: Sales growth and productivity. If a company pursues a growth strategy it looks for new products, markets and customers as well as the increase of the benefit to the client in order to increase turnover and profitability. An increase in productivity is primarily achieved through cost reductions and efficiency increases in existing activities or processes. From the customer's perspective it is considered which segment of the customer is to be dealt with, with the aim of achieving the economic success of the company. The core performance indicators are the following: Customer profitability; Customer satisfaction; Customer loyalty; Customer acquisition; Market share. Leading indicators for the lagging indicators listed before are derived from one of the following individual differentiation strategies: Strategy of product leadership; Strategy of customer loyalty; Strategy of operational excellence. Targets and indicators for the internal process perspective must be aligned with this strategy and reflected in the value proposition for customers. The sub-processes can be oriented towards the value chain "Innovation", "Operation/Production" and "Customer Service"(lagging indicator) with their key figures (leading indicator) regarding "Quality", "Costs" and "Time" such as "punctual deliveries", "ppm error rate", "yield", "material waste" or "process costs". In the learning and development perspective the required abilities/motivation of the employees as well as the necessary structure organizations and operational organizations are described which are needed for the objective accomplishment of the other perspectives specified before. Proposed are the categories "Employee Loyalty", "Employee Productivity" and "Employee Satisfaction" (lagging indicator) with the corresponding leading indicator "Working Climate", "Infrastructure" and "Personnel Potential". The SBSC as a further development of the BSC will be presented in the following Chapter 4.2.2

4.2.2 Sustainability balanced scorecard

Through its value-oriented view and the transformation of corporate environmental strategies and social strategies into operational parameters it represents a suitable concept and is able to answer the question of the measurability of sustainability in companies. The purpose of the SBSC is to integrate the three pillars of the sustainability concept into the respective strategy of the company whereby the company performance is to be increased through improved sustainability contributions of all criteria. Since the conventional BSC hardly takes environmental and social aspects into account the approach of an extension of the BSC by an additional perspective with regard to environmental and social aspects is recommended to be used to formulate an SBSC. A so called "Non-Market Perspective" is introduced ensuring that those environmental and social aspects that have not yet been considered are also integrated which appear essential for the successful implementation of the company strategy. The steps necessary to create an SBSC are explained as follows and can be obtained from Figure 4 [20, 21]. The first step is the selection of the strategic business unit for which an SBSC is to be created. A strategy for this unit must be available or formulated for this purpose. The second step is to determine the environmental and social exposure of the business unit. The aim is to draw up as complete as possible a catalogue of requirements for the strategically relevant aspects of the business unit. In the last main step the strategic relevance of environmental and social aspects is determined. The individual perspectives are examined starting with the financial perspective up to the non-market perspective and all identified environmental and social aspects are checked for their strategic importance. Three levels can be classified with regard to the strategic relevance of environmental and social aspects: a) Strategic core elements leading directly to key earnings figures. b) Ecological or social leading indicators. c) Hygiene factors [22] are defining the relevant aspects. By using the top-down approach of the individual perspectives the causalities of the various aspects can be mapped to the sustainable success of the company. The influence of the identified leading indicators and lagging indicators on the higher-level perspective must be examined consistently. Thus, all environmental and social aspects can be aligned either directly or indirectly with the financial perspective and finally with the economic performance of the company.

Figure 4
Procedure for Formulation of a SBSC

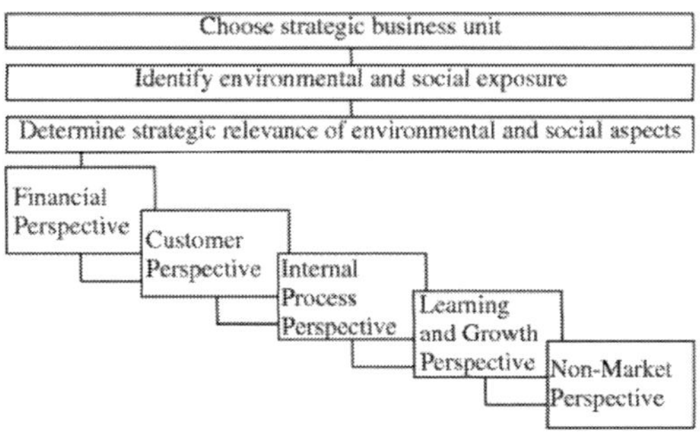

 https://doi.org/10.1051/shsconf/202111503002

5 Conclusion and recommendations

The vision of sustainable development is not a mayfly and is a topical issue for more than 70 % of large companies and more than 30 % of all companies in Germany. The challenges facing our society and the companies operating in it can be identified ecologically in the reduction of raw material and energy consumption, socially in the demands for transparency and accountability to stakeholders and economically in the growing need for value orientation in environmental and social management.

Here the validity of an SBSC as an integrative concept comes to light which overcomes the previous deficits by identifying environmental and social aspects relevant to success and establishing causal links with the company's performance. Furthermore the management of the allocated aspects according to relevance is made possible and an integration of environmental and social management into general management is feasible. This can be manifested in improvements related to environmental- and social- effectiveness and enhancements in environmental- and social- efficiency. The openness of the SBSC also makes it possible that sustainability is not limited to a small number of companies that are primarily active in ecologically and socially oriented industries. The concept is therefore well suited for all companies that want to improve their performance in all three dimensions of sustainability regardless of size.

The following recommendations can be summarized from the findings in order to overcome the trade-off between profit orientation and sustainability alignment with the aim of generating an important contribution to the overall well-being of society:

- Formation of strategic alliances [23].
- Development of a systemic framework: The literature shows a lack of consensus on how to establish a clear link between SBSC architecture and environmental performance outcomes. As a result, a holistic conceptual framework in which SBSC knowledge acts as a mediator and the presence of experts as a facilitator could provide a more consistent relationship between SBSC architecture and environmental performance outcomes [24].
- Review and assurance of supply chain materiality through the SBSC. Explicit targets are to be formulated around the selected SDGs, including setting performance indicators and monitoring progress [25].
- Cooperation of technical disciplines with "Finance" and "Controlling" has to be intensified;
- Long-term risks and opportunities need to be taken into account;
- Be "1. Mover" and introduce CSR -with SBSC- into SME's. This can prevent future obligations that SME's will have to fulfil in the near term;
- Rewarding entrepreneurial (sustainability) activities through investment policy and public procurement and ratings as extrinsic incentives;
- Establishment of training measures, e.g. in mentoring programmes and founding circles of industry experts that communicate and cooperate transparently in order to exchange experiences;
- Introduction of longitudinal studies which, for example, are accompanied by universities and accumulate discovered improvement potential as "lessons learned" and transfer it into practice.

References

1. Figge, F., Hahn, T., Schaltegger, S. & Wagner, M. Sustainability Balanced Score-card. Wertorientiertes Nachhaltigkeitsmanagement mit der Balanced Scorecard. Leuphana Universität Lüneburg, 6 (2001).
2. Lueg, K. Deconstructing corporate sustainability narratives: A taxonomy for critical assessment of integrated reporting types (2021).
3. BMU / BDI. Nachhaltigkeitsmanagement in Unternehmen; CSM; Universität Lüneburg, Schaltegger, Herzig, Kleiber, Müller (2002).
4. Hauff, Volker. Unsere gemeinsame Zukunft, Greven, 46 (1987).
5. Malay, OE; Aubinet, S. Improving government and business coordination through the use of consistent SDGs indicators. A comparative analysis of national and business sustainability indicators. *Ecological Economics*, 184, Art. No 106991 (2021).
6. Crossley, RM; Elmagrhi, MH, Ntim, C. G. Sustainability and legitimacy theory: The case of sustainable social and environmental practices of small and medium-sized enterprises. *Business Strategy and the Environment*, (2021).
7. Bieker, Thomas / Dyllik, Thomas/ Gminder, Carl-Ulrich / Hockerts, Kai. Management unternehmerischer Nachhaltigkeit mit einer Sustainability Balanced Scorecard. Forschungsmethodische Grundlagen und erste Konzepte, St. Gallen, 22, (2021).

8. Pierobon, Marianna. Nachhaltigkeitsbewertung mit der Ökoeffizienz-Analyse, NTA4 - Vierte Konferenz des „Netzwerks TA", Berlin, 22, (2010).
9. Gelbmann, U. & Baumgartner, R. J. Strategische Implementierung von CSR in KMU. In A. Schneider & R. Schmidpeter (Hrsg.), Corporate Social Responsibility: Verantwortungsvolle Unternehmensführung in Theorie und Praxis. Berlin, Heidelberg: Springer, (2012).
10. Europäische Kommission. Grünbuch. Europäische Rahmenbedingungen für die soziale Verantwortung der Unternehmen. Brüssel, 31, (2001).
11. Greiner, M., Sun, J. How corporate social responsibility can incentivize top managers: A commitment to sustainability as an agency intervention. *Corporate Social Responsibility and Environmental Management,* (2021).
12. Kober, R; Northcott, D. Testing cause-and-effect relationships within a balanced scorecard. *Accounting and Finance*, 61, 1815-1849 (2021).
13. Massingham, R. Improving integrated reporting; A new learning and growth perspective for the balanced scorecard, (2019).
14. Kaplan, R. und Norton, D. Using the Balanced Scorecard as a Strategic Management System. *Harvard Business Manager*, 75, (1996).
15. Dusch, Michael/Möller, M. Praktische Anwendung der Balanced Scorecard - Ein neuer Ansatz zur Fabriksteuerung in der Philips Bildröhrenfabrik Aachen. *Controlling*, 8, 119, (1997).
16. Horvath, Peter/Kaufmann, L. Balanced Scorecard - ein Werkzeug zur Umsetzung von Strategien. *Harvard Business Manager*, 20, (1998).
17. Ehrmann, H. Kompakt-Training Balanced Scorecard, 2. Aufl., Ludwigshafen, 50/186, (2002).
18. Kaufmann, L. ZP-Stichwort Balanced Scorecard. *Zeitschrift für Planung*, 8, 422 (1997).
19. Kaplan, R. und Norton, D. Balanced Scorecard. Strategien erfolgreich umsetzen, Stuttgart, 49, (1997).
20. Figge, F., Hahn, T., Schaltegger, S., Wagner, M. The Sustainability Balanced Scorecard – Translating Strategy into Value – Based Sustainability Management. Conference Proceedings of the 2001 Business Strategy and the Environment in Leeds, UK. ERP Environment, Shipley, (2001)
21. Hansen, E. G., Schaltegger, S. The Sustainability Balanced Scorecard: A Systematic Review of Architectures, (2016).
22. Herzberg,F. , Mausner. B., Snyderman, B. The motivation to work. 3 rd print., New Brunswick, NJ: Transaction Publishers, (1999).
23. Russo, A., Schena, R. Ambidexterity in the context of SME alliances: Does sustainability have a role? *Corporate Social Responsibility and Environmental Management*, 28, 606-615 (2021).
24. Jassem, S; Zakaria, Z; Azmi, A. C. Sustainability balanced scorecard architecture and environmental performance outcomes: a systematic review. *International Journal of Productivity and Performance Management*, (2021).
25. Claro, PBD, Esteves, N. R. Sustainability-oriented strategy and Sustainable Development Goals, *Marketing Intelligence & Planning*, 39, 613-630 (2021).

Human Resource Management in Healthcare

Jana Blštáková[1,] , *Jana* Palenčárová[1]

[1]Department of Management, Faculty of Business Management, University of Economics in Bratislava, Dolnozemská cesta 1, 852 35 Bratislava, Slovakia

Abstract. Contemporary healthcare systems face several challenges. The main management challenges include shortage and low satisfaction of health professionals. Satisfaction and motivation of medical staff is crucial for their stabilization and quality work-medical performance. The positive impact of HRM practices on increasing employee satisfaction and engagement is scientifically proven. Currently, HRM practices in healthcare are used in a limited way, often only at the level of HR or labour law and union bargaining. This is indicative of the low level of HR development. It is desirable to make full use of HRM practices in healthcare. The aim of this paper is to identify and define the challenges of human resource management in healthcare and explain the importance of implementing developed HRM practices to improve the quality of health care delivery. The research method used is a literature search. The investigation of the relationship between quality human resource management and healthcare delivery is considered important because the knowledge and competency to manage people are not necessary to hold a management position in most healthcare professions in Slovakia. In the world's best hospitals, we find developed HRM with a positive impact on employee satisfaction and medical indicators.

Introduction

Healthcare as a system is a major social, economic and political issue worldwide. According to Ozorovský and Vojtek, in many countries there is a drive to reform the health care system because for various reasons there is dissatisfaction with the level or manner of care provided [1]. As a result of the COVID pandemic, the need for health reforms continues to grow.

Several authors agree that EU health systems are increasingly interacting and have faced increasing common challenges over the last decade [2,3,4].

The sector suffers from a shortage and unequal distribution of health workers. The ageing population, coupled with the rise of chronic and age-related diseases, is leading to an increasing demand for healthcare. Health systems are facing rising healthcare costs due to the continuous development of technology and increasing demands on competences. As a result of technological and medical advances, the structure of procedures is also changing, highly specialised activities are being centralised, and demands for safety, quality and efficiency have increased, even for less specialised interventions. Further, there are inequalities in preventive healthcare and access to healthcare [2].

The aforementioned realities are particularly challenges for the management of health care facilities. It is necessary to seek strategies for performance management and medical staff development in line with technological and demographic trends in society. We consider it important to pay attention to human resource management tools and policies for managing people in healthcare. The application of modern human resource management tools in healthcare is one of the key responses to the challenges of healthcare.

2 Literature review

The development of human resource management follows the development of the economy and the advent of innovation during all 4 industrial revolutions. The first HR managers and the first HR departments appeared in manufacturing companies at the turn of the 19th and 20th centuries and their work was initially limited to administration, accounting and resource planning. For the purpose of knowing the level of quality of human resource management in healthcare, we consider it necessary to describe the developmental stages and to recognize the development of personnel management. The development of human resource management can be briefly described by 4 basic developmental stages [5].

The basic level of human resource management (HRM) is the personnel department, which provides personnel administration, payroll accounting and basic labor law. In the second tier, there is a specialization of HRM into selection, training, organizational design management, and compensation. The HR function includes HR service centres that provide services to employees and HR business partners that support managers at a strategic level. The third level is a superstructure in which HRM provides integrated talent management. New roles are succession planning, talent management, leadership support, and the development of a coaching culture. The most advanced departments of HRM are fully integrated with the business, are digitized and can predict future developments and deliver value through big data analytics. They are constantly expanding their knowledge and their impact [6]. They are not focused on what they do, but on what they deliver [7]. The impact of HRM activities on the performance of the organization is measured and evaluated.

The impact of the 4.0 era on HRM through changes in the values, roles, architecture and content of HRM is described by Blstakova's model [8], which, through conceptual issues of HRM, can form the basis for setting a developed HRM strategy and subsequent sub-strategies.

Several global studies have confirmed the relationship between the use of modern, developed HRM systems and the organizational performance of a company [9]. Similarly, Jankelova's research [9] in 44 Slovak hospitals confirmed the positive impact of the blended role (mature role) of HRM in healthcare facilities (hospitals) on their organizational performance. For example, West et al. reported that HRM systems in healthcare facilities directly influence the quality of healthcare provided and the mortality rate of patients in hospitals [10].

West et al. list specific practices of developed HRM that previous research suggests are most important to use - sophisticated management appraisal systems, staff safety, and the level of investment in people (as expressed by effective human resource management and, in particular, investment in training and development) [10]. Similarly, Townsend et al. state that, an HRM system must use advanced techniques (such as miracle question) to understand employee problems and solve them [11]. West points out that it is not enough to use advanced HRM techniques in isolation, but it must be a group of interrelated high-performance HRM practices [10]. Townsend et al. state that the growing importance of

HRM is a signal to employees that employees also have a strong position in the company [12]. The development of HRM strategies and tools is not homogeneous. The development of HRM is often positively influenced by the competitive environment, the form of ownership of the organization and its size. HRM develops fastest in highly competitive environments with high demand for quality labour (e.g., IT services, telecommunications sector). In sectors such as primary agricultural production, human resource management is sufficient to cover basic staffing activities in the long term.

The positive relationship between a well-developed HRM system and transformational leadership of managers in the healthcare sector, which in turn has a positive impact on employee satisfaction and loyalty, has been described in studies by several authors [9].

According to Joniaková and Jankelová, burnout and low job satisfaction are among the problems of contemporary healthcare [13]. Meanwhile, according to West et al. the satisfaction and motivation of medical staff is crucial for their stabilization (reduction of attrition to other departments, increased interest in studying and working in healthcare) and quality work performance (reduction of patient mortality, compliance with procedures and regulations, reduction of nosocomial infections, etc.). [10, 13]. Other studies show that job satisfaction is among the key factors for health professionals worldwide [14]. The positive impact of HRM practices on increasing employee satisfaction and motivation has been scientifically proven by several authors [10,13, 15, 16]. It is desirable to make full use of the current knowledge on human resource management in healthcare.

3 Methodology

The aim of this paper is to identify and define the challenges of human resource management in healthcare and to explain the importance of implementing well-developed human resource management practices to improve the quality of healthcare delivery. The research method is a literature search. The article is the result of analysis of published studies, analogy and comparison of relevant findings. The result is a synthesis and enrichment of the current knowledge about human resource management in healthcare in Slovakia. The article presents the author's personal experience with human resource management practices and procedures from the position of an HR leader of a new generation hospital in Slovakia. The research findings and discussion are enriched with her own empirical knowledge about the practical human resource management in the context of inpatient healthcare.

4 Research results and discussion

Based on an examination of the findings of published studies, we find a demonstrated link between the sophistication of human resource management systems and medical performance. The subject of the research published in the article was the current challenges for human resource management in healthcare in the Slovak Republic and the readiness of medical personnel with managerial competence to manage people. Examples of good practice of developed human resource management from world hospitals were part of the research.

4.1 The challenges of human resource management in healthcare

In order to define the main challenges facing human resource management in the Slovak healthcare system, it is important to know the broader context and development trends. A major influence is the persistent shortage of health workers in Slovakia. In 2019, the Health

Policy Institute found from statistical collection that there is a shortage of 2,900 doctors and 3,600 nurses [14].

In 2030, the Institute predicts that this will be nearly ten thousand nurses [14]. The 2017 census of doctors in Slovakia showed that there are more than four thousand doctors from Slovakia abroad, most of them in the Czech Republic, according to Haníková and Koník [18]. The shortage of doctors and nurses worries more than 82% of Slovak hospital directors involved in the HealthCare Institute's Healthcare Barometer 2020 survey, a non-profit organization focused on improving the quality of healthcare provided in hospitals [19]. As a consequence of staff shortages, the problem of overtime work of doctors and nurses is growing. More than half of hospital directors consider this to be a problem for physicians, and as many as two-thirds of directors consider it to be a problem for nurses [19]. Staff shortages also have a negative impact on the training of young doctors in clinics. In a state where there is a shortage of staff in the wards, doctors do not have time for med students [18]. Teaching others is seen, as a burden and does not support the transition to a learning organization in the future. Another management challenge in the health sector is relatively low competitiveness of remuneration, which results in health workers going abroad (doctors and nurses) or to other sectors where they receive higher pay for equally demanding work (especially nurses).

The historically stable healthcare environment has changed significantly over the last two decades, with an outflow of workers abroad and a decline in interest in the healthcare profession in general. Unaware of the seriousness of the situation, the HRM units have reacted to the changes in the environment mainly in a transactional manner. For example, they compensated for staff shortages by increasing financial compensation and benefits, often at the expense of deepening the organization's losses [17]. From author´s own empirical experience, we conclude that transformational tools of higher levels of HRM, such as defining the employer value proposition (EVP) for the employee, talent management, and the development of leadership skills of managers, can be observed only in their infancy so far.

At present, modern HRM tools are used in the healthcare sector in a limited way - often only at the level of wages and personnel administration, labour law and union bargaining. This indicates a low level of HR development [9, 10].

Empirical evidence shows that in Slovakia we can observe significant differences in the development of HR departments also between state, Slovak and foreign organizations. More often we find developed HR departments especially in organizations with highly competitive pressure and in foreign organizations that bring good practices from their parent headquarters. Furthermore, the possibility of specialisation of individual HRM staff increases with the size of the organisation and this allows for an increase in the quality of HRM processes.

For a long period of time, the Slovak health sector was a stable environment with absolute predominance of state ownership, while in the field of inpatient healthcare, individual hospitals belonged in size to medium-sized organisations with a regional monopoly. These characteristics did not create an environment in which there was pressure to develop the quality of HRM and, compared to other organisations (especially foreign ones in a highly competitive segment), therefore, human resource management in institutional healthcare facilities lagged behind.

The main role of the HR department in hospitals is usually recruitment, provision of mandatory statutory training and payroll. It is often part of the finance or legal department. This empirical evidence shows that in most hospitals in Slovakia the HRM department is at the first or second development level.

The fact that Slovak hospitals are generally not using developed HR processes nor are participating in the management of corporate culture is documented in the article by Raijani et al. [20], in which they describe in detail the managerial processes in Slovak hospitals, but with only a marginal mention of the current active role of human resource management. This view is confirmed by the results of the Healthcare Barometer survey, according to which the most popular measure to ensure sufficient staffing is benefits such as company meals and extra holidays for up to 91% of Slovak hospital directors [19]. However, these tools are transactional and do not address the causes of disinterest in the healthcare profession and a given hospital.

When interviewing health workers, we repeatedly encounter information that selections are not transparent, recruiters do not respond to CVs sent, adaptation of new employees is unmanaged and the management skills of supervisors are weak. Similar experiences are communicated by the media from time to time [18].

Low level of managerial skills has been also observed by author during selection procedures for managerial positions in nursing for the New Generation Hospital Bory. Candidates usually had excellent professional knowledge and skills, but their management style was predominantly directive, coupled with low self-awareness of its impact on the teams being managed. In doing so, Jankel and Joniakova [13] state that it is the management skills, especially the managerial style of the leader, that have a significant impact on employee satisfaction and motivation. They point out that organizational support (e.g., including HRM systems) has been identified as a predictor of employee satisfaction and manager empowerment. Alshamari emphasizes that organizational culture plays a significant role in enhancing organizational performance [14]. Several authors emphasize that an organization needs a well-developed HRM system to enable managers to be effective. Leadership can only be as good as the infrastructure that supports it [13, 21].

4.2 Readiness of managers to manage people in the health sector in the SR

Nowadays, the best medical experts often become managers in Slovak medical institutions. When they take up the position, they often have only basic managerial knowledge and learn the management of the entrusted department on the fly [22]. These first-line managers (senior physicians, station nurses, day shift managers) later move up the career ladder to middle management (chief physicians, head nurses) and up to hospital management (deputy medical-preventive service, deputy for nursing, hospital director).

Management and human resource management in the Slovak healthcare sector is not a prerequisite for all positions (e.g., for nurses, a minimum of 15 years of experience is sufficient). It is taught only marginally (1 semester) in undergraduate medical studies and includes the basics of healthcare management and leadership. Theoretical knowledge is not supported by practice in the workplace [23].

HRM is also taught within the management specialties that are the basis for the medical management position. HRM is usually only a small part of the content, which is surprisingly low for a sector with a high proportion of value added by human labour. The content presented is second level HRM, i.e., strategic HRM with a strong emphasis on selection, recruitment and training of staff and, from the organisational area, on the motivational skills of managers.

The question remains of translating what has been learned from the human resource management field into practice. According to research by Wild and Fulopova [24], graduates of the MPH program subjectively rated their theoretical knowledge and practical skills on average at the very good level. On the basis of more than 400 interviews completed with healthcare professionals over the last 2.5 years, we can conclude that the above knowledge often remains at the level of knowledge and is not translated into behaviour change of healthcare managers or into HRM processes.

Alongside the still prevailing model of the manager-expert (doctor, nurse, etc.), an independent role of the professional manager in healthcare has emerged in the Western Europe. These are managers who have graduated in healthcare management (MPH, MSc., MHA, or others) and their role in the healthcare facility is not medical, but managerial. The staffing of management professionals into senior healthcare positions is seen in countries such as the UK, Spain, and Germany, where the decentralization of healthcare has increased the need for management skills at the highest levels of healthcare facilities [22]. In Slovakia, a similar trend is gradually occurring - professional managers without a medical background take leading roles in hospitals (e.g., Ing. Miriam Lapuníková, MBA - Director of FNSP F.D. Roosevelt Banská Bystrica, or Ing. Marián Haviernik – Director of the New Generation Hospital Michalovce).

4.3 Advanced human resource management in hospitals - examples of good practice

HRM in the healthcare segment of the economy is not usually among the best the world of HR has to offer. According to Lüthy [25], there is still a negative attitude among hospitals towards trying to be an attractive employer. Hospital management often believes that working in a hospital is fundamentally very difficult and there is not much that can be done about it. However, according to Lüthy [25] this is not the right approach and suggests measures to improve the situation: professional leadership, motivation and rewarding of co-workers, further education and career planning, optimization of processes and avoiding bureaucratization, flexible working arrangements, work-family balance, and the formulation of codes of ethics to deal with internal and external stakeholders. For example, a study from Germany concludes that a hospital is not an attractive workplace, but can become one if it improves in particular personal development opportunities and the work life balance [26].

Below we give examples of good practice, of some hospitals that are successful in ensuring the above and their HRM departments are in the third to fourth developmental level of human resource management.

The Mayo Clinic network of hospitals is one of the best healthcare facilities in the United States. Their leadership and corporate culture rating are A to A+, according to Comparably.com, a site comparing multiple US employers over the long term [27]. HRM at Mayo Clinic supports the organization's strategy and measures the impact of HRM activities and policies on its performance, which is typical of fourth level of HRM. Mayo Clinic has continuous professional learning in its organizational DNA, but also strong support for diversity and adherence to the organization's values, with teamwork as a core value.

The largest private hospital network in Germany is Helios. Employing over 66,000 employees in 86 clinics throughout Germany and treating over 5.2 million patients annually, it addresses the topic of human resource management and productivity long term and professionally. According to Bornewasser et al. Helios managers point to several areas that enable high employee motivation and the associated continuous improvement of the productivity of their clinics [28]. High labour productivity brings the network long-term growing revenues, which are further invested in increasing healthcare delivery. The main employer levers at Helios include education, career opportunities, work-family balance and healthcare.

An example of a high level of HRM is the British NHS (National Health Service). The NHS is the largest public healthcare provider in the UK and the second largest public healthcare provider in the world (after Brazil). The NHS presents its human resource management to a high standard. It has elaborated in detail the different HR specialties (selection, training, remuneration, ...) and presents them openly externally through its web portal [29]. It makes significant use of elements of the third and fourth development levels of HR. A clear HR strategy is publicly communicated and accessible to all employees on the intranet and to job applicants and the public on the external web. For example, the employer's employee value proposition - description of what it is like to work for the organization - is communicated through the "People Promise" video. In it, individual employees talk about their experience in the NHS, communicating the culture of the organisation and how they work within it. The NHS actively responds to societal challenges and encourages new ways of working (flexible working for all roles, working remotely via technology - e.g., through video consultations with a doctor), new innovative roles (e.g., involving scientists in the treatment process), acquiring new skills (generalist skills from other areas) and learning through technology (a significant part of learning available 24/7 online). In the area of management and leadership, organization emphases the need for inclusive situational leadership, taking into account the diversity of the workforce and creating space for employee empowerment. It relies heavily on data collection, data analysis and modelling of future scenarios [30].

Conclusion

Based on current knowledge about human resource management in the healthcare, we find a proven relationship between developed human resource management and healthcare quality. Research findings highlight the need for a focused improving the functioning of relevant HRM systems in healthcare as one of the factors for improving patient care [10]. The quality of healthcare and the level of medical performance in Slovakia have long been undermined by a shortage of healthcare personnel and the poor readiness of health care managers to manage people. Research findings suggest that the key challenge of human resource management systems in healthcare in the Slovak Republic is attraction and retention. The way to build the attractiveness of the health profession and eliminate the departure of Slovak health professionals abroad is quality human resource management in its mature stage of development. Furthermore, defining and communicating the value of the employee experience of working in healthcare is important. According to Jankelova [9], it is leaders who play a key role in creating the organizational conditions in which a developed HRM system can be implemented. Here we see the role of management education of healthcare employees, which does not remain only at the level of the University curriculum, but continues further on the job. We see the quality of human resource management in hospitals as the foundation of quality healthcare.

Aknowledgement

Supported by the Scientific Grant Agency of the Ministry of Education of Slovak Republic and the Slovak Academy of Sciences VEGA Project No. 1/0412/19 Systems of Human Resources Management in 4.0 Industry Era.

References

1. V. Ozorovský, I. Vojteková a kolektív, Zdravotnícky manažment a financovanie, *Wolters Kluwer,* **2016**, 168-178, (2016) ISBN 978-80-8168-522-4

2. Európska Komisia, Biela kniha o budúcnosti Európy - Úvahy a scenáre pre štáty EÚ-27 k vývoju do roku 2025, *Európska komisia*, **2017**, (1. 3.2017) ISBN: 978-92-79-67631-4, DOI:10.2775/88238
3. P. Hroboň, Changing forms of care needed to ensure accessibility and quality - lecture during Prague International Health Summit, *intern document of Svet Zdravia – online conference recording*, **1.-2.10.2020** (2020)
4. W. Quentin, Transformation of national hospital networks - lecture during Prague International Health Summit, *intern document of Svet Zdravia – online conference recording*, **1.-2.10.2020** (2020)
5. J. Bersin, Talent Strategies in Asia: Do Asian Leaders Behave Differently?, *HR Management App*, (2019) visited 16.1.2021 on https://www.hrmanagementapp.com/talent-strategies-in-asia-do-asian-leaders-behave-differently/
6. E. Volini et al, The social Enterprise at work: paradox as path forward – 2020, *Deloitte Global Human Capital trends*, (2020)
7. D. Ulrich, Majrstrovské řízení lidských zdrojú, *Grada*, 29 (2009) ISBN 978-80-247-3058-5
8. J. Blštáková, Habilitation lecture - Readiness of Slovak enterprises to manage the new generation of employees in the conditions of Industry 4.0, *FPH EUBA*, 4-5 (2018)
9. N. Jankelova, The Key role of Strategically and People - oriented HRM in Hospitals in Slovakia in the Context of Their Organizational Performance, *Healthcare*, **Vol 9**, Iss 255, 31-33 (2021)
10. M. West, J. Guthrie, J.Dawson, C.Borrill, M. Carter, Reducing patient mortality in hospitals: The role of Human Resource Management, *Journal of Organizational Behaviour* **27**, 983-1002 (2006)
11. K. Townsend, A. Wilkinson, C. Allan, G. Bamber, All we need is a miracle: Using a solution-based approach to human resource management in hospitals, *Asia Pacific Journal of Human Resources*, **49 (2)**, 177 (2011) DOI: 10.1177/1038411111400163
12. K. Townsend, A. Wilkinson, C. Allan, Mixed signals in HRM: the HRM role of hospital line managers, *Human Resource Management Journal*, **Vol 22, no 3**, 267-282 (2012)
13. N. Jankelová, Z. Joniaková, Communication Skills and Transformational Leadership Style of First-Line Nurse Managers in Relation to Job Satisfaction of Nurses and Moderators of This Relationship. *Healthcare,* **2021**, 9, 346 (2021) https://doi.org/10.3390/healthcare9030346
14. S. Alshamari, Organizational Culture and Organizational Performance in the Primary Health Care Sector in Qatar : A proposed Theoretical Framework, *Cross-Cultural Management Journal*, **Volume XIX**, Issue 2/2017, 130 - 136 (2017)
15. K. Alfes, A. Shantz , C. Truss, E. Soane, The link between perceived HRM practices, engagement and employee behaviour: a moderated mediation model, *International Journal of Human Resource Management*, **Vol 24** , no. 2 , 330-351 (2013) DOI:10.1080/09585192.2012.679950
16. A Baluch, T. Salge, E. Piening, Untangling the relationship between HRM and hospital performance : The mediating role of attitudinal and behavioral HR outcomes *International Journal of Human Resource Management*, **September 2013**, 24(16) 3038-3061 (2013) DOI: 10.1080/09585192.2013.775027
17. Value for Money Unit Ministry of Finance of the Slovak Republic, Health Expenditure Review II - Final Report, **October 2019**, 11 (2019) https://www.mfsr.sk/files/archiv/44/revizia_zdravotnictvo_2.pdf
18. L. Haniková, J. Koník, A young doctor went to the Czech Republic, here she was asked in interviews when she wanted to have children and what the parents were doing, *Denník N*, **19.10.2020**, (2020)

19. Healthcare Institute, Healthcare Barometer - survey of Czech and Slovak healthcare directors, *HCI*, **2020**, 5 (2020)
20. I. Raijani, R. Bačík, R. Fedorko, M. Rigeslký, K. Szczepanska-Wosyczyna, The alternative model for quality evaluation of health care facilities based on outputs of management processes, *Polish Journal of Management Studies*, **17(1)**, 194-208 (2018), DOI: 10.17512/pjms.2018.17.1.16
21. M. Bender, Conceptualizing Clinical Nurse Leader Practice: An Interpretive Synthesis, *J. Nurs. Manag*er, 23-31, (2016)
22. T. Hudák, V. Šušňáková, Manažérske vzdelávanie zdravotníckych pracovníkov v Európe a USA, *Katolícka univerzita v Ružomberku*, Pedagogická fakulta, Katedra manažmentu, I180, visited 15.1.2021 on https://www.tvp.zcu.cz/cd/2014/PDF_sbornik/hudak%20susnakova.pdf
23. P. Ondruš, I. Ondrušová a spol., Manažment a financovanie v zdravotníctve – príručka zdravotníckeho manažéra, Matica slovenská, 9-19 (2017), ISBN 978-80-972535-9-2
24. K. Wild, A. Fulupova, Hodnotenie manažérskeho vzdelávania v odbore odborník na riadenie vo verejnom zdravotníctve – Master of Public Health (MPH), *Verejné zdravotníctvo (online)*, **ročník VII č.1,** (2010) ISSN 1337-1789, visited 15.1.2021 on www.verejnezdravotnictvo.sk
25. A. Lüthy, Mitarbeiterorientierte Personalpolitik: Wie Krankenhäuser attraktive Arbeitgeber werden, Krankenhausmanagement : Strategien, Konzepte, Methoden, Berlin : MWV Medizinisch Wissenschaftl. Verl.-Ges.,133-141 (2010) ISBN 978-3-941468-26-9
26. J. Hunák, Master thesis: From public to private: effects of hospital privatization on the work environment of physicians – a case from Slovakia, *Wirtschaftsuniversität Wien*, 17-20 (2018)
27. UKESSAYS, Strategic HRM at the Mayo Clinic business essay, November 2018, visited 15.1.2021 on https://www.ukessays.com/essays/business/strategic-hrm-at-the-mayo-clinic-business-essay.php?vref=1
28. M. Bornewasser, B. Kriegesman, J. Zuelch, Dienstleistungen in Gesundheitssektor – Produktivität, Arbeit und Management, *Springer Gabler*, (2014) ISBN 978-3-658-02957-9
29. NHS, Workforce plan, 2020, visited 15.1.2021 on https://www.nhsemployers.org/your-workforce/plan/medical-workforce/new-to-managing-doctors-a-guide-for-hr-practitioners)
30. NHS, We are the NHS – action for all of us, 2020, visited 15.1.2021 on https://www.england.nhs.uk/wpcontent/uploads/2020/07/We_Are_The_NHS_Action_For_All_Of_Us_FINAL_24_08_20.pdf

Consumer protection in the conditions of the Slovak Republic in line with the harmonization of legislation and consumer rights within the EU

Stanislava Deáková[1] * and *Sylvia* Bukovová [1]

[1]University of Economics, Faculty of Business Management, Dolnozemská cesta 1, 852 35 Bratislava, Slovakia

Abstract. The paper is devoted to analyzing the conditions and possibilities of assisting consumers in protecting their consumer rights. As Slovakia is a member of the EU, it is necessary to include the conditions resulting from harmonizing the law within the Union. This paper points out different possibilities of consumer protection documented in the research part by evaluating data available in statistical sources. We also point out the current state of the frequency of initiated individual proceedings, the analysis of which is used to formulate the conclusions and recommendations of the scientific study.

1 Introduction

Consumer protection is now an increasingly topical issue connected with the development trends of the global business environment in the form of globalization, intensification, intellectualization, humanization, greening, flexibility, and acceleration. The global business environment is part of the business environment and, therefore, significantly shapes the relations between businesses and the entities in their vicinity through its global reach. Another essential part is the general surroundings of the business entity, which relates to the entities and relationships built between them in the economy of the given state through elements such as the state, lower territorial unit, suppliers, customers, competitors, and the public. When building relationships, it is necessary to point out the ties according to their nature, i.e., authoritative or partners, through which they communicate. At the level of the specific surroundings, we perceive communication between entities in a specific space and time [1,2]. We analyze processes and tools, institutional and legislative provision of consumer protection in the Slovak Republic. We point out the possibilities provided by the legislation and individual organizations dealing with support and assistance to consumers within the framework of the rules of the Slovak Republic.

1.1 The current situation in the field of consumer protection in the Slovak Republic

* Corresponding autor: stanislava.deakova@euba.sk

As we mentioned in the introduction, current global trends contribute to the rapid development of new technologies and materials, providing the possibility of a free transfer of goods and providing services without being limited by the borders of a given country. Scientific and technological progress accelerates communication processes, which is also reflected in expanding internet services while providing goods and services and the subsequent payments for the goods provided. These facts are only a fraction of the impact of global trends. [3]

1.1.1 Advantages and disadvantages of the impact of business environment trends

On the one hand, the business environment creates wider opportunities to meet the demand for goods and services without unnecessary barriers for consumers and entrepreneurs. Consumers can choose from a much broader range of products, while at the same time, the internet space and some applications allow consumers to easily find and compare different criteria for the choice of goods. The fast availability of offered goods and services from virtually any place in the world is also tempting. The development of a sophisticated logistics system also contributes significantly to this. [4-5]

Negative impacts also accompany the opening of new opportunities and horizons for entrepreneurs and consumers. Among the most common are the possibilities of cyber attacks and the misuse of databases with personal data, access codes and passwords to collect money from paying consumers' accounts. The second group of risks is due to the possibility of choosing a wide range of products, where the customer is unable (despite access to information) to navigate through the quality of the goods and services provided, mainly due to the provision of false information on the Internet, which is very difficult to verify. Counterfeit goods of renowned brands, products made from harmful materials, or products made in countries where the labor is exploited, working under inhumane, often harmful conditions and just for a minimum wage that is not sufficient to cover even necessities are available on the market. In such countries, we are also faced with an undesirable impact of production on the environment because of missing legislation or control of compliance with environmental protection. [5-12]

The above shows that countries seek to harmonize legislation and trade rules within the cooperating countries. For us, as a country that is part of the EU, this area has the most significant impact. In the following section, we present a cross-section of the basic legal standards and institutions that help consumers ensure that consumer rights are protected.

1.1.2 Legislative and institutional protection of consumer rights in the Slovak Republic

The state is an essential institution ensuring the legal framework and compliance with consumer protection rules. Pursuant to § 19 sec. 1 sec. 2 letter e) of the Act no. 250/2007 Coll. on consumer protection and the amendment of the Act of the Slovak National Council no. 372/1990 Coll. on offences as amended, the Ministry of Economy of the Slovak Republic (MHSR) within its competence in the field of consumer protection informs consumers about their rights and obligations [13]. On its website, the MHSR provides clear information in the following categories: [13]

- consumer information
- European Consumer Center in the Slovak Republic
- consumer associations
- consumer protection legislation in English,
- consumer protection legislation,
- documents concerning consumer protection,

- safety of consumer products,
- EU consumer protection legislation,
- alternative resolution of consumer disputes,
- administrative action.

The Slovak Trade Inspection (SOI) is the state control body of the internal market regarding consumer protection with nationwide competence. It is subordinate to the MHSR. Act no. 128/2002 Coll. on state control of the internal market in consumer protection matters as amended (hereinafter referred to as the "Act") is a legal regulation determining SOI competencies. This law defines the rights and obligations of SOI in general, but also in relation to special regulations governing consumer protection in specific areas.
The core activities of the SOI are: [14]

- to inspect and supervise the compliance with obligations arising from 350 legal regulations,
- to impose on-the-spot measures, protective measures, interim measures and fines for breach of obligations arising from legal regulations within its supervisory competence,
- to propose, based on facts established by an inspection, the suspension or cancellation of trade licence,
- to provide professional and methodological assistance to district authorities, municipalities, consumer associations and other legal persons established for consumer protection,
- SOI provides advice to consumers,
- cooperates with state and public authorities, consumer associations, supervisory authorities of other Member States with the European Economic Commission for Europe and the relevant UN bodies in the exchange of information in the field of market surveillance,
- collects fines and receivables from business entities and natural persons,
- informs about the danger to the life, health and property of consumers or the environment by selling dangerous products,
- develops and renews a Manual on Fuel Consumption and CO2 Emissions according to the Government Regulation no. 384/2004 coll. on the availability of consumer information on fuel consumption and CO2 emissions in the sale and leasing of new cars,
- handles consumer claims and complaints;
- clarifies the infringements.

Given the limited scope of the paper, we do not provide more detailed information about the different categories mentioned in this section of the paper. Links to the detailed information are provided in the references.

2 Research design

The research aimed to point out, in theory, an overview of the basic approaches to the given issue. In the research section, we focused on statistical processing, assessment and analysis of the results of inspection activities, and evaluating cases that SOI dealt with in the monitored period from 2016 and 2019, emphasizing the development, structure of areas and inspection results. The research outputs are presented in a clear graphical and tabular form. We used mainly scientific articles in indexed journals and scientific monographs as sources of information. We used data from the available statistical sources of quoted institutions and SOI annual reports in the application section. We used general and specific scientific methods to process the research, such as analysis, synthesis, abstraction, induction, deduction, comparison, and standard mathematical and statistical methods.

3 Results and Discussion

The research part focuses on analyzing the outputs of SOI inspections from 2016 to 2019, with subsequent comparison and evaluation of changes in the cases in the monitored period.

3.1 SOI Inspection Activity

The SOI's inspection activities focus on those areas where the consumers cannot effectively protect their rights. These include, in particular, health and safety, misleading advertising, covert harm, discrimination, unsolicited advertising, the compliance of non-food products with technical requirements, the activity of travel agencies and accommodation facilities and the provision of services. In particular, emphasis is placed on protecting consumers' economic interests, their protection against unfair commercial practices or unacceptable terms in consumer contracts.

The development in the number of inspections and the results of the SOI inspection activities in the monitored period are shown in Tables 1 to 4:

Table 1. Inspections in establishments aimed at selling products and providing services to consumers in the internal market.

Year	2016	2017	2018	2019	Change Index 2019/2016	
Number of checks	16 867	16 846	17 636	16 720	-147	-0,87%
On the basis of the complaints	3 366	3 167	3 078	2 384	-982	-29,17%
Shortcomings	5 906	5 864	5 791	5 214	-692	-11,70%

Source: own table [15 -19]

Inspections in the establishments selling products and providing services to consumers in the internal market included the areas of claim processes, the right of withdrawal from the contract, the possibility of refunds on sales via the Internet, disagreement with expert judgement, flight delays, etc.

Most inspections were carried out in 2018, the least in 2019. The number of inspections decreased by 147 checks (almost 1%) in the monitored period compared to the base year. Based on input from consumers, the most inspections were carried out in 2016, the least in 2019. There was a decrease of 982 (29.17%) in the inspections performed based on submissions in the given period. The highest number of shortcomings was identified in 2016, the lowest in 2019. A positive trend is a decrease in the detected shortcomings in the number of 692 inspections (-11.70%).

Table 2. Results of inspections in establishments for the sale of products and providing services to consumers in the internal market.

Year	2016	2017	2018	2019	Change Index 2019/2016	
Suspended sale of goods types	7 386	10 708	8 525	9 922	+2 536	+34,34%
Value in Eur	688 496	1 486 419	3 075 058	1 572 350	+883 854	+128,37%

Source: own table [15 -19]

Based on the performed inspections and identified shortcomings, the sale of the various goods has been suspended. The most significant number was recorded in 2017, the lowest in 2016. The value of goods whose sales were suspended showed the lowest value in 2016, the highest in 2018. In 2019, compared to 2016, the number of types of suspended goods increased by 2,536 (34.34%), its value increased by EUR 883,854 (128.37%).

Table 3. The most represented assortment in inspected establishments, aimed at selling products and providing services to consumers in the internal market in %.

Year	2016	2017	2018	2019	Change Index 2019/2016
Travel agents	1,26	0,59	0,71	1,97	+0,71
Energy	3,76	4,55	4,00	3,56	-0.20
Non-food goods	60,50	59,89	58,95	62,89	+2,39
Hospitality facilities	8,64	9,75	10,68	10,98	+2,34
Food products	10,03	7,46	6,99	5,91	-4,12
Various goods	2,89	3,63	4,13	2,78	-0,11
Services	10,07	11,20	11,87	10,12	+0,05
Accomodation	2,84	2,93	2,68	1,80	-1,04

Source: own proceedings [15 -19]

Table 3 shows that the largest share of inspected operations in the monitored period was in producing non-food goods, the lowest share being travel agents. In fact, the structure of the inspected range does not change significantly in the monitored period; in 2019, compared to the base year, inspections decreased the most in the area of food products (4.12), the least in the area of various goods (0.11). The most significant increase was recorded in non-food goods (2.39), the smallest in services (0.05%).

Table 4. Energy inspections.

Year	2016	2017	2018	2019	Change Index 2019/2016	
Number of inspections	634	767	705	596	-38	-6,00%
Shortcomings	172	221	256	215	+43	+25,00%

Source: own table [15 -19]

In the area of energy inspection, the SOI considers it necessary to supervise the whole range of competencies, such as the energy efficiency of buildings, the area of biofuels and illegal gas consumption or the check of air-conditioning systems. The inspections were carried out mainly in the heating sector (Thermal Energy Act). The most negative findings were in budgeting the cost of heat and hot water for final consumers. The number of energy inspections was highest in 2017 (767), the lowest in 2019 (596). The resulting decrease was 38 inspections (6.00%). The highest number of shortcomings was identified in 2018, the lowest in 2016. The number of shortcomings increased by 43 (25 %) over the monitored period.

3.2 Addressing consumer complaints

The inquiries, suggestions and requests received from consumers included, in particular, the claim process, the right to withdraw from the contract, the possibility of a refund when selling via the Internet, disagreement with the expert judgement, flight delays, etc. Their structure remains almost identical over the monitored period.

The number of complaints and their structure over the monitored period is illustrated in Tables 5 to 8.

Table 5. Complaints received, requests for advice and notifications.

Year	2016	2017	2018	2019	Change Index 2019/2016	
Accepted	10 014	9 957	9 254	9 249	-765	-7,64%
Examined	3 366	3 167	3 078	2 384	-982	-29,17%

Justified	1 499	1 412	1 251	1 010	-489	-32,62%
Subject matter of the complaint unconfirmed	1 867	1 755	1 827	1 374	-493	-26,41%

Source: own table [15 -19]

The highest number of complaints was dealt with by the SOI in 2016, the lowest in 2019. The resulting decrease is 765 complaints (7.64%). During the monitored period, the number of investigated claims decreased by 982 (29,17 %), justified complaints by 489 (32,62 %) complaints with the unconfirmed subject matter by 493 (26.41%).

Table 6. Inputs received by region.

year	2016	2017	2018	2019	Change Index 2019/2016	
Energy	445	379	327	260	-185	-41,57%
services	2 451	2 557	2 446	2 460	+9	+0,37%

Source: own table [15 -19]

The number of complaints in services significantly exceeds the number of energy complaints every year in the monitored period. However, in energy, the number of complaints decreased significantly in 2019 compared to 2016 - by 185 (41.57%), while in services, it increased very slightly - by 9 (0.37%).

Table 7. Most represented areas of service by type in the complaints received.

Year	2016	2017	2018	2019	Change Index 2019/2016	
Services related to the use of the apartment	438	512	507	384	-54	-12,33%
Internet services	166	104	x	x	x	x
Hospitality services	x	319	498	406	x	x
Travel agents and agencies	157	138	154	226	+69	+43,95%
Construction services	106	199	133	174	+68	+64,15%
Air transport	111	165	151	290	+179	+161,26%
Advertising	97	x	x	x	x	x

Source: own table [15 -19]

During the monitored period, consumer complaints received by SOI were mainly related to the use of apartments. However, in the results, they decreased by 54 (12.33%). Hospitality services recorded significant growth in 2018. The complaints were also related to travel agency and agency services, construction services and air transport. The highest growth in 2019 compared to 2016 was in the number of complaints in air transport – an increase by 179 (161.26%), the lowest in construction services – an increase by 68 (64.15%), the decrease was recorded in the area of internet services and the advertising.

Table 8. The most represented areas of sale of non-food goods in the complaints received.

Year	2016	2017	2018	2019	Change Index 2019/2016	
Footwear	654	583	487	425	-229	-35,02%
Household articles	460	165	84	124	-336	-73,04%
Textile products	422	510	425	523	+101	+23,93%
Mobile phones	364	394	328	253	-111	-30,49%
Furniture	x	245	312	322	x	x
Household appliances	238	262	207	237	-1	-0,42%

Building material	196	162	168	133	-63	-32,14%
Sports equipment	70	70	x	x	x	x
Motor vehicles	183	157	169	153	-30	-16,39
Computing	173	185	130	117	-56	-32,37%

Source: own table [15 -19]

In non-food goods sales, the most significant increase in consumer complaints concerns the quality of purchased footwear, the claim procedure and the consequent disagreement with the conclusions of expert judgements, based on which the seller rejects consumer complaints. The most represented were the complaints or inquiries connected with claim procedure, internet sales, withdrawal rights. Numerous representation in the complaints belongs to filing the claims in footwear, textile products, mobile phones, household appliances, motor vehicles, and computing sectors. In the monitored period, the number of household articles complaints decreased the most – by 336 (73.04%), the least - by 1 (0.42%) the number of complaints in household appliances, no complaint was made in sports equipment and furniture. The increase in complaints was recorded only in textile products by 101 (23.93%).

3.3 SOI Punitive Activity

SOI is entitled to impose, on an inspected person with found shortcomings, based on the inspection results carried out at the point of sale or services provision, not only an obligation to take measures for their removal but also a fine. The amount of fine depends on several factors, such as the duration of the unlawful conduct, the seriousness of the shortcoming, or the consequences caused by such unlawful conduct. If repeated and serious shortcomings are identified in performing the business activities, the SOI may also submit a proposal to revoke or suspend the trade license.

The development and results of the SOI's punitive activities over the monitored period are documented in Tables 9 to 12 :

Table 9. Fines and their value.

year	2016	2017	2018	2019	Change Index 2019/2016	
Number of fines	3 442	3 745	4 269	3 570	+128	+3,72%
Value of fines in Eur	2 498 996	2 488 738	3 210 090	2 665 982	+166 986	+6,68%

Source: own table [15 -19]

In the monitored period, the highest number of fines with the highest value was in 2018, the lowest number of fines with the lowest value was recorded in 2016. The resulting change was an increase in the number of fines by 128 (3.72%) and an increase in the value of fines by 166,986 (6.68%).

Table 10. Pecuniary fines imposed on FO (natural persons) and PO (legal persons).

year	2016	2017	2018	2019	Change Index 2019/2016	
Number of FO	981	1 246	1 462	1 102	+121	+12,33%
Number of PO	2 314	2 363	2 559	2 275	-39	-1,69%
Value of fines in Eur for FO	428 218	588 754	745 214	547 075	+118 857	+27,76%
Value of fines in Eur for PO	1 836 183	1 681 752	2 061 156	1 804 688	-31 495	-1,72%

Source: own table [15 -19]

In the monitored period, the number of fines imposed on FO (natural persons) was significantly lower than that imposed on the PO (legal persons), the lowest in 2016 and the highest in 2018. However, the resulting difference increased by 121 fines (12.33%). The number of imposed fines on PO (legal persons) was the lowest in 2019, the highest in 2018, but fell by 39 (1.69%) in the end. A similar trend was also reflected in the value of fines - the value of fines imposed on FO(natural persons) in 2019 increased by EUR 118,857 (27.76%), the value of fines imposed on PO (legal persons) decreased by EUR 31,495 (1.72%).

Table 11. Fines for obstruction of inspection.

Year	2016	2017	2018	2019	Change Index 2019/2016	
Number	147	136	248	193	+46	+31,29%
Amount	234 595	218 232	403 720	314 219	+79 624	+3,39%

Source: own proceedings [15 -19]

The highest number of fines for obstruction of inspection (even with the highest value) was in 2018, the least - with the lowest value - in 2016. The number of fines increased by 46 (31.29%), their value by 79,624 (3.39%).

Table 12. Proposals for the cancellation or suspension of a trade license due to the detection of repeated and serious shortcomings in the performance of business activities.

Year	2016	2017	2018	2019	Change Index 2019/2016	
Number	8	17	6	14	+6	+75%

Source: own table [15 -19]

Most proposals for the cancellation of the trade licence were recorded in 2017, the least in 2018. The resulting difference represents an increase of 6 proposals, representing 75%.

4 Conclusion

By analyzing data from the annual reports of the Slovak Trade Inspection on the results of its inspections and punitive activities in Slovakia for the period 2016 to 2018, we found that the number of inspections carried out in the monitored period decreases, fewer inspections are carried out also based on fewer suggestions from consumers, and as a result, the number of detected shortcoming decreases. However, the suspended sale of goods and their value is increasing. A positive trend is a decrease in the complaints received; on the contrary, the decrease in the complaints investigated is negative. The structure of the areas to which consumer complaints relate does not fundamentally change. In the case of punitive activities, the negative development is deepening, the number of fines and their value is increasing, as well as the number of motions to cancel or suspend the trade license to conduct business activities.

Notes

The article is a partial result of research within the KEGA project no. 002EU-4/2019" Integrating and Systematizing the Results of Scientific Research Activities in the Field of Consumer Protection, with a Primary Focus on Food Security, In Order to Model Adequate Consumer Behaviour ".

References

1. A. Neumann et al., *Business Diagnostics.* (Bratislava : Iura Edition, 2012)

2. S. Deáková, *Diagnostic methods and procedures of the company*,(Bratislava: Economist, 2013)
3. M.A. Marinov, S.T. Marinova, J.A. Larimo, T. Lepovsky, Characteristics and Future Trends. *Internetional Business and Emerging Economy Firms* **1**, 1 (2020)
4. P. Hamilton, K. A. Getulio, T. Takeshy, Innovation and technology processes in micro and small business. *Cogent Business & Management*, **6** (2019)
5. G. K. Amoako. A conceptual framework: Corporate environmental management activities and sustainable competitive. *Management Of Environmental Quality*, **31**(2), 331-347. (2020)
6. T. Cernev, R. Fenner. The importance of achieving foundational Sustainable Development Goals in reducing global risk. *Research & Management Science*. **115**, 102492, (2020)
7. J. Gunawan, P. Permatasari, C. Tilt. Sustainable development goal disclosures: Do they support responsible consumption and production? Journal Of Cleaner Production, **246**, 118989. (2020)
8. J. D'Arcy, P. B. Lowry, Cognitive-affective drivers of employees' daily compliance with information security policies: A multilevel, longitudinal study. *Information Systems Journal*, **29**, 1, 43-69 (2019)
9. V. Bolek, A. Látečková, A. Romanová, F. KORČEK, Factors affecting information security focused on SME and agricultural enterprises. *Agris on-line Papers in Economics and Informatics*, **8**, 4, 37-50 (2016)
10. P. Balozian, D. Leidner, M. Warkentin, Managers' and employees' differing responses to security approaches. *Journal of Computer Information Systems*, **59**, 3, 197-210 (2019)
11. P. Ifinedo, Roles of organizational climate, social bonds, and perceptions of security threats on IS security policy compliance intentions. *Information Resources Management Journal*, **31**, 1, 53-82 (2018)
12. E. Hofmann, M. Rüsch, Industry 4.0 and the Current Status as Well as Future Prospects on Logistics. *Computers in Industry*, **89** (2017)
13. MHSR. Informácie pre spotrebiteľov, https://www.mhsr.sk/obchod/ochrana-spotrebitela/informacie-pre-spotrebitelov, accessed 2021/10/05
14. SOI. Postavenie, https://www.soi.sk/sk/SOI/Postavenie-SOI.soi, accessed 2021/25/05
15. SOI. 2016 Annual Report, https://www.soi.sk/files/documents/vyrocne-spravy/vs%202016.pdf, accessed 2021/10/05
16. SOI. 2017 Annual Report, https://www.soi.sk/files/documents/kcinnost/vs%202017.pdf, accessed 2021/25/05
17. SOI. 2018 Annual Report, https://www.soi.sk/files/documents/vyrocne-spravy/vs%202018.pdf, accessed 2021/10/05
18. SOI. 2019 Annual Report, https://www.soi.sk/files/documents/vyrocne-spravy/2019/vs%202019.pdf, accessed 2021/10/05

Negative Effects caused by COVID-19 on Critical Path of Construction Projects

Oliver Haas[1,*], *Peter* Markovič[2]

[1] Faculty of Management of the Comenius University, Odbojárov 10, 820 05 Bratislava 25, Slovakia
[2] Department of Corporate Finance, Faculty of Business Management, University of Economics in Bratislava, Dolnozemská cesta 1/b, 852 35 Bratislava, Slovakia

Abstract. Begun in 2020 COVID-19 disease spread all over the globe and negatively affected all areas of human activities and societies inclusive engineering and construction projects. This paper analyses how COVID-19 implications can be used as a general excuse by a construction company to relief from contractual obligations, e.g., liquidated damages. To answer this question, we compared existing case law, previous court decision and project management literature with potential global and force majeure arguments based on COVID-19 cases. Our results showed that a general "free out of jail" card using COVID-19 argument is not applicable and COVID-19 cannot be seen per se as a force majeure in general and for every situation. However, if the effects from COVID-19 implications are beyond the parties' control and evidence shows that the contractor is actually disrupted and severely delayed the contractor is entitled for an extension of time or a delayed delivery of his performances. The paper concludes among other details that a COVID-19 argument can only be based on existing contractual provisions, accurate records of the as-built situation, proper notices of all kind of impacts and cause-and-effect considerations. Despite of any warning and if a contractor decides to rely on a global COVID-19 argument only it is a very risky endeavor and may backfire on him later on.

1 Introduction

The outbreak of COVID-19 disease officially announced by WHO on March 11th 2020 negatively affected engineering and construction projects around the globe as much as it affected the societies in different ways and magnitudes. Labor and skilled worker shortages due to travel bans and temporarily or permanent self-isolations reduces the ability for a return to the engineering and construction projects and therefore created disruptions and delays to deliver the project in time to a certain extent.

The same effect happened to the interrupted supply chain for construction materials, plants and equipment is relevant for the progress of construction project, which stuck either due to closed country borders or due to a lockdown of established transportation and traffic routes. All of those shortages of labor and goods in general slowered or even stopped the progress and timely performance of a contractor while the contractor may bar and tight up to a concrete and fixed delivery time for the project and liability for delays associated with a contractual liquidated damage provision.

That creates a problem for contractors, but finally as well for the clients, owners, all other subcontractors and the society involved in the same engineering and construction project. The goal of this paper is to assess and verify whether a contractor affected by COVID-19 implications can be relieved of contractual obligations, e.g., liquidated damages and can use contractual Force Majeure or other excusable disruption and delay provisions to seek for an extension of time (EOT) and additional compensation for such disruptions and delays. The proof of cause by COVID-19 and its direct effect on the contractor programme is important and a precondition to successfully place and being paid for a disruption or a delay claim. Further, it is mandatory that COVID-19 implications can be distinguished into excusable or inexcusable reasons for disruptions and delays.

In addition, if the contractor cannot deliver interim or final project milestones the contractor is not able to achieve interim or milestone payments or maintain sufficient cash flow. Ultimately, the

* Corresponding author: haas9@uniba.sk

contractor's ability to perform and progress will be restricted or even collapse as he might not be possible to issue and submit pro-rata invoices for his performances and run out of liquidity to run his overall business.

That is the reason why this paper and its outcome matters to construction projects.

2 Relevant Background Knowledge of COVID-19 Implications on Construction Projects

Without doubt events to cause delays and disruption often if not always happens to any small, middle or large-scale construction project worldwide. Events for such delays and disruption have different nature and associated with different circumstances, e.g., disorganization or mismanaging of a party, wrong or late or even no design, late procurement, wrong plant or machinery, wrong skills of labor, miscalculation of the construction sequence etc. Most of such incidents can be seen as "internal" causes with the contractor responsibility for delays and disruption. This paper seeks to list and identify related issues to frame the COVID-19 implications and the delay and disruptions effects on construction projects hereafter. [1-3] COVID-19 as a virus can be seen as an "external" cause for delays and disruptions.

2.1 The Problem

An essential element of a construction projects is the time to deliver the project. With time in global it is meant the duration of individual or numerous activities in sequence and/or one specific date to hand over or to bring the project into a stage where an owner or a client can take over or use the project for his purposes.

Very often "time is of the essence"-clauses or alterations of it are very common in construction contracts. By use of "time is of the essence"-clauses in general it is described that duration of activities and/or start or interim or death lines and substantial completion to specific dates are of the parties most interests. "The time is of the essence"-clause could also be seen as contractual obligation of the parties involved and even more, the contractor fails to complete in time it is seen as a material breach of contract.

If the contractor fails to deliver such project or substantial completion in time and to a specific date, the contractor is exposed to (but not limited to) such delay effects:

- damages at large respectively liquidated damages,
- further contractual and legal consequences,
- disputes,
- loss of reputation,
- arbitration,
- proceedings.

Another essential element of construction project are contracts associated with a lump sum prices for the works. In other words, the price is fixed according to works specified in the tender and procurement stage or bill of quantities (BOQ). In combination with a fixed end date to deliver a project the contractor is exposed to a high level of risk by the means of the contract. The contractor has to manage and control the risk in relation to time and works. Works can also be seen as equivalent of money spent for the works carried out, e.g., labor, material, equipment, plant etc. and quality of the works, e.g., quality of the material etc. delivered.

2.2 The Research Question (simplified)

As COVID-19 is a worldwide pandemic with effects to almost all of any aspect of the society we derivate the following research question:

- Can a contractor affected by COVID-19 implications be relieved of death lines and associated liquidated damages and/or other delay effects?
- Can a contractor use contractual Force Majeure or other excusable disruption and delay provisions associated with COVID-19 to seek for an extension of time (EOT)?
- Can a contractor seek for additional compensation for such disruptions and delays caused by COVID-19?

2.3 Examples of Contract Clauses dealing with Time

In almost all of the construction projects and if a contract is in place, contract clauses are dealing with time in large or specifically. Beside dates of the commencement or the sign off of the contract agreement itself, the start of the project activities, the duration of one or more project activities and/or end date to occupy or utilize the project we can find additional contract provisions as follows (but not limited to):

- Construction Change Directives
- Emergencies
- Safety of Persons and Property
- Suspension by the Owner-for Convenience
- Termination by the Owner of Convenience
- Evidence of the Owner's Financial Arrangements
- Change Orders, e.g., Design, Extensions, Upgrades, Adjustments etc.
- Adjustments for Changes in Legislation
- Delays caused by Others
- Delays caused by Authorities
- Delays and Extensions of Time (i.e., Impossibility of Performance, Force Majeure, Act of God).

The list is not comprehensive but frames the battery limits to contractual provisions where time is or can be a matter to deal with from the parties involved. In the light of COVID-19, one of the potential contractual provisions that a contractor may call to apply for an EOT is Force Majeure or similar provisions like Act of God. As any of the aforementioned contractual provisions implies different legal and contractual consequences, we need to explore more on the disruption or delay matter hereinafter.

2.4 Disruption vs Delay

Disruption and delay are not the same. Both events can concur the same time or each after another or independently from the other. Disruption and/or delay are not a prerequisite for the other.

Delay is an event when the as-build performance or end date of project delivery is extended or prolonged compare to the original as-planned programme or schedule (timely as-build-situation does not meet as planned programme). Delays coming always with an overrun of costs or budget of one or more parties in comparison what was planned and bid associated with the question if those costs are compensable from either party to the other party. Throughout the research for this paper and under consideration of the authors' experience in this field it becomes obvious the construction and project management industry by use of the term delay implies that those delays are critical for the overall project completion. Interim delays or challenges with minor subsequent delays are not meant as critical if lost time can be regained or does not have an effect for an EOT or a prolongation of the overall project completion date. Missing the overall project completion date and responsible for those delays associated with a contractual LD provision ultimately leads to the damage for the delaying party.

Disruption is an event that finishes the project on time, but created an overrun of the original budget cost in comparison what was planned and bid (timely as-build-situation does meet as-planned-programme). Disruptions are very often used in different terms and KPI's, e.g., loss of productivity of labor and equipment and/or inefficiency occurred in conjunction with measured lower production rate of works completed within a certain time window. Those effects can be assessed as damages for the parties. Generally speaking, disruption can be measured by metrics or units where labor is converted to cost and further those costs are associated with a certain amount of work completed in a certain time window. Typical examples to measure assess and identify losses and inefficiencies are

- Less performances of cubic meters of pouring concrete per hour, shift or day or
- Fewer square meters of laying bricks of a certain wall thickness of e.g., 24cm or 36.5cm of masonry blocks.

The question whether a contractor is entitled to receive a compensation of such losses and inefficiencies relies on the root causes of the events of disruption, the proof of it as well on the contractual provisions.

2.5 Disruption

Disruption can be categorized into two major groups:

- Loss productivity of labor [4] respectively equipment and
- Inefficiency occurred in conjunction with measured lower production rate of works completed within a certain time window.

Examples for disruption are (but not limited to): [5]

- Absenteeism
- Acceleration
- Beneficial or joint occupancy
- Bid error
- Crowding
- Defects with different causes, e.g., fabrication, design, transportation etc.
- Defective plans and specifications
- Dilution of supervision
- Extended overtime
- Fatigue
- Higher quality standards
- Idle time due to bad or severe weather
- Inadequate labor pool
- Inexperienced personnel
- Insufficient equipment
- Labor relations
- Labor strikes
- Labor turnover
- Lack of coordination
- Lack of protection from elements
- Late deliveries of material, plant, equipment, design, construction plans, and information required for permits, construction drawings, specifications etc.
- Learning curve
- Limited site access
- Manpower changes
- Materials, tools, equipment shortages
- Multiple shifts
- Numerous changes with start-stop-effects
- Out-of-sequence work and additional work orders
- Over-inspection
- Overmanning
- Performing field work planned for shop
- Poor morale and lack of motivation
- Project Management factors (tactical or strategical nature)
- Rework and errors
- Schedule compression
- Seasonal weather conditions
- Shift work
- Simultaneous operations
- Site conditions
- Skill dilution
- Slow response on requests for information
- Suspended work
- Trade stacking
- Underbidding work
- Undermanning
- Unusually severe respectively unfavorable weather
- Work restrictions

2.6 Delay

On high-level, delays can be categorized into five major groups:

- Excusable Delays ("ED"): [5]
  - Beyond the control of Client and Contractor, e.g., Force Majeure (see below) etc.
  - In general, entitlement to Extension of Time (EOT), but not to additional reimbursement.
- Non-excusable Delays ("NE"): [5]
  - Contractors' responsibility, therefore no relief of contractual obligation, e.g., shortage of material, plants and labor, breakdown of equipment & plants, accidents, loss of own productivity, discoordination, lack of management etc.
  - No entitlement for Extension of Time (EOT) and not additional reimbursement.
- Compensable Delays ("CD"): [5]
  - Clients' responsibility, therefore relief of contractual obligations, e.g., access to construction site, clients permit, mismanagement of client, inaction, no decision of client, non-Collaboration by the customer, late approval, changes of clients, reworks in design and already executed works, soil risks, differing site conditions etc.
  - Entitlement for Extension of Time (EOT) and additional reimbursement based on proof and evidential facts.
- Concurrent Delays: [5, 6]
  - Two or more delays occur at the same time and delaying the construction project.

- Concurrent delay comes often in combination of various delay types.
- Difficulties are
  - to distinguish to whom a delay is objective and finally attributable;
  - to which extent and affect a delay grows on later;
  - fragments of delays are apportioned to different causes, various parties and different events, which then later is or is not compensable.
- Critical respectively Non-Critical Delays: [6, 7]
  - Critical Delays:
    - At large and in simplified words critical delays on a construction project extend the overall or final date of the expected completion for the parties involved or by using, any other contractual definition where the parties believed is the date of being "substantial" finished and the "project" can come in use or operation.
    - Per definition, a critical delay massively modifies and changes mainly negatively the overall or final completion date and lies on the critical path and is therefore itself a critical path activity. The term critical path will be explained later in this document.
    - A critical delay consists of a scope of work or an activity of work which is mandatory for the project to continue and to deliver but cannot be skipped or waived "off the (project) table". This scope of work or activity of works requires time by nature to solve the technological constraints which the project does not have.
    - A critical delay always hurts in terms of time, money and quality or in combination of it. A critical delay cannot be discussed away or be ignored or forgotten. A critical delay has at least a temporary negatively sometimes permanently negatively consequence of the entire project completion. No doubts, if a critical delay cannot be recovered or eliminated the success of the construction project is in danger or even can be the cause of a ruined project.
  - Non-Critical Delays:
    - In simplified words, non-critical delays are delays with are of the opposite of the aforementioned definition for critical delays.
    - Non-critical delays consist of scope of work or one or more activities of work, which is not on the recent critical path.
    - Per definition a non-critical delay does not modify and change the overall or final completion date yet and does not lie on the critical path and is therefore itself a non-critical path activity. The term critical path will be explained later in this document.
    - The deceptive challenge to understand non-critical delays is that those delays still have the potential for delay damages once their negative effect radiate to the completion date. In addition, and over the course of a construction project previously former non-critical delays may leap up on the critical path and therefore grow to critical delays per definition. Suddenly low risk elements and non-critical delays turns in harmful critical delays for construction projects.

2.7 Overview of specific Terms and their Definitions in Relation to COVID-19 Implications

As previously listed different types of delay and disruption may affect the completion date of a construction project. Although construction projects are always unique and specific the individual area, environment, society and time-budget constraints in the relation between the contractor and client, if we want to clear sight and focus what COVID-19 brought to construction, in sum the following set of main root causes can be assessed:

- Impossibility of Performance:
  - An event which made performance for a contractor impossible, e.g., governmental shutdown or without any alternative options, but exhaust any of reasonable measures for performance, e.g.
  - Specific level of skilled labor needed.
  - Limited contractor in venue of construction project available.
  - Shortage of skilled labor in conjunction of the needed numbers of contractor.

- Impossibility of Performance shifts the completion date of the contractor. Difficulties for this argument is whether the event for Impossibility of Performance was objectively unexpected. The following key questions needs to be considered.
- Before February 2020 COVID-19 and its magnitude was not expected.
- Courts may rule COVID-19 as pandemic at this magnitude was not expected.
- Future construction contracts should specifically describe COVID-19 affects resp. pandemic in general.

- Force Majeure:
  - Force Majeure means an exceptional event or circumstance (based on FIDIC Fédération Internationale des Ingénieurs Conseils > other construction contracts may use similar definitions);
  - which is beyond a Party's control (one or the other);
  - which such Party could no reasonably have provided against before entering into the Contract (one or the other);
  - which having arisen, such Party could not reasonably have avoided overcome (one or the other);
  - which is not substantially attributable to one or the other Party;
  - e.g., hurricane, earthquake, volcanic activity, floods, storms, societal or governmental actions, war, invasion, strikes, civil unrest, war, hostilities, invasion, act of foreign enemies, rebellion, terrorism, revolution, contamination by radio-activity (list is not comprehensive);
  - New COVID-19 considerations: epidemics, pandemics, quarantines, preventive health protection. [8]
- Act of God: [5]
  - Act of God has a similar but at the same time different logic as for the term Force Majeure, but mainly focus on unexpected, unique and extraordinary force of nature. While e.g., rain, heavy rain and snow blizzards can be regularly assessed as events with comes with certainty or expected frequencies if those events led and result to shear failures of soil, flooding at large or large rock falls such consequences may be deemed as Act of God. Similarities to Force Majeure are applicable, e.g.;
  - which is beyond a Party's control (one or the other);
  - which such Party could no reasonably have provided against before entering into the Contract (one or the other);
  - which having arisen, such Party could not reasonably have avoided overcome (one or the other);
  - which is not substantially attributable to one or the other Party;
  - beyond a Party's control (one or the other).

2.8 The Critical Path (CP)

Understanding of the CP and alterations of the CP are essential and the only objective way to determine if a delay or disruption event really occurred to the construction project and its schedule and ultimately led to contractual and legal consequences.

The importance of the understanding of the CP led further to the understanding of the concept of Critical Path Methodology (CPM). [9] CPM is every time applied if a delay and disruption analysis is required and/or at least one (1) party seeks to be compensated from the other parties, which denies the existing delays and disruptions events. CPM is mainly carried out by appropriate programming called Program Evaluation Review Techniques (PERT) and supported by standard software applications. Although those mechanisms are mature disciplines, experience of the professionals and experts in this field is much more crucial. Finally, the outcome of CPM is assessed by those experts and the input data are interpretated differently. Common sense shall be applied, but facts and evidence viewed from a different angle may produce different results. Hence, the outcome of a proper and accurate executed CPM may vary from expert to expert and from case to case. This is no surprise as it reflects life experience from other fields.

CP per definition is the longest continuous sequence of activities through the entire construction project. The CP therefore connects the Start and End Date directly. At the same time and as those sequence of activities has no individual time buffer or float left, a delay or disruption has a direct impact, mainly negative, to the completion date of a construction project. In other words, any minor or larger alteration of the CP will change the entire project in the same affect. The project permanently breathes along the CP.

Over the course of a construction project, the CP may change through the sequences of activities as well. The CP can hop on or off of different set of activities and also switch by looking at different time windows. CP is dynamic and requires constant monitoring and control for the greater good of all stakeholders of a construction project.

CPM, Delay, and disruption analysis methods needs to the best-in-class methodology to prove or disprove delay and disruption claims and their delay and disruption damages, e.g. (but not limited): [9, 10]

- Formatting Impacted as-planned
- Time impact analysis
- Time slice windows analysis
- As-planned versus as-built windows analysis
- Retrospective longest path analysis
- Collapsed as-built analysis
- Equivalence method by Gutsche
- Construction process model by Toffel
- Approach by Thode
- Multi-stage approach by Lang
- Entitlement determination by Heilfort

The differences of each of aforementioned CPM's are not forming part of the conclusion of this research paper, but should once more emphasize the complexity of the assessment of CPM once delays and disruptions occur.

Again, a construction project is a complex and complicated integrated endeavor with thousands of thousands of construction activities, which are interconnected. Which CPM is finally chosen depends on numerous matters, for instant contract provisions, proceedings and formal agreed set up within arbitration and litigation, level of records, whether the analysis shall be prospective (looking in the future) or retrospective (looking the historic) and documentation and expert evaluation of the use of CPM.

Upon the execution of sophisticated and mainly cost intensive, CPM has to be discussed and agreed by the involved parties. Otherwise, the parties fail to finally agree on the outcome of the assessment of such CPM. This requires time and the involvement of experts in that field. [11]

2.9 COVID-19 hypothetical Case Study

For the purpose of clarity, a hypothetical Case Study with illustrations were rendered, compiled and simulated to demonstrate the negative effects by COVID-19 implications on the critical path of construction project by use of Program Evaluation Review Techniques (PERT).

The Case Study start to show the as-planned programme first as arrow diagram and then translated into a bar chart format for clarity. This is also known as the baseline schedule which best represents how the contractor intends to carry out the scope of work based on the contractual provisions and technological project specific. Baseline schedule mainly exchanged during the procurement process of the parties and undergo various updates and changes before sign off the contract.

Afterwards a COVID-19 delay respectively disruption element was inserted on and beside the critical path activities to show the effect on the completion date.

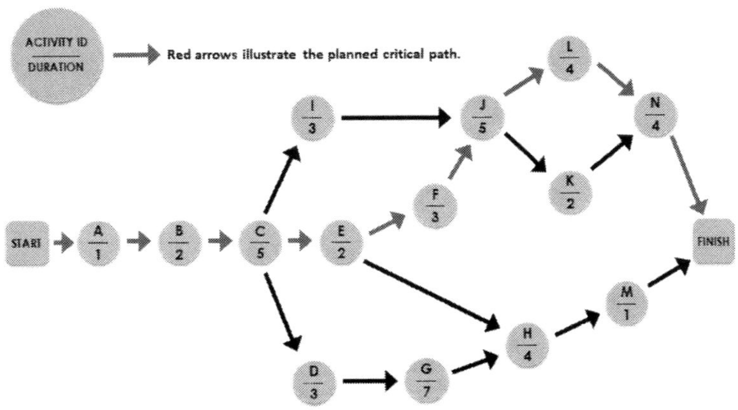

Fig. 1. Arrow Diagram of the as-planned Critical Path of the case-study project

Figure 1 shows numerous construction activities (A to M) and a definite start and finish milestones of the hypothetical case-study project. Each of the activities comes with a definite as-planned duration (shown in the denominator of the activity ID text field). Unit of duration could hours (h), days (d), weeks (w), months (m) or any other unit specified for construction project. For the ease of this paper, we use the term "time unit".

Red arrows illustrate the planned critical path where as black arrows represents planned non-critical path activities of the hypothetical case-study project. Those non-critical activities lying beside the critical path activities.

Based on the programme of the hypothetical case-study project the total project duration is twenty-six (26) time units and the critical paths runs through activities with ID numbers A, B, C, E, F, J, L and N.

Translated into more familiar chart techniques and network schedule of the same hypothetical case-study project the bar chart looks like:

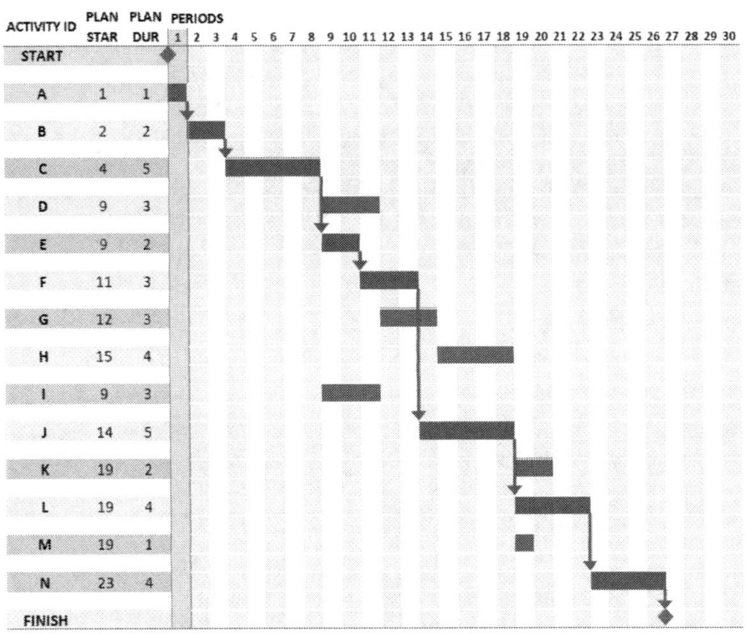

Fig. 2. Bar Chart of the as-planned Critical Path of the case study project

Figure 2 shows a bar chart exactly the same logic and intelligence of the hypothetical case study like in the aforementioned arrow diagram Figure 1, but with more clarity to understand.

For the exercise of this paper, we insert now a COVID-19 delay and disruption element and name it with activity ID "O" and an assumed duration of three (3) time units. The new bar chart of the hypothetical case study looks like:

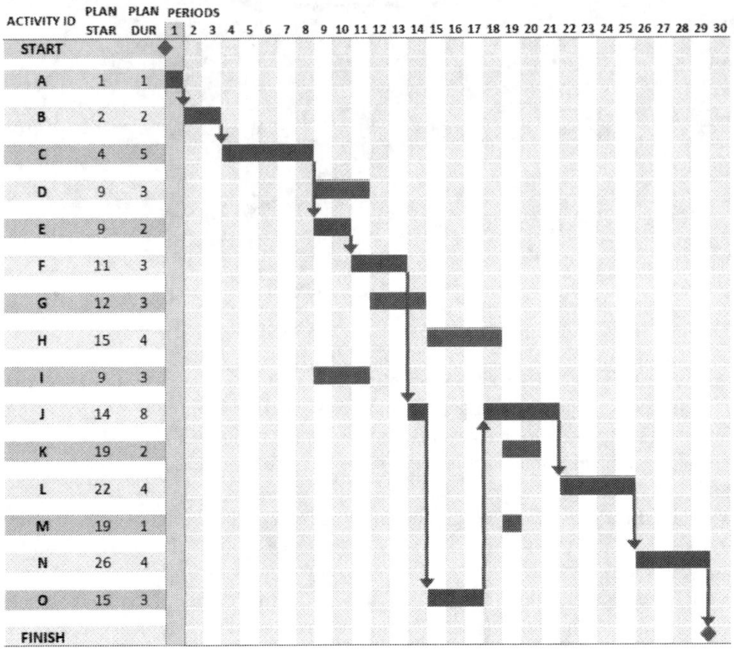

Fig. 3. Bar Chart with COVID-19 delay effect (activity ID "O") of the case study project

Figure 3 shows the new programme of the hypothetical case study delayed and disrupted by COVID-19 with activity ID "O" and its related duration of additional three (3) time units. The total project duration is now twenty-nine (29) days and the critical paths runs through activities with ID numbers A, B, C, E, F, J, 0 (the COVID-19 delay and disruption), back on J, L and N. Hence, the COVID-19 delay and disruption activity ID "O" shifted the as-planned completion date of additional three (3) time units. The non-critical activities lying beside the critical path activities are not yet affected by the COVID-19 delay and disruption activity ID "O".

As the effect of COVID-19 hits the entire construction world at once and without change for preparation, the outcome the research questions raised in this paper is relevant for all stakeholders involved in construction projects.

2.10 Case Law and Court Decisions

In the absence of recent High Supreme Court Verdicts in relation to delay and disruption claims caused by COVID-19 we have to assess the recent status:

- Knowledge and competence how to deal with delays and disruption in construction projects are driven purely by facts and applicable law (contract law, civil code). Recent case law and High Supreme Courts Verdicts dealing with delay and disruption claims and the appropriate use of the "correct" analysis methods set out the frame to assess the current new COVID-19 situation.
- The new experience and aspect with negative impacts caused by the COVID-19 pandemic for the lawmakers, judges, experts, contractors, clients etc. can be simplified framed as follows:
  - apart from a slightly different delay, starting from Wuhan, China, the entire construction world was hit more or less at the same time, but with different magnitude of its consequences;
  - some of the construction project using the same supply chain for labor, material, plants and equipment. One construction project may hit harder by the shortage of this supply chain than other construction projects;

- no preparation time for all involved stakeholder in the worldwide construction industry was given. The same is applicable for all lawmakers, judges, experts, contractors, clients etc. involved;
- COVID-19 pandemic and its knock-on-effect to the societies of the world still continues and still affecting not only the construction world with unknown concrete date once it is over;

- Although different Universities, Faculties and Societies from different countries published specific articles and produced valuable documents in this area [14-16], it seems that the previous high supreme courts dictate and leads the standard needed for the applicable and accurate delay and disruption analysis method and influenced most of the methodology to prove or disprove delay or disruption claims by contractor or client resp. owner.
- In Germany those required standards are associated with the High Supreme Court Verdicts of, e.g. (but not limited):
  - VII ZR 286/84 dated 20.02.1986;
  - VII ZR 245/94 dated 23.05.1996;
  - ZR 185/98 dated 21.10.1999:
  - VII ZR 224/00 dated 21.03.2002;
  - VII ZR 141/03 dated 24.02.2005;
  - ZR 225/03 dated 24.02.2005.
- The same logic and line of arguments of aforementioned High Supreme Court Verdicts from Germany applies more or less in other countries and jurisdictions. Although different countries and jurisdictions weighs and calculate specifics events and their effects in other way, the majority of such very specific alterations can be neglected for the purpose of this paper. In other words, the final decision of a delay and disruption case is based and assessed by reference to common law principles in every jurisdiction.
- The ability of a claiming contractor seeking an extension of time in a construction project by the client is associated to his ability of providing facts and evidence concrete enough to show COVID-19 is objectively affecting the critical path of construction project. Insufficient evidences led always to the situation that the ruling judges do not understand the objective as-built and as-is construction situation.
- Executive Summary:
  - Entitlement under the contract and/or recent changes in regional or federal legislation due to COVID-19 measures has to been demonstrated and recited by the contractor. This is a matter linked to applicable and existing law even if the legislation or regulations changes during COVID-19 period.
  - Burden of proof for any disruptions & delays events and related costs caused by COVID-19 lies on contractor. It is a question of fact and can be supported with proper documentation submitted by the contractor or client resp. owner. The same logic applies for the proof of detailed Quantum resp. damages. [12] This is a matter of contractor's abilities and way of project management or monitoring his activities and events preventing such activities.
  - Cause and Effect [13] has to be in forensic, reasonable and reflect-reality-level ("as-built"). The specific methodology chosen for the disruption and delay analysis and the accurate link that such COVID-19 event had a negative effect on the completion date of the construction project is crucial to proof such negative effects. As construction project are very complex endeavors the proof of cause and effect [13] is of complex nature. Even if the contractor or client were capable to provide a holistic delay analysis, it would be seen as a substantiated, but subjective assertion of one Party. Therefore, this is a matter to involve an independent, but mainly objective expert.
  - The claimant of delay and disruption claim shall deeply prepare and maintain accurate all kind registers that records and supports those delays and disruptions. This comes together with time-stamped photographs, precise written site diaries, supervisions of construction internal and external construction experts with experience in that technical field.
  - The same logic for an extension of time applies for the rules of compensation for such disruption and delay occurred. If extension of time due to COVID-19 reason is grant and gets a contractual and legal merit the claimant may be compensated for some or all damages occurred. However, if the judging court rules the case as Force Majeure and/or as Act of God the claimant, e.g., contractor will be compensated most likely for the loss he or she suffered. It may be seen that it is beyond any party involved and therefor common law principle applies.

3 Conclusion

On first sight, it seems relatively easy for a construction contractor to use COVID-19 implications as an accurate and valid argument to prove disruptions and delays and related costs.

However, this paper shows that COVID-19 per se not Force Majeure. [17 - 23] The contractor should instead carefully examine and assess the contractual provisions linked to the pandemic disease COVID-19 and act accordingly. This includes the provision of proper notices of impact or notices of disruption respectively delay to the client or subcontractors involved. It also includes accurate evidential facts and proper documentation to demonstrate any negative effects caused by COVID-19 on the critical path of a construction projects.

In the absence of contractual provisions, the contractor may have to find statutory or common law equivalent and evaluate further legal options.

The contractor has to consider in his overall assessment to prove and demonstrate non-liability, causation and damages for his disruption and delay claims. [13] In the absence of recent high supreme court decision for COVID-19 cases he has to consider previous high supreme court decisions dealing with disruption and delay claims for what the contractor was not responsible and therefor an excusable delay with or without compensation can then be determined. The contractor has to assess and prove disruption and delays by accurate records [18], cause-and-effect considerations based on comparison of as-planned vs as-built construction programmes, use proper CPM and PERT. Finally, the contractor has to demonstrate how he carried out mitigation. [24, 25]

In the absence of recent higher or highest court decision the contractor is well-advised not to rely on a general COVID-19 "free of obligation" thinking. The individual and specific effects due to COVID-19 implication on disruption and delays without any proper records and reasonable way of doing might be differently seeing by the court and judges at a later stage.

COVID-19 related disruption and delay implications may or may not be seen as excusable reason to apply for an extension of time under a construction contract. It depends on the concrete contract provisions, on the specific impacts and situation the construction project is in.

A "one COVID-19" argument or a kind of "general wild card" fits for an all disruption and delay reason for a contractor to relief from one or more contractual obligations is a risky endeavor.

As any construction projects varies from size and duration, it true that is a complex and an individual process of different parties involved and specific and individual activities occurred. Hence, the outcome of the view on disruption and delay of a construction period will also vary from construction project to another.

I have no doubt that long after the COVID-19 pandemic is settled and entirely under control contractors, clients, different societies, experts, courts and judges are forced to continue to deal with the implications of COVID-19 on construction projects. The COVID-19 implications on construction have just begun. [3]

Acknowledgment

The article is an output of research project VEGA (1/0240/20) "Financial Aspects of Sustainable Business – Enterprise Succession Solution for Small and Medium-sized Enterprises".

References

1. FIDIC International Federation of Consulting Engineers. *The FIDIC Golden Principles*. https://fidic.org/sites/default/files/_golden_principles_1_12.pdf. (accessed 06.01.2021).
2. A. Burr. *Delay and Disruption in Construction Contracts*. 5th ed. Milton: Taylor and Francis, (2016).
3. A. Burr. *Delay and Disruption in Construction Contracts: First Supplement*, 1st ed. Milton: Taylor and Francis, (2017).
4. W. Schwartzkopf. *Calculating lost labor productivity in construction claims*. 2nd ed. New York, NY: Aspen Publishers, (2004).
5. B. B. Bramble and M. T. Callahan. *Construction delay claims*. 3rd ed. Gaithersburg, Md.: Aspen Law & Business, (2000).
6. B. Eggleston. *Liquidated damages and extensions of time: In construction contracts*. 3rd ed. Chicester: Wiley-Blackwell, (2009).

7. N. Javed et al. Managing risk in green building projects: toward a dedicated framework. *Smart and Sustainable Built Environment* **9**, 2 (2020).
8. R. Gibson. *A Practical Guide to Disruption and Productivity Loss on Construction and Engineering Projects*. 1st ed. Wiley-Blackwell (2015).
9. Maclean, Ewen, Partner and Blackrock Expert Services. *COVID-19 - Stay alert (to lost productivity and delay).* https://www.blackrockx.com/wp-core/wp-content/uploads/2020/05/COVID-19-Stay-alert-to-lost-productivity-and-delay.pdf (accessed 06.01.2021).
10. P. Netscher. *Construction claims: A short guide for contractors*. Subiaco [Western Australia]: Panet Publications, (2016).
11. Society of Construction Law. *Delay and Disruption Protocol*. Society of Construction Law Delay and Disruption Protocol, 2nd ed. 234 Ashby Road, Hinckley, Leicestershire, LE10 1SW: Society of Construction Law (UK), (2017).
12. A. Balleer, S. Linke, M. Menkhoff, P. Zorn. *Demand versus supply: Price adjustment during the Covid-19 pandemic*. VOX, CEPR Policy Portal." https://voxeu.org/article/demand-versus-supply-price-adjustment-during-covid-19-pandemic (accessed 06.01.2021).
13. N. G. Bunni, ed. *The FIDIC forms of contract: The fourth edition of the red book, 1992 ; the 1996 supplement ; the 1999 red book ; the 1999 yellow book ; the 1999 Silver book*. 3rd ed. Oxford: Blackwell Publ, (2005).
14. W. Wang. The concept of sustainable construction project management in international practice. Environment Development and Sustainability DOI: 10.1007/s10668-021-01333-z (2021).
15. P. Amirtash et al. Integration of project management services for International Engineering, Procurement and Construction projects. *Built Environment Project and Asset Management* DOI 10.1108/BEPAM-06-2020-0106 (2021).
16. W. S. Dale, R. M. D'Onofrio. *Construction schedule delays*. 2018th ed. [Eagan, MN]: Thomson Reuters, (2018).
17. N. Sergeeva, S. Ali. The Role of the Project Management Office (PMO) in Stimulating Innovation in Projects Initiated by Owner and Operator Organizations. *Project Management Journal* **4,** 51 (2020).
18. J.-H. Kues, F. S. Thomas. *Covid-19-Krise-Die-Behandlung-von-Störungen-des-Bauablaufs-aufgrund-einer-Ausnahmesituation.* https://www.leinemann-partner.de/wp-content/uploads/2020/07/Covid-19-Krise-Die-Behandlung-von-St%C3%B6rungen-des-Bauablaufs-aufgrund-einer-Ausnahmesituation.pdf (accessed 06.01.2021).
19. L. Foggie, Turner, Townsend Inc. *COVID-19 Your project(s) may have stopped. But, you shouldn't! What you can do now to remain productive and in control.: covid-19-managing-project-delays.* https://www.turnerandtownsend.com/media/4980/covid-19-managing-project-delays.pdf (accessed 06.01.2021).
20. P. T. Nguyen et al. Exploring Critical Risk Factors of Office Building Projects. *Journal of Asian Finance Economics and Business* **8,** 2 (2021).
21. A. Mahdavian et al. Hybrid Genetic Algorithm and Constraint-Based Simulation Framework for Building Construction Project Planning and Control. *Journal of Construction Engineering and Management* **146**, 12 (2020).
22. R. Thomas. *Construction contract claims*. Houndmills, Basingstoke, Hampshire: Palgrave, (2001).
23. H. P. Thanu et al. Developing a building performance score model for assessing the sustainability of buildings. *Smart and Sustainable Built Environment* DOI: 10.1108/SASBE-03-2020-0031 (2020).
24. F. A. Ghansah et al. Exploration of latent barriers inhibiting project management processes in adopting smart building technologies (SBTs) in the developing countries. *Construction Innovation-England* DOI: 10.1108/CI-07-2020-0116 (2021).
25. R. F. Cushman. *Proving and pricing construction claims*. 3rd ed. Gaithersburg, MD: Aspen Law & Business, (2001).

SHS Web of Conferences **115**, 03006 (2021) https://doi.org/10.1051/shsconf/202111503006
Current Problems of the Corporate Sector 2021

Assessment of the competition level in the sector of bicycle manufacturers in the Slovak Republic

Vladimír Hojdik[1*], *Monika* Raková[1]

[1]University of Economics in Bratislava, Faculty of Business Management, Department of Business Economy, 852 35 Bratislava, Slovakia (EU)

Abstract. Bicycle industry has recorded significant boom recently, mainly as a consequence of global pandemic situation, but also as a result of green policies of countries. The popularity of cycling is on its all-time high, which also transferred to rising sales of bicycles worldwide. This paper evaluates current situation of bicycle industry in Slovak republic, mainly with the objection on performance of Slovak bicycle producers. With use of appropriate indicators, the paper aims to assess the rate of competitiveness among Slovak bicycle producers. For this purpose, relevant indicators as market share, concentration ratio CR_3, CR_5, and Herfindahl index are calculated and evaluated. The aim of this paper is to analyze financial results of Slovak bicycle industry producers, then measure the degree of concentration of whole industry in order to explain industry competitiveness in context of Slovak economy.

1 Introduction

The bicycle industry (known as cycling industry) is the sector of economy, which is concerned by bicycles and cycling. Within the industry, many different subjects execute their activities. The industry includes especially following groups of business entities [4] :

- bicycle manufacturers,
- component/part manufacturers,
- accesory manufacturers.

However, not only these subjects are interested in the operation of bicycle industry. Very important for the overall industry functioning are also other business groups [4] :

- distributors,
- retailers,
- bicycle organizations,
- bicycle event promoters,
- bicycle service providers.

Businesses operating in this sector are cooperating together and create complex supply chains. Such cooperation is very important and any disturbance in the supply chain will

* Corresponding author : Ing. Vladimír Hojdik, PhD. : vladimir.hojdik@euba. ORCID : 0000-0001-6955-1677.

result in the lack of final products (bicycles) on the market – the phenomenon that is present during the times of COVID-19 pandemic.

Sector of bicycle industry has recorded growing tendency in Slovak economy. Bicycle manufacturing complies to industry manufacture and it is also subcategory of manufacture of other transport equipment. It is registered by SK NACE as autonomic business activity under the statistical code SK NACE 30.92.0, which represents not only bicycle manufacture, but also wheelchair manufacture. According to data of Statistical Office of Slovakia, following key indicators are related to subjects classified as SK NACE 30.92.0 businesses [21].

Table 1. Basic information about SK NACE 30.92.0 industry (Manufacture of other transport equipment) for years 2017-2019

	2019	2018	2017
Number of businesses	19	19	17
Total revenues (mil. EUR)	514,7	406,6	312,3
Number of employees	3872	3651	3385

By the end of 2020, seven businesses in Slovak Republic were engaged in bicycle manufacturing. These businesses differ significantly in many aspects. This research analyses differences especially from revenues perspective, and then it discusses how it influences market structure and industry competition level. All in all, this paper aims to enrich the knowledge base of this research area with new findings from Slovak bicycle manufacturing industry.

2 Methods

The aim of the paper is to assess the dynamics of Slovak bicycle industry, with special focus on bicycle manufacturers. Particularly, it measures industry concentration, based on relevant indicators. During the research process, we used the combination of general and specific research methods. General methods (especially analysis, comparison or synthesis) were predominantly used for purposes of theoretical explanation of the topic in the introductory part of the paper. Introductory part had mainly descriptive character as it outlines current situation among Slovak bicycle manufacturing sector. However, the use of specific methods was the main part of the research to achieve the aim of the paper.

Among the specific research methods, following were used:

- mathematical and statistic methods,
- data processing with use of MS Excel interface,
- calculation and implementation of key indicators relevant for the research: concentration ratio and Herfindahl index.

Mathematical and statistical methods enabled the quantification of indicators that measure the competitiveness level of business sector – in this case bicycle manufacturers sector. First, market share of businesses (expressed in %) was calculated, based on total revenues of reviewed subjects. From market share values we later proceeded to quantification of another important indicator – concentration ratio.

Concentration ratio depends on market shares of businesses within the industry, and most often is identified for three (CR_3), five (CR_5) and ten (CR_{10}) businesses with highest market share (also expressed in %). In our research, we used CR_3 and CR_5 values.

The final indicator was Herfindahl index (also known as Herfindahl-Hirschmann index) which measures absolute concentration of industry. Concentration ratio and Herfindahl index are two most frequently used indicators which evaluate the level of industry competitiveness [3]. Results obtained from CR_3, CR_5 and Herfindahl index quantification became the basis for evaluation of competitiveness among bicycle manufacturers in Slovak republic. In the following table, there are basic information about the research sample – Slovak bicycle manufacturers.

Table 2. Basic information about Slovak bicycle manufacturers

Business name	Brand of bikes	Number of employees	Base
Kenzel, s.r.o.	Kenzel	20-24	Hurbanovo
ISTRO – HGA, spol s r.o.	Vedora	5-9	Hurbanovo
DEMA Senica, a.s.	Dema	25-49	Senica
Belve, s.r.o.	CTM	50-99	Stará Turá
LIBERTY TRADE, s.r.o.	Mayo, Liberty	2	Trenčín
KELLYS BICYCLES, s.r.o.	Kellys	150-199	Veľké Orvište
Cyclision s.r.o.	Cyclision	1	Piešťany

3 Results and discussion

3.1 Current situation among Slovak bicycle manufacturers

Considering the Slovak bicycle market as a whole, there is wide variety of bicycle brands available for potential customers. These bicycle brands are made especially by manufacturers from Germany, USA, Italy, Czech Republic or China. However, only seven manufacturing companies which are currently operating on the Slovak bicycle market are from Slovakia.

With the use of website finstat.sk, we summarized financial data needed to execute the research. For the purposes of the paper, business revenues represented the most important financial metric. In the Table 2, we present total revenues of bicycle manufacturers during the five-year period between 2016-2020. Among the manufacturers with the highest revenue we can include Belve, Kellys Bicycles and Dema Senica [8]. For fiscal year 2020. data were not yet available for all companies, so in the research we later worked with estimated values.

The data in the Table 3 show the development of revenues among the observed businesses. Total revenues of Belve for 2020 have not been reported for all year (12 months), but only for nine months (*). Three manufacturers: Istro – Hga, Kenzel and Liberty Trade have not reported financial data for 2020 yet.

Table 3. Revenues of Slovak bicycle manufacturers in years 2016-2020 (EUR thousand)

Business name	2020	2019	2018	2017	2016
BELVE, s.r.o.	17 433*	11 155	10 799	9 890	9 993
Cyclision s.r.o.	568	245	171	-	-
DEMA Senica, a.s.	11 880	15 537	10 492	9 647	11 192
ISTRO – HGA, spol s r.o.	230**	230	284	316	372
KELLYS BICYCLES s.r.o.	44 655	32 094	34 657	30 305	2 674
Kenzel, s.r.o.	712**	712	504	532	459
LIBERTY TRADE, s.r.o.	939**	939	993	937	1 057

* estimated value, revenues reported only for 9 months

** revenues for 2020 still not reported, so we used values from previous year

The revenues progression of bicycle manufacturers remains steady during analyzed period, without any radical change in the dynamics. Kellys Bicycles holds leading position in the whole industry asi it controls approximately half of the market. For the dataset in the Table 3 we also present descriptive statistics in the Table 4 for all of observed years.

Table 4. Descriptive statistics for total revenues in years 2016-2020

Variable: revenues (EUR thousand)	2020	2019	2018	2017	2016
N	7	7	7	6	6
Mean	10 917	8 702	8 271	8 605	4 615
Standard deviation	16 364	12 023	12 585	11 544	5 480
Minimum	230	230	171	316	372
Maximum	44 655	32 094	34 657	30 305	11 192

3.2 Key indicators for the research (chapter processed according to [14])

Market competitiveness and its changes indicates what is the quality of competition in the economy. Measurement and assessment of market concentration indicators give institutions an opportunity to monitor the competitiveness of businesses which operate in specific sector of economy [16].

The literature is home to numerous methods that evaluate the degree of competitiveness within the industry. To follow the aim of this paper, indicators of absolute concentration were calculated (these indicators quantify the degree of concentration in all subjects). Among absolute concentration indicators we include concentration ratio and Herfindahl index [10]. Both indicators are based on the calculation of market share. Therefore, the fundamental data for quantification of selected indicators are total revenues of individual businesses.

3.2.1 Market share

Market share is the basic indicator of concentration ratio. It represents the share of the individual enterprise revenues on the revenues of whole sector. Market share allows to determine, how the manufacturer performs in comparison to the other market participants – other competing businesses. Analytically, it is possible to quantify market share by the following formula [10]:

$$r_k = \frac{q_k}{q} = \frac{q_k}{\sum_{j=1}^{n} q_j} \tag{1}$$

where n – number of companies operating in the sector,
q_k – revenues of k company, $k = 1, ..., n$,
q – total revenues for whole sector,
r_k – market share of k company.

Market share can be expressed as relative or as percentage result. It takes values from the interval <0; 1> for the first, and <0; 100> for the latter. Extreme values indicate either full concentration of the industry – monopoly for r_k=1, or the absence of market operators in the sector for $r_k = 0$ [10].

3.2.2 Concentration ratio (CR)

Concentration ratio is an indicator of the market share sum held by few of the strongest industry businesses. Formula of the concentration ratio quantification is as follows [10]:

$$CR_\psi = \frac{1}{q} * \sum_{k=1}^{\psi} q_k \tag{2}$$

in which ψ lays in the interval <1;n> and $CR\psi$ takes values from range <0;1> when calculated in relative numbers, or <0,100> if we calculate percentage results. In practice, the concentration ratio is commonly quantified for three, five or ten strongest companies in the industry (quantification of indicators CR_3, CR_5 and CR_{10}). In some other studies, authors prefer indicators CR_4 and CR_8 quantified for four and eight strongest subjects [13, 17, 18].

3.2.3 Herfindahl index

Herfindahl index measures the absolute concentration in the industry. In the literature we may also find the concept of the Herfindahl-Hirschmann index [11]. It is widely used for sector analysis and it is well-known and commonly accepted indicator of market competitiveness [6]. It explains, how financial power (business revenues) of individual subjects impacts the concentration of whole industry. Herfindahl index quantifies the square of the market share of each subject in the industry and sums up the resulting numbers [7].

Analytical expression of Herfindahl index is as follows [10]:

$$H = \sum_{k=1}^{n} \left(\frac{q_k}{q}\right)^2 \tag{3}$$

and the meaning of the variables remains identical, as in the previously defined formulas.

Herfindahl index H takes values from <1/n, 1>. Extreme low value H = 1/n means that every company has equal market share in the industry. For extreme upper level, for which H = 1 there is a pure monopoly on the market – one company produces 100% of whole industry production [5].

In practice, the Herfindahl index may not be used as a number of the interval mentioned above – because in sectors with a higher number of enterprises we would operate with very low numbers. Therefore, the value of H tends to be adjusted by the appropriate multiplier, most commonly used multiplier is 10,000. Depending on what is the value H after conversion, we can classify the degree of concentration of the industry, according to the following criteria defined by the US Federal Trade Commission (FTC) :

https://doi.org/10.1051/shsconf/202111503006

- for *H* <1500, the industry is not concentrated,
- for *H* is in the range of <1500; 2500>, the industry is moderately concentrated,
- for *H*> = 2500, the industry is concentrated [22]. Herfindahl index values also impact how government sets and enforces competition policy [16].

The Herfindahl index as used currently was firstly applied in 1950 in dissertation thesis of O. C. Herfindahl. Later, in 1959 he used it in a study about international copper industry [12]. The similar concept (sum of squared market shares), however is known since 1940 when A. O. Hirshmann executed a trade study. Hirschmann in his paper suggests that original idea of index belongs to him [13]. Therefore, in the literature we can find concepts of Herfindahl index or Hirchmann-Herfindahl index, but both of them represent the same indicator.

Herfindahl index has been used in multiple studies to consider competition level in variety of industries, for example in banking, insurance, automotive or creative industry [1, 5, 9, 14, 15, 19, 20, 23]. Its construction, however, has some limitations – it can measure concentration only on geographically isolated markets, but cannot be applied for international companies, as the results may be misleading [2].

3.3 Quantification of selected indicators

This chapter presents and discusses calculated results, which are listed in following tables separately for different indicators – market share, concentration ratio and Herfindahl index. Table 2, Table 3 and Table 4 complete all data needed for the research purposes.

3.3.1 Quantification of market share

According to revenues data listed in Table 3, it was possible to calculate the market share of observed businesses. Market share is expressed in % and following table (Table 5) illustrates the relative strenght of businesses based on the amount of revenues.

Table 5. Market share of Slovak bicycle manufacturers during observed period 2020-2016 (%)

		2020	2019	2018	2017	2016
1.	KELLYS BICYCLES s.r.o.	58,43	52,69	59,86	58,70	-
2.	BELVE, s.r.o.	22,81	18,31	18,65	19,16	43,31
3.	DEMA Senica, a.s.	15,54	25,51	18,12	18,69	48,51
4.	LIBERTY TRADE, s.r.o.	1,23	1,54	1,72	1,81	4,58
5.	Kenzel, s.r.o.	0,93	1,17	0,87	1,03	1,99
6.	Cyclision s.r.o.	0,74	0,40	0,29	-	-
7.	ISTRO – HGA, spol s r.o	0,30	0,38	0,49	0,61	1,61

The power distribution between bicycle manufacturers reflects that three businesses (Kellys Bicycles, Belve and Dema Senica) keep the distance from other four (Liberty Trade, Kenzel, Cyclision and Istro-Hga), as the leading trio control almost 97% of the market. The leading company – Kellys Bicycles clearly holds top position during observed period with approximately 58% of market share.

Second strongest business is Belve, as it covers almost 23% of the market, while Dema Senica holds third place with more than 15% of market share. Data also show changes in power balance within the industry. In 2016, only five manufacturers were on the market and Dema Senica possessed almost one half of the market. However, during next four years

Kellys Bicycles and Belve gained momentum and Dema Senica was subsequently squeezed from the first to the third position.

Among the four weakest players, there are not any significant changes between 2016 and 2020. All businesses together – Liberty Trade, Kenzel, Cyclision and Istro-Hga are currently accounting for approximately 3% of the market, therefore their impact on the sector is almost negligible. Trinity of top manufacturers clearly dictate conditions of the Slovak bicycle manufacturing sector. Production among other four businesses is significantly fragmented and none of them comes close to the revenues level of top three producers.

3.3.2 Quantification of concentration ratio

Concentration ratio of industry is dependant on market share values. As already mentioned before, concentration ratio in the industry is most often calculated for three, five and ten strongest companies in the industry. As we analyze the sector with only seven businesses, CR_{10} quantification is not relevant. Table 6 contains already calculated concentration levels CR_3 and CR_5 for the bicycle manufacturing industry in Slovakia, based on revenues achieved by enterprises in the relevant period.

Table 6. Concentration Ratio CR_3, CR_5 and of Slovak bicycle manufacturing industry (%)

	2020	2019	2018	2017	2016
CR_3	96,78	96,51	96,63	96,55	96,4
CR_5	98,94	99,22	99,22	99,39	100

The results show stable development of concentration ratio during the period. Concentration ratio CR_3 represents the share of three strongest manufacturers which constantly control almost 97% of local bicycle manufacturing industry. Their position remained absolutely dominant even during COVID-19 recession and probably anything can threaten it in the near future. Moreover, their hegemony is not short-term at all – top three businesses dominate the market throughout whole monitored period .

Concentration ratio CR_5 reaches similar values as CR_3. This is caused by the fact, that fourth and fifth strongest business (Liberty Trade and Kenzel) account only for 2,16% of the market. As we stated above, the dynamics of the whole industry is influenced by three biggest players and four remaining companies have only minimal impact on the market operation.

Concentration ratios CR_3 and *CR5* did not record any fluctuations between years 2016 and 2020. Comparison of both indicators makes it possible to state that Slovak bicycle manufacturing industry shows signs of oligopolistic market structure – with three very strong and four very small companies operating within the sector.

3.3.3 Herfindahl index quantification

A first step before Herfindahl index quantification, a key metric evaluating a degree of industry concentration, we determined the number of businesses in the sector to 7 (n = 7). Despite the number of industry participants is relatively low, it is still possible to execute Herfindahl index quantification. Based on the business revenues and market share, values of Herfindahl index were calculated and then collected in Table 7.

Table 7. Herfindahl index of Slovak bicycle manufacturing industry in years 2016-2020

	2020	2019	2018	2017	2016

Herfindahl index H	4179,9	3766,2	4263,1	4166,6	4390,6

Herfindahl index as a complex indicator assesses the degree of industry concentration. During the analyzed period, values of Herfindahl index stay in the range of <3766,2; 4390,6>. According to generally accepted methodology of the Federal Trade Commission (FTC), the industry can be classified as concentrated (because values of Herfindahl index sigbificantly exceed 2500) [22]. This suggestion is supported by other criterion where Herfindahl index for concentrated industry takes values which are higher than 1800 [5, 10].

Herfindahl index values calculated for Slovak bicycle manufacturing industry classify it as concentrated industry. Based on the quantified results, it could be also defined as highly concentrated industry. Currently, only seven subjects are operating within the industry while three of them have dominant position in the sector as they account for 97% of the total industry revenue, what underlines the thesis of a highly concentrated industry. Three strongest entities possess such a power that makes it very difficult for other competitors to expand their market activities.

4 Summary/Conclusion

Results obtained from the research prove that Slovak bicycle manufacturing industry is almost exclusively dependent on performance of three strongest producers. Analysis and explanation of quantified indicator values enable us to state that sector of bicycle manufacturers in Slovakia undoubtly shows signs of highly concentrated industry. According to methodology for Herfindahl index calculation, Slovak bicycle manufacturing industry ranks in the category of highly concentrated industries what means relatively low competition level and unsuitable conditions for enterpreneurship within the sector.

However, the industry in the Slovak Republic has recorded succesful development in past years, which was caused mainly by increased demand for bicycles. Moreover, sales of bicycles also increased during the COVID-19 crisis and global bicycle industry was one of few sectors not affected by economic slump.

As global trends favor a healthy lifestyle, ecological solutions, green mobility and environmental sustainability, further expansion can be expected for bicycle industry, even in the conditions of Slovak Republic.

Acknowledgement

This paper is one of the partial outputs of the research grant no. I-21-107-00 entitled "Analysis of factors influencing the development of green mobility in the context of business sphere of the Slovak Republic", led by Ing. Vladimír Hojdik. PhD., University of Economics in Bratislava, Faculty of Business Management, Department of Business Economy.

References

1. F. M. Aijde, J. O. Ajileye, Market concentration and Profitability in Nigerian Banking Industry : Evidence from Error Correctiion Modeling. *International Journal of Economics, Commerce and Management* **3**, 1-12 (2015)
2. P. Aghion, N. Bloom, R. Blundell, Competition and Innovation : An Inverted-U Relationship. *Quarterly Journal of Economics* **120** (2), 701-728 (2005) DOI : 10.1093/qje/120.2.701
3. J. A. Bikker, K. Haaf, Competition, concentration and their relationship : an empirical analysis of the banking industry. *Journal of Banking and Finance* **26**, 2191-2214 (2002) DOI :10.1016/S0378-4266(02)00205-4
4. T. Blondiau, B. van Zeebroeck, H. Haubold, Economic Benefits of the increased cycling. *Transportation Research Procedia* **14**, 2306-2313 (2016) DOI : 10.1016/j.trpro.2016.05.247
5. I. Brezina, A. Oršulová, J. Pekár, Analýza absolútnej koncentrácie vybraného odvetvia pomocou Herfindahlovho-Hirschmannovhio indexu, *Ekonomický časopis* **57** (1) 77-94 (2009)
6. I. Brezina, J. Pekár, Z. Čičková, M. Reiff, Herfindahl-Hirschmann index level of concentration values modification and analysis of their change, *Central European Journal of Operations Research* **24**, 49-72 (2016) DOI : 10.1007/s10100-014-0350-y
7. M. C. Cavalleri, A. Eliet, P. McAdam, Concentration, market power and dynamism in the euro area. *ECB Working Paper Series, Discussion Paper No. 2253*, European Central Bank (2019) DOI :10.2866/379250
8. Database of Slovak Businesses and Organizations. (2021) Available at <https://www.finstat.sk >
9. J. Fedderke, D. Naumann, An Analysis of Industry Concentration in South African Manufacturing 1972-2001, *Applied Economics* **43**, 2919-2939 (2011) DOI : 10.1080/00036840802631835
10. M. Fendek, E. Fendeková, *Mikroekonomická analýza* (Iura Edition, Bratislava, 2008)
11. G. Grullon, Y. Larkin, R. Michaely, Are US industries becoming more concentrated ? *Forthcoming, Review of Finance* (2017)
12. O. C. Herfindahl, *Copper Costs and Prices : 1870-1957* (Baltimore, The John Hopkins Press, 1964)
13. A. O. Hirschmann, The Paternity of an Index, *American Economic Review* **54**, 761-762 (1964)
14. V. Hojdik, Evaluation of Slovak Automotive Industry Competitiveness Based on Market Concentration Indicators. *SHS Conferences* **83** (2020) : Current Problems of the Corporate Sector 2020 : 17th Scientific Conference. – Paris : Edition Difussion Presse Sciences. DOI : 10.1051/shfconf/20208301022
15. M. Kozáková, M. Kmety Barteková, Analysis of market concentration in creative industry. *SHS Conferences* **83** (2020) : Current Problems of the Corporate Sector 2020 : 17th Scientific Conference. – Paris : Edition Difussion Presse Sciences. DOI : 10.1051/shfconf/20208301022
16. P. MacKay, G. M. Phillips, How Does Industry Affect Firm Financial Structure ? *The Review of Financial Studies* **18** (4), (2005) DOI : 10.1093/rfs/hhi032

17. E. Nauenberg, K. Basu, H. Chand, Hirshmann-Herfindahl Index Determination under Incomplete Information, *Applied Economics Letters* **4** (10), 639-642 (1997) DOI : 10.1080/758533291
18. S. Rhoades, The Herfindahl-Hirshmann index, *Federal Reserve Bank of St. Louis* **52** (3), 188-189 (1993)
19. Z. E. L. Rolim, R. R. Oliveira, H. M. Oliveira, Industrial Concentration of the Brazilian Automobile Market and Positioning in the World Market, *Working Paper* (2019) DOI :10.13140/RG.2.2.11210.11207
20. M. N. Al Arif Rianto, T. B. Awwaliyah, Market share, concentration ratio and profitability : Evidence from Indonesian Islamic Banking Industry. *Journal of Central Banking Theory and Practice* **2**, 189-201 (2019) DOI : 10.2478/jcbtp-2019-0020
21. Statistical Office of Slovak republic, Yearbook of Industry of the SR 2019 (Bratislava, 2019)
22. U.S. Department of Justice and Federal Trade Commision, Horizontal Merger Guidelines, (2010)
23. D. M. L. Williams, P. Molyneux, J. Thornton, Market Structure and Performance in Spanish Banking. *Journal of Banking and Finance* **18**, 433-443 (1994) DOI :10.1016/0378-4266(94)90002-7

Current measures and challenges to strengthen consumer protection

Jana Kissová[1*], *Gabriela* Dubcová[1]

[1] University of Economics in Bratislava, Faculty of Business Management, Dolnozemská cesta 1, 852 35 Bratislava, Slovakia

Abstract. Over a short period of time, the individual countries of the world must face a common problem that affects them and adversely affects the lives of individuals. In the context of the current emergency related to the coronavirus pandemic, fundamental changes and enormous impacts can be seen in the social or economic dimension. The article aims to provide an overview of the current situation in selected countries and compare the system of measures in the Slovak Republic and the Czech Republic to stabilize or retain workers or eliminate impending damage.

1 Introduction

For a long time, the topic of consumer protection has been brought to the attention of individual EU countries in terms of ensuring the highest possible safety and elimination of any problems. The institutions involved are making considerable efforts not only to solve problems but also to prevent them and prevent potential risks. Within individual countries, efforts are made to harmonize the rules while respecting their differences or specific features.

The article aims to provide an overview of the current situation in consumer protection with the aim to improve the position and strengthen consumer protection.

1.1 Theoretical starting points related to consumer protection

Consumer protection is ensured through the European single market as one of the largest markets in the world. The main task of EU consumer policy is to help consumers make the most of this market's benefits. The key objectives of EU consumer policy are to strengthen their position and effectively protect their security and economic interests, intended to provide an effectively and adequately functioning single market. At the same time, it presupposes protecting consumers' rights in relation to traders and providing enhanced protection for vulnerable consumers.

Consumer policy is one of the most thoughtful policies in the EU. The consumer protection system is based on several key concepts and application principles. [1, 2, 3]

* Corresponding authors: jana.kissova@euba.sk, gabriela.dubcova@euba.sk

Dzurová states: *"Consumer policy is a set of measures aimed at informing and protecting consumers. It, therefore, focuses on consumer protection and seeks ways to prevent harming it.* " [4]

According to the Great Economic Encyclopedia, consumer policy is defined as "*social, economic, commercial, legislative and other measures taken in the consumer's interest.*" [5]

Consumer rights are defined by legislation, in our country, they are a part of the Act on Consumer Protection, which defines "*consumer rights and obligations of producers, sellers, importers and suppliers, the competence of public authorities in the field of consumer protection, the status of legal entities established or established for consumer protection.* " [6]

The Great Economic Encyclopedia defines the concept of consumer law as "*a right that expresses the position of the buyer in relation to sellers. It shall assist in ensuring consumer protection on the market.*' [5]

Under legislative consumer protection, consumers are entitled to claim the following rights: [7,8]

- **the right to meet basic needs** – includes the right to food, clothing, housing and shelter, health care and education,
- **the right to product safety** – this right is to protect an individual from products and services that may pose a risk to health or life,
- **the right to information** – means sufficient access to information, protection against unfair advertising and labelling of goods,
- **the right of choice** – the right to access a variety of products and the right to choose goods and services from a wide range with a guarantee of quality that satisfies the need of the individual,
- **the right to be heard** – the right of the consumer to pursue their consumer interests in the development and implementation of government policy,
- **the right to** compensation – includes handling of complaints, relevant responses to complaints, compensation for purchase shortcomings (e.g. for scrap goods and unsatisfactory services),
- **the right to education** – the right to acquire knowledge and skills for the correct and reliable choice of goods and services based on high-quality, credible information,
- **the right to a healthy environment** – improves the quality of life and means living in an environment that does not endanger current or future generations.

1.2 Overview of the priorities and achievements in the consumer protection area

As part of consumer protection, consumer policy was implemented in individual EU countries through the European Consumer Program 2014-2020. Its objectives focused primarily on four areas: [9]

- monitoring and safety compliance
- information and education initiatives
- legislation to strengthen consumer rights and effectively compensate consumers
- enforcement of rights beyond borders

EU policy ensures a high level of consumer protection in many areas. High safety standards apply to toys, electrical appliances, cosmetics, medicines, food, lighters, personal protective equipment, machinery, and recreational craft. [10, 11, 12]

EU consumer policy is intended to help citizens enforce their consumer rights and thus actively participate in the functioning of the Single Market. The total budget of the consumer program (2014-2020) is EUR 188.8 million (about 5 cents per consumer/year). Despite the modest budget, the impact of this program on consumers is truly tangible. Consumers in all Member States under EU projects will have high-quality and affordable protection from European consumer centers and national state authorities if their consumer rights are violated. [9,13]

Key achievements of consumer policy to date include: [9]

- **Product safety** – only safe products can be placed on the EU market. The rules include the requirement to use 'CE' conformity marking on several product categories - the manufacturer's declaration that the products meet the basic EU safety criteria. Safety rules are often extended or modified to cover new products or take into consideration the latest scientific knowledge.
- **EU Rapid Data Exchange System RAPEX** - informs inspectors in all Member States of the measures taken by a particular Member State concerning a dangerous non-food product. This triggers a chain reaction and removes the product notified by the Member State from other national markets in the EU. It has been in operation since 2004.
- **Food safety** – EU rules guarantee consumers a high level of protection for food and food products sold in the EU at all stages of production and distribution. They regulate hygiene conditions for food products and storage, animal health and welfare, plant health and the risk of food contamination
- **Cosmetic safety** – The EU guarantees it through standard rules on the composition, labelling and packaging of cosmetic products. In 2013, the rules were extended, leading to the end of testing cosmetics on animals and placing such products on the EU market.
- **Prohibition of coercive sales or aggressive trading methods** – misleading advertising and unfair commercial practices have been banned in the EU since 2005, e.g., no hidden costs, tricks, hidden claims, misleading information, and advertisements aimed at children. If it is stated that something is free, it should be so.
- **Distance contracts** - EU rules protect consumers when buying products or services remotely, whether via a mail order service or the Internet. Thanks to them, the seller is obliged to provide sufficient information before purchase; consumers have the right to withdraw from the contract within 14 calendar days without any consequences or penalties. The regulations also protect them from misuse of the payment card.
- **Product repair** – EU rules provide legal protection for consumers if they have paid for defective goods or goods that do not have declared characteristics. A two-year warranty covers all products they buy. If the purchased product ceases to function during this period, the consumer has the right to request the repair of the defect present at the delivery time. If it is not possible to get the product repaired within a reasonable time, he/she has the right to request a refund of the purchase price or a discount. Commercial guarantees provided by the seller do not replace the minimum 2-year free legal guarantee but may supplement it.

- **Labelling and packaging of products** - the labelling of products and services as an information tool guarantees quality, statements and services are subject to strict rules of certification and control. Most products placed on the EU market (national or European or both) are subject to technical regulations to ensure a minimum level of health and safety for humans, animals, and the environment. Eco-labelling of products is a voluntary tool of environmental policy and standards. The product labelling system guarantees quality for both consumers and manufacturers.
- **Advice and assistance** - if a new computer breaks down if the consumer does not get the goods ordered via the Internet, or he/she wants compensation for a night spent at a hotel because of a flight cancellation, he/she is entitled to certain guarantees and support under EU consumer policy. When a trader refuses to recognize his/her rights, they can contact one of the EU consumer organizations – the European Consumer Centers Network (ECC - NET). It is represented in every EU country, Iceland, and Norway. It provides free assistance and advice to consumers on their rights to purchase goods and services from traders based in another EU country. Complaints are dealt with through a settlement between the consumer and the trader; organizations help contact the trader and find an acceptable solution.

2 Objective

The article aims to provide an overview of the current situation in consumer protection with the aim of improving the position and strengthening consumer protection. The priority is to bring together current measures and challenges to strengthen consumers' rights and interests and thus increase their level of protection.

3 Results and discussion

This part aims to approach and map an overview of current measures and challenges in the field of consumer protection in order to strengthen their position and protection.

3.1 The system of existing consumer protection measures

Empowering consumers and effectively protecting their security and economic interests have become key objectives of European policy. Although EU consumer policy has some of the strictest consumer-protection rules, there are still cases showing that even the current regulations and their enforcement have their limits. To this end, in April 2019, the EU adopted a so-called **'New Consumer Agreement',** which contains stricter rules that can be better and more effectively enforced. The most important measures include greater transparency for consumers when shopping online, more effective punitive actions and transparent rules for tackling dual product quality in the EU. [14]

It followed the **'Strategy for the Digital Single Market'** of 2015, in which the EC implemented several initiatives adjusting consumer protection rules to the online world, such as abolishing roaming charges or unjustified geo-blocking. The modernized Consumer Protection Cooperation Regulation, adopted in 2017, seeks to improve enforcement by public authorities and cross-border cooperation between consumer protection authorities.

3.2 A System of Current Consumer Protection Measures and Challenges

Continued efforts to strengthen the position of consumers, their rights, safety, and security are shaped in the form of the **New Consumer Agenda** (*Strengthening Consumer Resilience for Sustainable* Recovery*)* **for 2020-2025.** This program follows the completed European Consumer Agenda (2014-2020) and is also based on the 2019 New Consumer Agreement. It presents a vision of consumer policy in the forthcoming five-year period. It aims to flexibly address the needs of consumers that many of them face because of the Covid-19 pandemic and thus increase their resilience. These pandemic-related issues affect consumers' daily lives and are not related only to the availability and accessibility of products and services but also to travel within and out of the EU. [15]

The program covers five main priority areas: [15]

1. green transition,
2. digital transformation,
3. corrections and enforcement of consumer rights;
4. the specific needs of particular groups of consumers;
5. international cooperation.

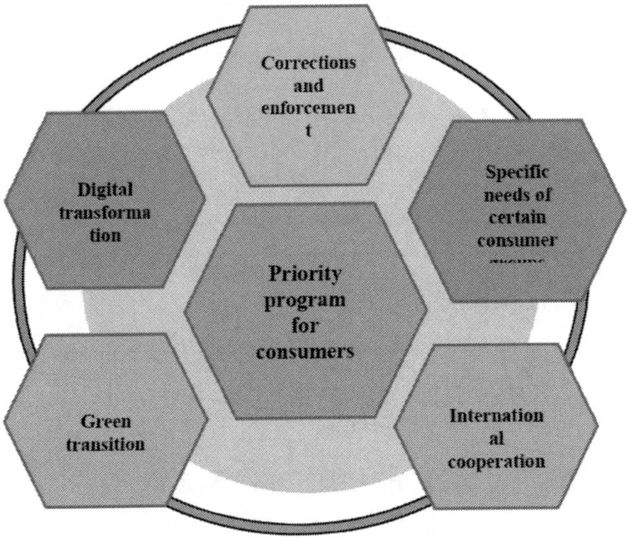

Fig. 1. Main priority areas

Source: processed according to [16]

Table 1. Overview of priorities for consumer protection measures

PRIORITIES OF THE PROGRAMME	CONTENT OF THE MEASURE
1. Green transition	▪ raising consumer awareness of the environmental sustainability options of products, such as their durability or reparability ▪ protecting consumers from certain practices, such as environmentally misleading advertising or early obsolescence, and ensuring that they have access to credible information ▪ promoting commitments of businesses concerning sustainable consumption measures that go beyond the legal obligations ▪ product repairs and support for the purchase of more sustainable and recyclable products → revision of the Directive on the sale of goods after 2022
2. Digital transformation	▪ strengthening consumer protection and security in the context of the digitization of retail financial services and → revision of the Consumer credit directive and the Directive on the provision of financial services to consumers at a distance ▪ addressing new product safety challenges caused by new technologies and online sales → revision of the General product safety directive ▪ ensuring consumer protection and protecting the fundamental rights in relation to AI and → proposal for a horizontal legal act setting out requirements for AI ▪ addressing online business practices that do not respect consumer rights → update guidance documents on the Unfair Commercial Practices Directive and the Consumer Rights Directive, and analyze whether additional legislation or other measures are needed
3. Corrections and enforcement of consumers' rights	▪ coordinating and supporting law enforcement activities carried out by the Consumer Protection Cooperation Network (CPC) to combat illegal practices ▪ supporting authorities by providing standard and innovative electronic tools for online investigations ▪ assessment of the set of tools available to authorities to address infringements of EU consumer law → evaluation of the CPC Regulation
4. Specific needs of certain groups of consumers	▪ improving the availability of debt advisory services in the Member States ▪ improving the safety of childcare products -> Revision of the General Product Safety Directive and new safety requirements for standards for childcare products ▪ supporting local initiatives to provide advice to consumers, including in remote areas
5. International cooperation	▪ developing an action plan with China to intensify cooperation on the safety of products sold online ▪ offer regulatory support to EU partner countries

Source: New Consumer Agenda [16]

The adopted program aims to meet consumer expectations regarding the benefits of the single market, make informed decisions, and be an active part (participation) in the green and digital transformation. At the same time, the intention is to ensure free access to goods and services throughout the EU and protect consumers' rights regardless of emerging challenges.

The program is a comprehensive set of different EU policies, focusing on consumers. The program also complements other EU initiatives, such as the European Green Agreement, the

EU Circular Economy Action Plan, and the Communication on shaping Europe's digital future. It also builds on and supports international concepts such as the UN 2030 Agenda for Sustainable Development and the UN Convention on the Rights of Persons with Disabilities. [15]

4 Conclusion

In the long term, it is possible to observe how the European Union has implemented a comprehensive set of policies and rules over the last five decades, enabling consumers to ensure a high level of protection and solid rights. Building on the results and successes of consumer policy to date, it continues with a new program for consumers, setting out how to promote these rights while protecting consumers and empowering them in the current context of the pandemic, as well as in the coming years.

The article provides an overview of the measures taken to improve and empower consumers by presenting a new consumer program aimed at responding to consumer needs during and after the COVID-19 pandemic. It includes a long-term vision for 2025, setting out concrete measures in five key areas to protect consumers and enable them to take an active part in shaping the green and digital transformation.

Acknowledgement

This paper is an outcome of the project: KEGA MŠ SR č. 002EU-4/2019 "Integrating and systematizing the results of scientific research activities in the field of consumer protection, with a primary focus on food security, to model adequate consumer behavior" funded by the University of Economics in Bratislava; led by doc. Ing. Mgr. Gabriela Dubcová, PhD. - project share is 100%.

References

1. D. Vokounová, Generation Y – Health Related Attitudes and Behaviour. Central and Eastern Europe in the Changing Business Environment. International Joint Conference. Central and Eastern Europe in the Changing Business Environment: Proceedings of 18th International Joint Conference: May 25, 2018, Bratislava, Slovakia and Prague, CzechRepublic,467-476 (2018)
2. J. Gunawan, P. Permatasari, C. Tilt. Sustainable development goal disclosures: Do they support responsible consumption and production? *Journal Of Cleaner Production*, **246**, 118989. (2020)
3. M. Pechota, M. Matuskova, L. Medlenakova, The trend of cost of universal services provided by national postal operator and correlation between price of letter mail and amount of sentlettermails in Slovakia. 12TH INTERNATIONAL SCIENTIFIC CONFERENCE OF YOUNG SCIENTISTS ON SUSTAINABLE, MODERN AND SAFE TRANSPORT, **192**, 574-579 (2017)
4. M. Dzurová, et al. *Consumer protection*. Bratislava: ECONOMIST, 2011. 279. ISBN 978-80-225-3180-1
5. D. Seibble, et al. *Great economic encyclopedia*. 1. ed. Bratislava: SPRINT, 2002. 967 p. ISBN 978-80-89085-04-0
6. Act No. 250/2007 Z.z. on Consumer Protection and on Amendment of the Act of the Slovak National Council No. 372/1990 On Offences, as amended
7. Hunt, Shelby D. – Scott, J.V. 1986. A General Theory of Marketing Ethics. *Journal of Macromarekting* 6 (Spring): 5-16. Reprinted in Ethics in Marketing. Eds. N. Craig Smith and John A. Quelch. 1993. Honeywood: Irwin, 757-774

8. M. Dzurová et al. *Consumer theories and realms*. 1. ed.. Bratislava: Economist, 2007. ISBN 80-225-2327-1
9. European Union policies. 2016. Consumers. [online]. Publications Office of the European Union. 2016. Available on the Internet: http://www.europedirectsnv.eu/wp-content/uploads/2013/11/Spotrebitelia.pdf [accessed 02.05.2021].
10. B. H. Clark, A Summary of Thinking on Measuring the Value of Marketing. Journal of Targeting. *Measurement and Analysis for Marketing*, **99**, 357 -369 (2001)
11. M. E. Silva, M. D. Figueiredo. Practicing sustainability for responsible business in supply chains. *Journal Of Cleaner Production*, **251**, 119621. (2020)
12. T. Ambler, F. Kokkinaki, S. Puntoni, Assessing Marketing Performance: Reasons for Metrics Selection, *Journal of Marketing Management,* **20**, 475-498 (2004)
13. V. S. C. Tunn, N. M. P. Bocken, E. A. van den Hende, J. P. L. Schoormans, Business models for sustainable consumption in the circular economy: An expert study. *Journal of Cleaner Production*, **212**, 324-333. (2019)
14. A new agreement for consumers. 2018. Representation in Slovakia Available on the Internet: http://ec.europa.eu/news/new_deal_for_consumers_sk [accessed 02.05.2021].
15. European Commission: Communication from the Commission to the European Parliament and the Council. A new program for consumers. Available at: https://www.mhsr.sk/uploads/files/MxnYjl9a.PDF [accessed 02. 05.2021]
16. European Commission: Factsheet. November 2020, New Consumer Agenda, New Consumer Program. Measures to protect European consumers 2020-2025. Available at: https://ec.europa.eu/info/sites/default/files/novy_program_pre_spotrebitelov_-_informacny_prehlad.pdf [accessed 02. 05.2021]

Patents as an indicator of the future development of individual region

Peter Kubík

TARIQ Company s.r.o., Křižovnická 86/6, Staré Město, 110 00 Praha 1, Czech Republic

Abstract. In today's world full of changes, whether the advent of new technologies that are authoritative and will be a trend for many economic sectors or impact Covid epidemic whose effects are still unaware can be sure that the changes in consumer preferences and aggregate demand and increasingly more competition between the largest economies in geographical terms, all this puts more importance on research and development. It is expected a new impetus in the field of patents that will respond to new impulses from the market.
Monitor the development of patents over the last year in the broader perspective implies growth or decline in individual industries and economies, are interesting differences in the intensity of new know-how even within different regions of the world.

1 Introduction

The long-term the nature of research and development regarded as an essential element of the economic condition. Patent as an innovative tool is one of the most important elements of creating a competitive advantage in the market. It is protected by legal norms that are implemented in every developed country. The creation of patents is involved in many businesses, such as industrial enterprises, universities, research agencies, technology parks, public research institutes. The issue of patents in individual countries and regions is managed by individual public organizations. These organizations accept individual patents, issue decisions and protect the rights associated with them. They headquarter information on patents, which often offer additional opportunities for their interpretations and may be used for subsequent decisions at different levels of the economy.

1.1 Theoretical starting points related to patents

Competitive advantage and competitive position are used to indicate the company's ability to compete with each other. Dynamic and unexpected environment business environment requires companies to be prepared to use innovative and effective competitive strategy to build a competitive advantage in the market. [1] Then have achieved competitive advantage impact on the competitive position of which can be described by the terms of its profitability. [2] All these phenomena are under a general term that describes the company's ability to compete in the market. [3]

Investment in intangible assets (patents, know-how, trademarks, designs, software) is still under the pressure of global competition in recent decades, a significant source of competitive advantage. [4]

The term invention is understood to mean a set of activities or processes that lead to a particular discovery. The inventor can register his patent with the Industrial Property Office in each country. A patent is an official document that is awarded by the government of the country. The rights associated with this document allows innovators to prohibit another person to perform certain activities related to the patent without his consent.

What is important is the basic characteristics of the features of a patent or improvement proposal, which include [5,6,7]:

- **Novelty** - is examined using state of the art at the date when the applicant has the Right of Priority. The Right of Priority is understood the date on which the applicant filed the first application for a given invention, regardless of the country in which the application was filed. To this date, it examines the state of the known technology as the novelty is understood to mean the absolute world novelty,
- **Inventiveness** - is understood as an activity of the inventor, which reached an invention that could be patentable. The examination of the application analyzes whether the invention is the result of creative activity. When assessing inventive step at the Industrial Property Office tends apply fictional person "expert" with average qualifications and general overview of the art to which the invention was made,
- **industrial applicability and technical nature** - is understood as the utility of the invention in all areas of the economy, regardless of whether they are industry, agriculture or services.

Due to the historical context of patent law, the issue of intellectual property relatively frequently examined through the collection and processing of statistical data. Patents or inventions are directly regulated by national and international legislative standards are accurately recorded and then the source outputs of varying quality are freely available on the websites of national authorities, such as the Slovak Industrial Property Office of the Slovak Republic or transnational joint authorities, such as a. within the European Union European Patent Office.

There are a large number of studies and surveys in measuring innovation. Along with using the input data, such as spending on research and development and human resources in research, patents have become the most common measure of innovative products in the survey. The aforementioned feature comfortable patents is that they are available relatively exact number of inventions in accordance with various aspects. An addition, patents have been well documented, especially in recent years, thanks to extensive online information that can be conveniently organized into databases. Another advantage of patents that may combine different characteristic indicators. [8,9,10,11]

According to Kucka, *the types of patent searches include:* [12]

- ***patent analysis and statistics,***
- ***patent monitoring.***

Patent analysis and statistics are methods that address the latest trends in science and technology. Also they allow to identify the possibilities of development in the various fields of science and technology worldwide. These analyzes help to get an overview of the authors and owners patent activity in scientific disciplines for the relevant time period or in the field of science and technology, where we can monitor the number of registered and published patent applications and also the number of granted patents for individual years.

Patent analysis and statistics provide important information for research institutions, the company management in decision-making and not least investors. All are made from commercial or non-commercial tools. Patent monitoring can be understood as the continuous collection of information about patents, according to certain attributes such as continuous or obtain information about the status of published applications and granted patents.

Start-ups play an important role in innovation in every region, because applied leadership has a decisive influence on the success or failure of a start-up. Team work and leadership in innovation process are necessary for the success of any start-up. [13,14] The general transformation of the global world creates new opportunities for many countries. For example, the role of some of the constraints on the labor market, such as geographical location, underdevelopment and other similar activities, is declining.[15,16]

A successful innovation process is an important part of the business model, which consists of 4 elements, which are interconnected:

- contribution to customer value,
- profit formula,
- key resources
- key processes. [17]

1.2 Patent law

Within the industrial property right it is regarded the protection of inventions as the oldest institute of protection. It was the laws for the protection of inventions and subsequently patents that were among the first to be adopted in modern times.

Patents are the national rights granted by national patent offices. There is no patent, which would be granted protection across the EU. One application for a European patent in the European Patent Office (EPO) provides national protection in European countries chosen by the applicant but must be members of the EPO. International application is simplified under the Patent Cooperation Treaty (PCT). Trademark rights and design can register at national level or at Community level. At Community level, are registered at the Office for Harmonization in the Internal Market, located in Alicante (Spain). These rights are exclusive and apply throughout the EU. [18]

Intellectual property rights are still protected by national laws rather than EU law. Their protection in each EU country separately can be complicated and costly, so further harmonization is needed.

In the Slovak Republic, patents are administered by the Industrial Property Office of the Slovak Republic (Úrad priemyselného vlastníctva Slovenskej republiky). The entire process for patents on their characteristics, the method patenting, use and protection governed by the following legal regulations in force in the Slovak Republic: [5,6,19]

- Act no. 435/2001 Z.z. on patents, supplementary protection certificates and on amendments to certain acts,
- Act no. 402/2002 Z.z., amending and supplementing Act no. 435/2001 Z.z.. on patents, supplementary protection certificates and on amendments to certain acts,
- Act no. 84/2007 Z.z.., amending and supplementing Act no. 618/2003 Z.z.. on copyright and rights related to copyright (copyright act) and on amendments to certain acts,
- Act no. 517/2007 Z.z. on utility models and on amendments to certain acts (Article III),

- Act no. 495/2008 Z.z. on the fee for maintaining the patent, the fee for the maintenance of the European patent with effects for the Slovak Republic and the fee for maintaining the validity of a supplementary protection certificate for medicinal and plant protection products and on amendments to certain laws (Article III)
- Decree no. 223/2002 Z.z., which implements Act no. 435/2001 Z.z. on patents, supplementary protection certificates and on amendment of certain acts.
-

2 Objective

The article aims to provide an overview of the current situation in consumer protection with the aim of improving the position and strengthening consumer protection. The priority is to bring together current measures and challenges to strengthen consumers' rights and interests and thus increase their level of protection.

3 Results and discussion

The aim of this section is to present an overview and mapping patents and patent holders in their development and status in the world to compare the differences in the economies of individual regions.

3.1 Global developments in patents

Prediction of future economic developments in individual countries and regions is also dependent on innovation activity in the market. The hallmark of innovation activity is the number of patent applications. The high volume of patents for certain time period makes the hypothesis of a healthy and sustainable growth, macroeconomic indicators and good condition of businesses in the market. It also allows a fall in the amount of time to prepare the properly stimulus measures by public authorities to reverse the condition. It is important to know the structure of the applied patent, its characteristics, to know the holder of the patent application. This information is concentrated in national and multinational organizations, which manage patents and create databases.

The World Intellectual Property Organization (WIPO) owns the world's largest patent database. From its databases it is possible to present developments in this area over the last 40 years. [19]

Region	Patent applications
Africa	9 596
Asia	1 200 024
Europe	1 219 399
Latin America	21 620
North America	1 213 077
Oceania	52 718
Others	4 507
Total	**3 720 941**

Table 1. Patent applications filed per region, 1978–2018

Source: processed according to [20]

Since 1978 the number of filed patent applications increases year on year and in 2018 it reached 253,000. In 2009, a decrease of total patent applications, when there was a global economic downturn. A similar course is expected in 2020-2021 due to the Covid epidemic, perhaps larger than in 2009. The reason may be a barrier to work, where many companies used home office and researchers did not have access to many of the tools needed for innovation. Yet in 1978, Europe and North America together accounted for about 88% of the total administration of patents. In 1995 there was a turning point, the combined market share is decreasing in favor of Asia. In 2009, Asia had a larger volume of patent applications than North America, and in 2010 it overtook Europe, becoming the world's largest patent-pending region. In 2018, more than half of the total number of applications came from a single region - Asia (50.5%).[20,21,22]

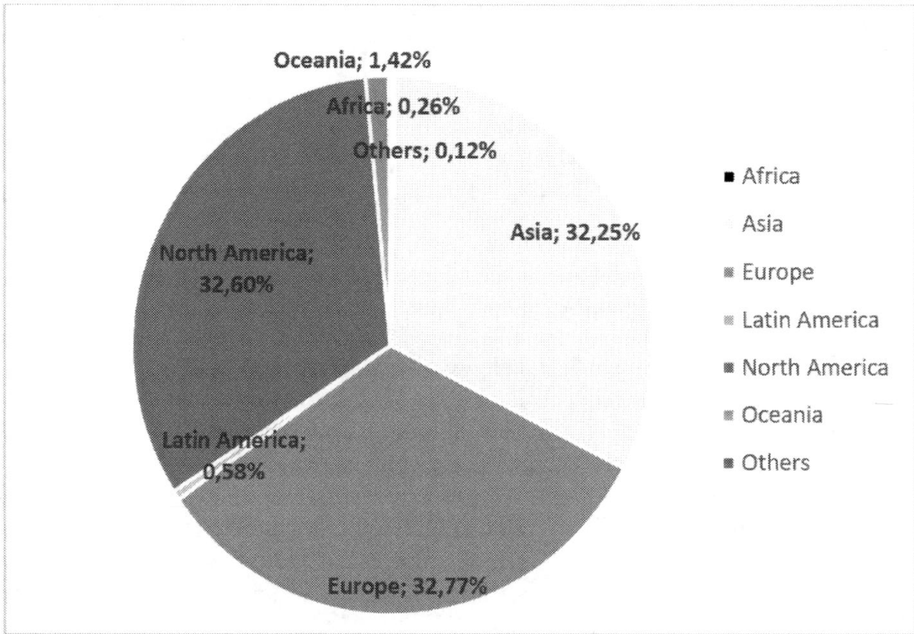

Fig. 1. Region share patent applications for 1978 - 2018

Source: processed according to [20]

In the last 40 years, Europe, North America and Asia have applied for patents at about the same volume - Europe (1,219,399) North America (1,213,077) and Asia (1,200,024). These three regions constitute 97.6% of the total administered in the period 1978-2018. Oceania (52,718), Latin America and the Caribbean (LAC) (21,620) and Africa (9,596) recorded in the same period a much lower volumes. The view of individual countries is also interesting. Applicants for patent applications from the United States of America (USA) have filed 31.1% of the total number of applications in the last 4 decades. The second largest number of applications was recorded in Japan (17.3%). If we add the applicants from Germany (10.4%), China (8.1%) and the Republic of Korea (4.6%), the total sum of the top five countries together accounted for 71.4% of the total number of patent applications during this period.

In 2018, the list of the most successful countries did not change, but their share increased to 77.5%. The largest increase in the proportion of experienced China, which grew from the cumulative 8.1% over the last 40 years to 21.1% in 2018. This is especially remarkable given the fact that between 1978 and 1993 filed by Chinese applicants only five patent applications.

The majority of applications submitted in Africa between 1978 and 2018 came from South Africa (76.6%), in Asia by Japan (53.3%), in North America and Oceania by the USA (95.3%) and Australia (86,1%). Less than half share in the continent achieved the most successful countries in Europe and LAC, where Germany and Brazil represents 31.8% and 41.6% of the total. In terms of the nominal number of patent applications for 40 years dominated by the USA (1,156,062), Japan (641,903), Germany (387,338) and China (303,103). [20,21,22]

The view of the Top 30 patent applicants in the world is also interesting:

Applicant	Country of Origin	Patent applications		Ranking in time			
		1978-2018	1978-2018	1990	2000	2010	2018
PANASONIC IP MANAGEMENT	Japan	34 081	1	23	6	1	12
HUAWEI TECHNOLOGIES	China	33 899	2		6 896	4	1
KONINKLIJKE PHILIPS ELECTRONICS	Netherland	32 783	3		2	5	18
ROBERT BOSCH CORPORATION	Germany	27 654	4	2	5	6	10
SIEMENS AKTIENGESELLSCHAFT	Germany	27 403	5	5	1	12	14
ZTE CORPORATION	China	25 746	6			2	5
QUALCOMM INCORPORATED	U.S.	24 858	7	1 510	16	3	4
TELEFONAKTIEBOLAGET LM ERICSSON	Sweden	22 429	8	98	3	9	9
MITSUBISHI ELECTRIC CORPORATION	Japan	19 342	9	72	19	14	2
INTEL CORPORATION	U.S.	17 963	10		22	62	3
LG ELECTRONICS INC.	Korea	17 349	11		378	7	8
SHARP KABUSHIKI KAISHA	Japan	16 593	12	833	261	8	16
SAMSUNG ELECTRONICS	Korea	15 680	13	1 510	37	17	6
3M INNOVATIVE PROPERTIES COMPANY	U.S.	14 746	14	195	13	16	30
BASF	Germany	14 654	15	22	7	13	34
SONY CORPORATION	Japan	14 563	16	127	17	28	13
TOYOTA JIDOSHA KABUSHIKI KAISHA	Japan	14 201	17	1 510	223	11	236
PROCTER & GAMBLE COMPANY	U.S.	13 603	18		4	27	36
NEC CORPORATION	Japan	13 282	19		122	10	22
MICROSOFT TECHNOLOGY LICENSING	U.S.	12 870	20		30	20	11
HEWLETT-PACKARD DEVELOPMENT	U.S.	12 095	21	1 510	160	18	15
NOKIA TECHNOLOGIES	Finnland	11 749	22		61	15	35
FUJITSU	Japan	10 049	23	34	27	19	43
HITACHI	Japan	9 854	24	84	18	26	27
MOTOROLA	U.S.	9 666	25	6	12	116	329
E I DU PONT DE NEMOURS	U.S.	9 352	26	8	11	21	460
FUJIFILM	Japan	8 736	27			48	21
UNIVERSITY OF CALIFORNIA	U.S.	8 096	28	21	23	38	38
INTERNATIONAL BUSINESS MACHINES	U.S.	7 706	29	195	62	22	100
BOE TECHNOLOGY GROUP	China	7 443	30				7

Table 2. TOP 30 patent applicants 1978–2018

Source: processed according to [20]

Top 30 patent applicants come from eight countries. On top of the list are the USA with 10 applicants, followed by Japan (9), Germany (3) China (3) South Korea (2), Finland (1), Netherlands (1) Sweden (1). Seven of the 10 best candidates in the last 40 years appeared among the 10 best candidates in 2018. The exception was Panasonic (12th place in 2018), Philips Electronics (18) and Siemens (14). Each active since 1978 10 best candidates in the electronics industry. [20,21,22]

List of 30 best covers only businesses, the best university is the University of California (8096) to 38th place.

Ranking	University	Origin	Published PCT 2018	Ranking All Applicants
1	UNIVERSITY OF CALIFORNIA	U.S.	501	38
2	MASSACHUSETTS INSTITUTE OF TECHNOLOGY	U.S.	216	108
3	SHENZHEN UNIVERSITY	China	201	119
4	SOUTH CHINA UNIVERSITY OF TECHNOLOGY	China	170	145
5	HARVARD UNIVERSITY	U.S.	169	147
6	UNIVERSITY OF TEXAS SYSTEM	U.S.	158	158
7	SEOUL NATIONAL UNIVERSITY	Korea	137	181
8	TSINGHUA UNIVERSITY	China	137	181
9	LELAND STANFORD JUNIOR UNIVERSITY	U.S.	121	211
10	CHINA UNIVERSITY OF MINING AND TECHNOLOGY	China	114	219
11	OSAKA UNIVERSITY	Japan	105	242
12	JOHNS HOPKINS UNIVERSITY	U.S.	99	259
13	KOREA ADVANCED INSTITUTE OF SCIENCE	Korea	94	273
14	UNIVERSITY OF TOKYO	Japan	92	280
15	HANYANG UNIVERSITY	Korea	89	294
16	TOHOKU UNIVERSITY	Japan	87	300
17	KYOTO UNIVERSITY	Japan	86	304
18	UNIVERSITY OF MICHIGAN	U.S.	81	324
19	OXFORD UNIVERSITY INNOVATION LIMITED	U.K.	79	332
20	UNIVERSITY OF FLORIDA	U.S.	79	332

Table 3. TOP 20 patent applicants by universities in 2018

Source: processed according to [20]

3.2 Patent examination and approval

The average time duration of the process from application to approval patent in each patent offices varies.

In 2018 the largest national and multinational patent offices needed for the approval process following time period:

https://doi.org/10.1051/shsconf/202111503008

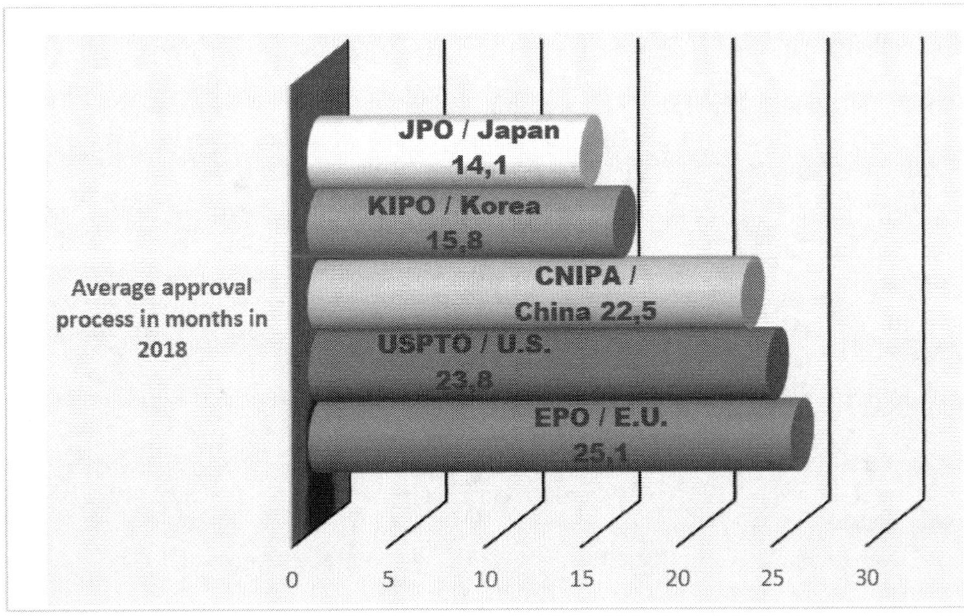

Fig. 2. Average approval process for patents in 2018

Source: processed according to [23]

The time required for the entire patent assessment process is the shortest at the Japan Patent Office (JPO) - 14.1 months. Almost twice as long this process takes at the European Patent Office (25.1 months). All TOP 3 leaders come from Asia.

JPO focuses mainly on accelerating internal processes within its organization. It employs more than 1,400 specialists / examiners who assessed the various patent applications, work closely and actively search for information on patents that are already established in international patent organizations. Speed process, however, must comply with established quality criteria of the organization. [23]

4 Conclusion

In the long run, it is clear that innovation efforts are shifting more to Asia. On the contrary, Europe has recently been losing innovation potential, which is very dangerous for its future economic development. The electronic sector is a key accelerator of innovation, which is developing rapidly year by year.

An important factor is also the support of innovation / patents in terms of legislative and approval processes in individual countries. It is certainly motivating for many applicants the shortest possible response to the submitted patent from the patent office. Innovative support from academic institutions is the basis not only for the emergence of new innovative solutions, but also the basis for the education of new innovators. Mentioned attributes should be the minimum for the European Union a major challenge to ensure its future.

References

1. S. Farhikhteh, A. Kazemi, A. Shahin, A., M.M. Shafiee. How competitiveness factors propel SMEs to achieve competitive advantage? *International Business Journal* **1**, 2-13 (2020).
2. M. Kharub, R. Sharma 2020, An integrated structural model of QMPs, QMS and firm's performance for competitive positioning in MSMEs. *Total Quality Management & Business Excellence*, **31**, 3-4 (2020).
3.]. E. Akben-Selcuk. Factors affecting firm competitiveness: Evidence from an emerging market. *International Journal of Financial Studies*, **2**, 1-10 (2016).
4. Š. Hittmár, M. Hrnčiar, V. Lendel. Management of innovation processes in the enterprise. Žilina: EDIS, p. 15-25 (2017). ISBN 978-80-5541-414-0.
5. M. Kropaj et al. Basics of intellectual property rights for economists. Bratislava: Wolters Kluwer , p. 12-45 (2017). ISBN 978-80-8168-706-8.
6. Z. Adamová. Intellectual property law. Bratislava: TINCT , p. 25-32 (2020). ISBN 978-80-9735-440-4.
7. V. Mihálová. The term "plausible" in the decision-making practice of the European Patent Office. *Journal Intellectual property*, **3**, ISSN 1339-5564.
8. B. Hall, A. Jaffe, M. Trajtenberg. The NBER Patent Citations Data File: Lessons, Insights and Methodological Tools. NBER working paper no. 8498 (2001).
9. M. Trajtenberg, G. Shiff, R. Melamed. The "Names Game": Harnessing Inventors' Patent Data for Economic Research, NBER WP no. 12479 (2006), Cambridge, MA.
10. D. Harhoff, K. Hoisl. Institutionalized incentives for ingenuity—patent value and the German Employees' Inventions Act. *Research Policy*, **36**, 1143–1162 (2007).
11. A. Jaffe, M. Trajtenberg, R. Henderson. Geographic localization of knowledge spillovers as evidenced by patent citations. *Quarterly Journal of Economics*, **63**, 577–598 (1993).
12. Ľ. Kucka. Types of patent searches - Patent analysis and statistics, patent monitoring. *Transfer technológií bulletin*. [online]. 2014, Available on the Internet: http://ttb.cvtisr.sk/buxus/docs/ttb_2_2014.pdf, ISSN 1339-2654 [accessed 20.05.2021
13. Zäch, S., Baldegger, U. 2017. Leadership in start-ups. International Small Business Journal: Researching Entrepreneurship, 35, pp. 157-177, ISSN 0266-2426.
14. J. Bussgang. Are you suited for a start-up? *Harvard Business Review*, **6** (November-December 2017), 150-153, ISSN 0017-8012.
15. H. Lasi, P. Fettke, H. G. Kemper, T. Feld, M. Hoffmann. Industry 4.0. *Business & Information Systems Engineering*, **6**, 239-242 (2012), ISSN 2363-7005.
16. K. Ozawa. Managers' experiences as recipients: impact on organizational change. *Journal of Organizational Change Management*, **33**,143-156 (2019), ISSN 0953-4814.
17. M. W. Johnson, C. M. Christensen, H. Kagermann. *Reinventing Your Business Model*. Harvard Business Review, **86**, 2008, ISSN 0017-8012
18. European Patent Office 2021, European Patent Register. [online]. Available on: https://www.epo.org/searching-for-patents/legal/register.html, [accessed 18.05.2021].
19. Industrial Property Office of the Slovak Republic 2021, What are patents. [online]. Available on: https://www.indprop.gov.sk/?co-su-to-patenty, [accessed 18.05.2021].
20. World Intellectual Property Organization 2020, Patent Cooperation Treaty Yearly Review 2020. [online]. Available on: https://www.wipo.int/edocs/pubdocs/en/wipo_pub_901_2020.pdf, [accessed 20.05.2021].
21. European Patent Office 2021, Patent Index 2020 – European Patent Application. [online]. Available on:

 https://doi.org/10.1051/shsconf/202111503008

https://documents.epo.org/projects/babylon/eponet.nsf/0/59B9A0C416D02E75C1258699004AA62F/$File/patent_index_2020_european_patent_applications_en.xlsx, [accessed 18.05.2021].

22. United States Patent and Trademark Office 2020, Annual report 2019. [online]. Available on: https://www.uspto.gov/sites/default/files/documents/PPAC_2019_Annual_Report.pdf, [accessed 18.05.2021].
23. Japan Patent Office 2020, Status report 2020. [online]. Available on: https://www.jpo.go.jp/e/resources/report/statusreport/2020/document/index/all.pdf, [accessed 18.05.2021].

Strategic management tools and their application in innovation processes

Iveta Kufelová[1], *Monika* Raková[2,*]

[1]University of Economics in Bratislava, Faculty of Business Management, Department of Business Economy, Dolnozemská cesta 1/b, 852 35 Bratislava, Slovak republic
[2]University of Economics in Bratislava, Faculty of Business Management, Department of Business Economy, Dolnozemská cesta 1/b, 852 35 Bratislava, Slovak republic

Abstract The development of new technologies and innovation processes lead companies to improve their processes, resulting in new products all the time. Throughout the product life cycle, a firm incurs various costs related to the activities carried out in the pre-production, production and post-production phases. The essence of strategic management tools is to optimize costs throughout the product life cycle, focusing on the different stages. Such tools include target costing, kaizen and life cycle costing

1 Introduction

The increasing importance of knowledge as an economic driver has major implications for innovation management. The systemic approach to innovation recognizes that innovation and knowledge generation take place as a result of a variety of activities, many of them outside the formal research process. In the current economic context, growth must mainly originate from increasing the productivity of knowledge work, and increasing this productivity is the most important contribution management can make. [15] The complex current managerial dynamics requires, not only immediate actions to ensure timely decisions in the present, but also a process of anticipation that allows to visualize possible scenarios in the future, through the application of methodologies and tools for the analysis of trends that facilitate the forecast of future situations or through the exercise of prospective in the construction of a desirable future. [14]

2 Current state of the solved problem at home and abroad

Cost management is the effort to provide the necessary information for the efficient use of resources in the products production that can compete in international markets in terms of time, cost, quality and functionality, thereby maximizing the profit of the enterprise and minimizing the costs. Cost leadership is to produce the same product at a lower cost than competitors without sacrificing quality. [12] There are several

* Corresponding author: monika.rakova@euba.sk

ways how to optimize costs during innovation process. In this topic it has been defined three strategic management tools: life cycle costing, target costing and kaizen.

The product life cycles theory describes the general development of certain performance indicators of a product over its life time such as the development of turnover, profit rate as well as imports and exports. This development is typically divided and described in four phases–introduction, growth, maturity and degeneration. Phase one refers to the early exploratory stage which is characterized by a low market volume, uncertainty regarding the technological specification required by the market, simple product design and hardly any automatization in the production process. Phase two is specified as a growth phase with a large market expansion rate in which the dominant design emerges. Phase three marks the stage of maturity. Market growth slows down, innovations are less relevant and the number of companies decreases while market shares stabilize. [9] The product life cycle is significantly curtailed by rapid innovation and development in technology. It is a main driver of life cycle of products. In technology based industry, the lifecycle is shorter than before. [8] Companies focus on product-oriented in-novation during the growth stage of the industry life-cycle, and on process-oriented innovation during the maturity phase of industry evolution. [10]

Strategic cost management should provide cost information for both development and implementation of strategies, creating tactics to implement them, as well as development and implementation of control mechanisms for monitoring and reaching strategic objectives. Target costing was created during the 1960s in Japan by Toyota. Target costing is perceived as a strategic management system designed to manage costs of goods and services, seeking to identify new materials, at lower costs, for replacing components of the product, aiming to improve competitiveness. Target costing requires the participation of people from all sectors of the company, such as management, marketing, production engineering, cost accounting, and sales. [7] Target costing is not a costing system as such; rather it is an activity which is aimed at reducing the life-cycle costs of new products, while ensuring quality, reliability, and other customer requirements, by examining all ideas for cost reduction at the product planning, research and development process. [3] Target costing uses customer preference analysis to identify the implementable price and future sales volume. The product's cost structure is then optimized by matching drifting costs. The costing mechanism helps assign reasonable costs to products, matching customers' respective utility. This optimization assumes the independence of customer preferences resulting in an additive utility function. [2]

Kaizen as a lean management philosophy preaches continuous and progressive improvement of activities to create more value while reducing waste. It can be expanded to all areas of an organization. [11] Kaizen costing adopts a cost reduction system and assumes continuous improvement in production. The purpose of Kaizen costing is to reach cost reduction targets. Kaizen costing is a cost reduction method that helps to eliminate activities that do not add value to the product in the production process by continuous improvement. [12] Kaizen aims to promote continuous improvement, through multifunctional teams training and productive activities planning, with the objective of identifying gaps and developing new methods to solve possible issues. [13]

3 Research Design

The main objective of the paper is to clarify the strategic management tools that can be used during the different phases of the product lifecycle as well as during innovation, with respect to cost optimization.

Such strategic tools include target costing, kaizen and lifecycle costing with a focus on individual stages - pre-production, production and post-production.

In the preparation of the paper, sources on the subject from domestic and foreign authors were used. The obtained data were processed on the basis of analysis and synthesis, mathematical calculations and schemes.

The paper is rather theoretical in nature with the possibility of recommending how to use individual cost optimization tools in a company that wants to stay on the market in the long term and has a tendency to innovate.

4 Result of the Paper and Discussion

4.1 Cost optimization during the product life cycle

During the individual phases of the product life cycle, the company incurs various costs, that are related to the activities performed. First, they are the costs of activities in the pre-production stage, then the costs of production and finally the costs of processes in the post-production stage (dismantling of old technology, etc.). We refer to these costs in terms of drawing on the company's resources as the costs actually incurred.

The essence of strategic management tools is the optimization of costs during the entire product life cycle with a focus on pre-production stages.

Increasingly, many managers are inclined to say that increasing operational efficiency is not enough. That there is something else that will allow the company to become more successful.

In order to find resources and reserves to increase performance and profitability, it is necessary to monitor the entire value chain. The production and sale of products are associated with costs associated with all phases of the life cycle, not only during production. From an operational point of view, we get to strategic cost management, managerial accounting, which monitors the costs and revenues of the company from a long-term perspective. If a company looks at it from this point of view, it can give it a competitive advantage in the innovations. In many sectors, more emphasis needs to be placed on the importance of the pre-production stages of the product life cycle than on the production phase itself. These are mainly technologically and structurally demanding industries with an extensive phase of research and development (e.g. the automotive industry). On the one hand, the life cycle of products or services will be shortened, which means that it works at a time when products generate profit. On the other hand, the technological and organizational complexity of non-production stages is growing. Complications also include the fact that it is difficult to interfere with technological parameters during production, as well as the fact that it is not possible to influence cooperating entities at this stage.

It follows from the above facts, that a large part of the costs (sometimes up to between 80-90%) is decided in the pre-production stages. If the company starts production, it has only small options for managing costs and purposefully influencing them.

The goal of strategic management and pricing decision-making is to focus on the costs in the pre-production stages and look for solutions to optimize them, as well as on the costs incurred during all stages of the product life cycle. One of the possible solutions to this problem is the use of the *target costing method.*

4.2 Essence and implementation of the target costing method

Target costing is currently one of the most widely used management accounting methods and an essential part of pricing decision-making processes. The method was developed at Toyota in 1965 and has been widely applied by companies in Japan since the 1970s. However, the principles of the method were already applied in the Baťa shoe factories in the 1920s in Czechoslovakia. The method is suitable for the production of more complex products, in research and development intensive production and in highly competitive industries, as well as for the innovation of already existing products. The production of complex products requires searching for new possibilities concerning the use of technology or construction solutions. Competition, in turn, forces companies to reduce costs.

Target costing method - is an activity aimed at verifying all possibilities of cost optimization in the phase of research, development and prototype preparation. The result of this activity should be a product design that meets all customer requirements and expectations, and the cost of its production and price will provide the company with the desired projected profit.

The basic steps of the target costing method include:

1. Determining the *market price* based on the market - customer requirements for the planned product.
2. Cost planning, which is aimed at influencing costs at the pre-production stage in an effort to achieve the lowest possible production costs.
3. Communication and cooperation at all levels (processes, departments, subcontractors, construction team, etc.)

The basis of this method is searching for the answer to the question: *"What costs can the company spend on the product at a given market price?".*

Cost calculation using the target costing (TG) method

The process of pricing using the target costing method is divided into the following stages:

1. Determination of target price and target cost.
2. Determination of the company's framework objectives -target profit.
3. Dividing the product into individual parts -functional units, levels.
4. Adjustment of planned costs to target costs and target price.

The TG method is extensively used in the automotive industry, as shown in the following example:

Suppose a model of a certain car is to be sold in the market for €16,900. This amount represents the target selling price that customers are willing to pay for a given type of car. The firm aims to achieve a profitability of 10 % of the selling price.

Target - selling **price** 16 900 €
Required profit surcharge 10% (of the selling price)
Target profit 1 690 €
Total - target cost 15 210 €

The calculated target cost of €15,210 should cover all the costs that the company has to incur to produce the car, which are:
- direct materials and semi-finished goods (semi-finished products),
- direct wages and other direct costs,
- overhead costs,
- a safety reserve (in case of unexpected circumstances),
- sales costs (matched with sales and distribution)
- marketing costs,
- research and development costs.

In terms of calculation of costs, the target costing method generally proceeds from the most general costs groups of a strategic character (research, development, marketing) to variable unit costs, which are deducted from the total cost.

The company further assumes that sales costs, research and development costs and strategic marketing costs will represent 38 % of total costs. Overheads are assumed to be 30 % of production costs. At the same time, it foresees a risk surcharge of 5 % of the total cost of production. It expects wage costs and other direct costs to account for 25 % of variable costs.

Calculation of variable costs per unit:

Total cost	15 210,00 €
- sales, research, development and marketing costs (38%)	5 779,80 €
Production costs (to cover overheads and variable costs)	9 340,20 €
- overheads	2 802,06 €
- risk surcharge (safety margin)	467,01 €
Variable costs per unit	6 071,13 €
- wage costs and other direct costs	1 516,78 €
Direct materials and semi-finished goods	4 554,35 €

Variable unit costs represent the maximum allowable level of costs within which a company must fit. They contain the costs of direct materials, raw materials, semi-finished goods, semi-finished products, direct wages, other direct costs such as packaging costs, pallets, etc. When procuring the above-mentioned items, the company asks itself the question "which variable costs should we focus on, in order to meet the calculated amount?".

With fully automated machines or standardized performances, the possibilities to reduce direct labor costs are very limited. The right way can be found in reducing the costs of direct materials or raw materials and semi-finished goods (e.g. by diversifying suppliers).

The following figure shows *the process of dividing the cost of a product into the individual components for which the company is making decisions with respect to their purchase price*. The company's objective in this procedure is to find the level of direct material costs that corresponds to the desired cost level. For the construction of the entire product is responsible the construction team, which supervises for meeting the required costs, based on which it determines the costs for the individual components.

Figure 1 Implementation of the target costing method [5]

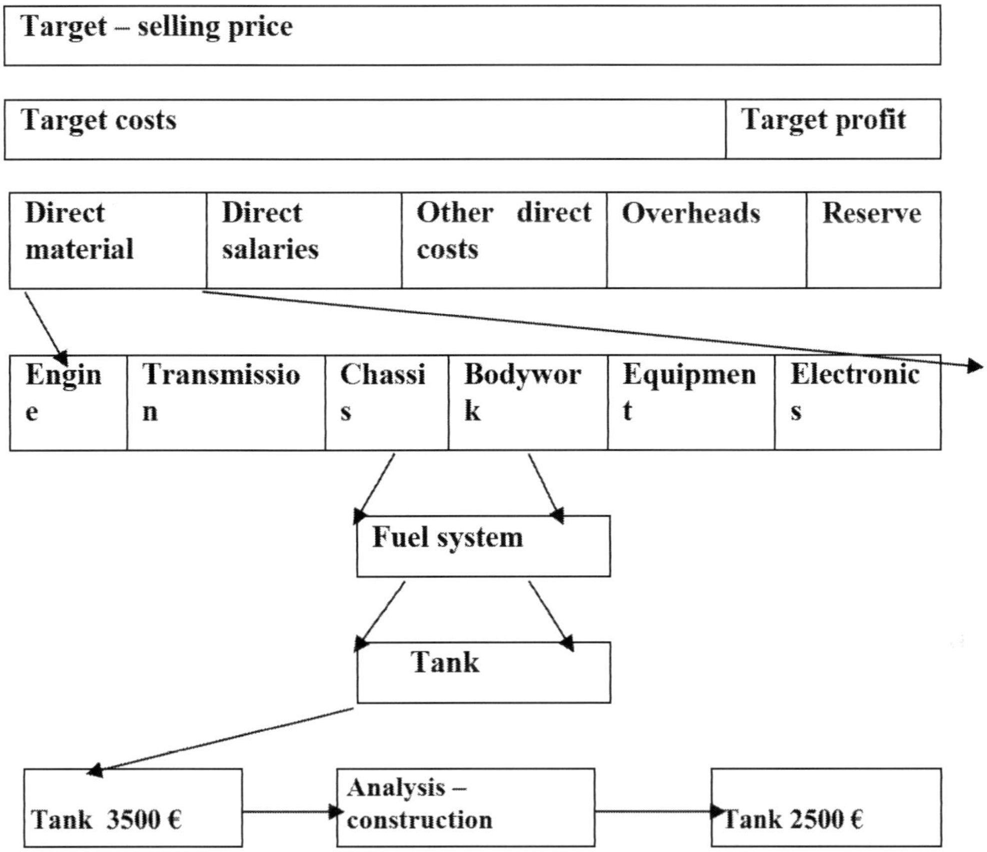

One of the main benefits of the target costing method is the strengthening of cooperation between individual departments of the company as well as professions, which have to cooperate efficiently and flexibly in the preparation as well as in the production of the product. The team involved in this must be composed of designers, technology and production staff, supply and marketing staff. The aim of all those involved is to design a product that will have all the features and functions required by customers, at a cost that will provide the company with the required profit.

Although the target costing method generally leads to cost optimisation, this reduction should not be associated with a reduction in quality or a reduction in the customer's perception of value. Another limitation of this method is that it works with estimated cost variables as well as estimated production volumes. Trying to get closer to the cost estimates may increase the time taken to prepare the product, which may lead to changes in input prices as well as in the customer preferences on which the target price was based and as a result, may have an impact on the projected - target profit.

4.3 The kaizen costing method and the use of life cycle costing

The kaizen method represents the introduction of smaller - continuous changes during the production phase of the product life cycle, in contrast to the TG method which focuses on the pre-production phase. Cost optimisation using kaizen is the most often achieved by increasing the efficiency of the production process through incremental changes in design and manufacturing processes, rather than undertaking major innovation projects.

The development of new technologies and innovation processes lead companies to improve their processes, resulting in new products all the time. Every product goes through a certain life process - the so-called life cycle - which they influence:
- the length of the life cycle,
- estimation of sales volume throughout the life cycle,
- estimation of the costs of each phase of the life cycle,
- analysis of the factors influencing costs and revenues during the life cycle.

Each stage is characterised by a different marketing situation, which also influences the choice of marketing mix strategy. During the life cycle, it influences its course:

Economic cycle - when economic changes in different areas need to be taken into account.

- Prosperity period - during this period the demand for a product increases. Prices are easier to increase, although costs also tend to rise. Costs rise more slowly than product prices. The difference between costs and prices affects the level of price, which influences firms' decisions on how to increase production capacity.
- Period of economic recession - in this period, buyers place emphasis on price. Firms reduce costs to a minimum level in order to keep prices low.

Distribution - is a factor that influences pricing decisions especially in small firms. Large firms are able to create their own market by spending large sums on advertising. Timely delivery of the product to the final customer requires important activities to be carried out which also affect the price. Smaller firms with limited resources have to use different distribution channels. They need to cooperate with them, but at the same time they need to take account of their reaction to their own pricing policy and be aware of the items that affect the price.

Product type

- Fashion product - the price of such a product is set at the discretion of the producers in the first period, as they know that there will always be customers willing to pay the price in order to show their serious interest in fashion. These firms therefore practise a policy of 'cream skimming', followed by a general reduction in prices after a more or less short period of time.
- Durable product - such product is not a source of impulse purchases. In deciding on the price, the firm relies on consultations at several levels and comparisons with competitors' products. In particular, product pricing must respect competitors' prices and the quality-price relationship of rival products.
- luxury product - some companies like to review the prices of luxury products. Luxury products can evoke a strong desire to have something special. If a firm launches such a new, lucrative product within its product range, it may be able to use its temporary, near-monopoly position to set the price. However, this pricing policy must also be expected to generate more competition, which will be attracted by higher profits.
- a basic demand product - there is no manipulative or manoeuvrable price range for these products. If buyers become accustomed to price reductions, they will reject any price increases and go to competitors.

Psychological factors

Every marketing (pricing) manager should monitor consumer buying behaviour, which is influenced by psychological factors. These are important to know in relation to pricing, in order to provide better, higher quality service to customers.

The following figure shows the relationship between target costing, kaizen and life cycle costing.

Figure 2 Life cycle costing [6]

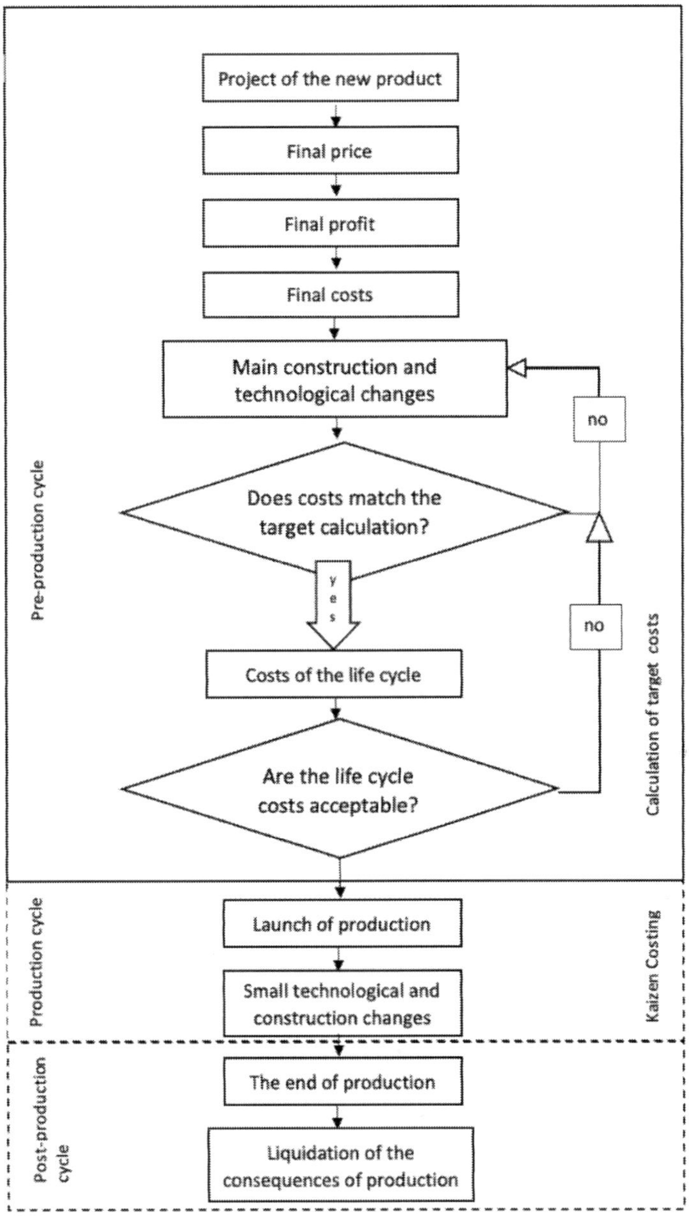

5 Conclusion

In business, almost all decisions are economic in nature. Thus, it is about comparing the benefits of business activity with the costs incurred, making a profit and maximising the market value of the business. In this respect, every company has to plan, calculate, record and manage costs mathematically. These activities enable managers to thoroughly understand and know the cost issues and, based on this, make decisions that are consistent with the company's goals and vision for the future.

It is about combining management, cost and financial accounting into a common cost management system, where the company can use cost optimization tools and thus compete in a fierce competitive battle. The emergence of these tools and related methods has been driven in particular by the development of information technology and the software tools available today, which are now a normal part of business and everyday life.

These strategic tools are aimed at optimising costs and achieving profitability through the volume of production (not by increasing output). As already mentioned, they are also suitable for innovation and can be considered as a competitive advantage of the company.

The article is an partial output of research project VEGA no. 1/0646/20 „Difúzia a dôsledky zelených inovácií na trhoch nedokonalej konkurencie, "in extent 50%, and VEGA No. 1/0708/20: "Socioeconomic determinants of sustainable consumption and production in terms of impact on business performance and competitiveness" in extent 50%.

References

1. E. Fendeková, I. Kufelová. *Ceny a cenové rozhodovanie.* Wolters Kluwer 2020, ISBN 978-80-7598-743-7. (2020)
2. C. Homburg, A. Hoppe, R. Schick, A. Braul. Accounting for preference dependency in target costing – a note. *Review of Quantitative Finance and Accounting*. doi:10.1007/s11156-021-00962-9. (2021)
3. K. Yutaka. Target costing support systems: lessons from leading Japanese companies. *Manage Account Res*, **4**, 33–47. doi: https://doi.org/10.1006/mare. 1993.1002. (1993)
4. J. Kita, a kol. *Marketing*. Jura Edition 2010. Prvé vydanie. 411 s. ISBN 978-80-8078-327-3. (2001)
5. B. Popesko, *Moderní metody řízení nákladu*. Praha: Grada Publishing, 233 s. ISBN 9788024729749. (2009)
6. L. Šoljaková. *Manažerské účetnictví pro strategické rizení,* Management Press 146 s. ISBN: 80-7261-087-2. (2003)
7. A. A., Cunha Callado, A. L. C. Callado, E. T. Do Bomfim. Target Costing in Micro and Small Companies from The IT Sector. *Revista Evidenciação Contábil & Finanças*, **8**, 125–141. doi:10.22478/ufpb.2318-1001.2020v8n3.50245. (2020)
8. Y. Sun, H. H. Cai, R. Su, Q. Shen. Advantage of low quality in short life cycle products. *Asia Pacific Journal of Marketing and Logistics,* **32**, 1038–1054. doi:10.1108/apjml-03-2019-0148. (2019)
9. M. O'Sullivan. Industrial life cycle: relevance of national markets in the development of new industries for energy technologies – the case of wind energy. *Journal of Evolutionary Economics,* **30**, 1063–1107. doi:10.1007/s00191-020-00677-5. (2020)

10. M. Cucculelli, V. Peruzzi. Innovation over the industry life-cycle. Does ownership matter? *Research Policy*, **49**, 103878. doi:10.1016/j.respol.2019. 103878. (2020)
11. E. E. Jeunon, F. de Oliveira Júnior, L. da Cunha Duarte, E. H. R. Guimarães, E. Lean manufacturing implantation impacts: a study on a large logistics operator. *Revista Gestão & Tecnologia*, **20**, 305-319. (2020)
12. D. Kelesbayev, H. Kozhabayev, G. Mombekova, R. Yermankulova, B. Dandayeva. An application for reduction of the non-value activities by Kaizen costing method in the Kentau Transformer Plant. *Entrepreneurship and Sustainability Issues*, **8**, 182. (2020)
13. L. S. Valamede, A.C.S. Akkari. Lean 4.0: A New Holistic Approach for the Integration of Lean Manufacturing Tools and Digital Technologies. *International Journal of Mathematical, Engineering and Management Sciences*, **5**, 851–868. doi:10.33889/ijmems.2020.5.5.066. (2020)
14. EE. Minguet-Carvajal. Management prospective. Background, currents and prospective worldviews at the philosophical and praxeological. *Vision GERENCIAL*, **19**, 61-74. (2020)
15. A. Hidalgo, J. Albors. Innovation management techniques and tools: a review from theory and practice. *R&D Management*, **38**, 113–127. doi:10.1111/j.1467-9310.2008.00503.x. (2008)

Education as the basis for the career development

Marta Matulcikova[1], *Anna* Hamranova[2*], *Tatiana* Hrivikova[3]

[1]University of Economics in Bratislava, Faculty of Business Management, Department of Management, Dolnozemska 1, 852 35 Bratislava, Slovakia
[2]University of Economics in Bratislava, Faculty of Business Management, Department of Information Management, Dolnozemska 1, 852 35 Bratislava, Slovakia
[3]University of Economics in Bratislava, Faculty of Applied Languages, Department of Intercultural Communication, Dolnozemska 1, 852 35 Bratislava, Slovakia

Abstract. A career can be seen as a sequence of employments, jobs and positions an individual fills in throughout their professional life. The choices made within the educational system seem to influence its successful start. The formal educational system is nowadays a preparatory stage for both professional life and lifelong learning. Retaining one's employment and career development goes, therefore, hand in hand with lifelong learning while employers play a significant role in all forms of further professional development. Our purpose for this paper is to analyse the transition from education to employment and subsequent lifelong learning. The paper aims to assess the importance of the achieved educational level in further career development and the attitude of Slovak companies towards further training and development. Another aim is to formulate proposals and improvements for a successful transition to the professional sphere, as well as progress in the preparation for a career and lifelong employment. We used standard methods of scientific research such as analysis, synthesis, comparison, and selection, but a questionnaire survey was used as our major research method. To evaluate the results, descriptive statistical methods were implemented together with pivot tables. Eventually, we verified our hypotheses by statistical methods and coefficients appropriate for the evaluation of associations between nominal or dichotomic parameters: chi-square and Cramer's V. For validation, PSPP statistical software was implemented.

1 Introduction

Career represents a lifelong process of self-creation, it begins with decisions about the field of formal education, constant improvement of one's potential and its adaptation to the dynamic conditions of the labour market. The requirements related to work performance grow ever higher and the intellectualisation of work requires constant improvements in qualification [1-2]. The qualification background needed for work performance is constantly changing, it is influenced by the necessity of permanent innovations in

* Corresponding author: anna.hamranova@euba.sk

production and services, the use of new technologies and above all, by the advancement of information and communication technologies [3-6]. All educational activities need to adapt to these trends:

- In the formal, i.e., primary, secondary, and tertiary education as part of school curricula, but also in further education [7-8], that is intended for individuals preparing for their profession within the framework of informal education which supplements the formal system.
- In adult vocational education which creates conditions for formal adult education within the framework of the school system leading to higher qualification with formal credentials of qualification. During that period of life, further adult vocational training is central [8] as it is closely related to work performance and accompanies one's career development. This includes:
  - Adaptive training.
  - Further qualification development focused on professional growth and improvement of one's existing qualification [9]:
    - Innovative education and qualification refreshment
    - Extending and deepening the achieved qualification.
  - Requalification – change of profession [9]:
    - Requalification in a related field – it can take the form of formal or informal requalification (more often the result of informal education).
    - Targeted requalification – change of existing qualification by acquiring knowledge, skills, and habits necessary for a new profession, usually aimed at a specific new working position. It is often acquired within the formal educational system.

The choice of formal education (secondary and tertiary ones) and their successful completion conclude the preparatory phase for a career. Completing the preparatory phase has an impact on the beginning of the first employment. We can often witness very difficult transitions from school to real working life or even move from formal education to unemployment which is an alarming fact requiring due attention. But we did not deal with it in our paper, as we included only those respondents into our sample who were at that time employed. The next relevant group consists of those who do not work in their area of study. This leads to the amortisation of finances invested into their education by society. All these factors had an impact on our empirical analyses. We collected information provided by experts and specialists in the area who explain the reasons for the existing discrepancy between education and subsequent employment. Their opinions – results of fundamental and empirical research can be characterised as follows:

- Personal perception of an individual: motivation, needs, faculties, talent, preferred values, willpower, intellect, emotional intelligence etc. [10-11].
- Attitudes of legal guardians (parents), their judgement of the aspirations and motivation for education and subsequent career prospects [12-14], considerations of performance, motivational characteristics, dynamic properties etc.
- Interpretation of an individual's personality by educational counsellors [15], Interpretation of an individual's personality by career counsellors [16-17], Interpretation of an individual's personality by human resources specialists [18-19].

In the context of the Slovak Republic, the decision-making power about one's education lies in the hands of the individual willing to study and his/her legal guardians. Unfortunately, the choice of an educational institution is often random, based on the opinions of parents, relatives and acquaintances and it does not take into consideration the personality of the individual who is being steered towards a particular career. Frequently, some extreme approaches can be registered when a school is chosen because of its social prestige or conversely, it is easy to pass. It is often overlooked that formal education takes time, and the graduates enter the labour market with a few years delay. Therefore, the choice of school should reflect the presumed conditions and employment prospects for the near future when the student graduates in the specific field of study. It is important to consider the number of graduates in various branches that the national economy is capable to employ. This requires the involvement of educational and career counsellors, specialists who are familiar with the trends in various professions on the labour market. This could create favourable conditions for future career success [20-21]. This way, situations could be eliminated when a graduate cannot find a job, becomes unemployed, needs to requalify, or works in a different field.

Dynamic conditions of the labour market, emergence of new and disappearance of some previous professions and the increasing demands for high work performance turn the attention to education as a central prerequisite for lifelong employment [22]. Therefore, formal education is currently more a preparatory phase for lifelong learning rather than just for one's professional life.

2 Research Framework and Methods

The main aim of the paper is to study the importance of the achieved qualification for career development and the approach of various companies operating in Slovakia towards further education. Based on a questionnaire survey, identify the problems in placing employees into the working process according to their qualification and assess the interest for and the scope of further education. The paper intends to interpret the results in a wider context. We wish to point out the practical needs from the employers' perspective which should initiate changes in formal education. We assume that if the education system is expected to become more adapted to the requirements of the professional sphere it needs to increase the cooperation with the entrepreneurial sphere.

The research consisted of three main stages. In the first stage, we focused on the study of scientific publications dealing with two important topics; the importance of achieved qualification for entering the labour market and further education and training in companies operating in Slovakia.

During the second stage, we formulated our research assumptions and research hypotheses. Simultaneously, a research model was created (model of research indicators) for the validation of the hypotheses. Table 2 shows the research model with the partial results.

The third stage included the statistical verification of the hypotheses and the formulation of conclusions. Adaptation of education at all levels to the requirements of economic practice reflects the future vision of the society.

2.1 Research assumptions and research hypotheses

We specified the following assumptions for the research:

1 Individuals with higher education are more easily placed in the working process according to their education and they are more interested in further education or training in comparison with people with a lower level of education.

2. Large enterprises pay more attention to achieving accord between employee placement and their qualification and are more interested in further training than small and middle-sized companies.

3. The sectors of the economy in which a company operates has no impact on employee placement neither on their access to further education.
4. Individuals working in managerial positions are more interested in further education than people working in non-managerial positions.

To validate the research assumptions, four sets of research hypotheses were formulated as the null and alternative hypothesis.

1. Set of hypotheses:

1 H_0: The achieved level of education of the respondents does not have a statistically significant effect on their work placement or their interest in further education.
1 H_1: The achieved level of education of the respondents has a statistically significant effect on their work placement and their interest in further education.

2. Set of hypotheses:

2 H_0: The approach towards employee placement based on their achieved qualification and towards further education is not different in micro, small, middle, or large companies.
2 H_1: The approach towards employee placement based on their achieved qualification and towards further education is different in micro, small, middle, and large companies.

3. Set of hypotheses:

3 H_0: The sector of the economy, in which the respondents work, does not have a statistically significant effect on their work placement or interest in further education.
3H_1: The sector of the economy, in which the respondents work, has a statistically significant effect on their work placement or interest in further education.

4. Set of hypotheses:

4 H_0: The working position (manager/non-manager) does not have a statistically significant effect on their approach towards further education.
4 H_1: The working position (manager/non-manager) has a statistically significant effect on their approach towards further education.

2.2 Methodology

We used standard methods for scientific research such as analysis, synthesis, comparison, and selection, but a questionnaire survey was used as our major research method. The questionnaire survey was carried out in the period between December 2019 and December 2020 with a sample of 192 companies operating in Slovakia. The questionnaire was based on our research model and consisted (besides setting of the parameters P1, P2, P3, P4 – Table 1) of four research indicators (VZ1, VZ2, VZ3, VZ4 – Table 2) focused on the work placement of the respondents based on their achieved qualification level and their approach towards further education with only one possible answer chosen from the alternatives. To evaluate the results, descriptive statistical methods and contingency tables were implemented and the hypotheses were verified using statistical methods and coefficients appropriate for evaluation of associations between nominal or dichotomic parameters: chi-square and Cramer's V [23]. For validation, PSPP statistical software was implemented.

3 Results and Discussion

The results are presented according to the following structure: the research sample and its parameters, evaluation of the indicators, and statistical verification of the hypotheses.

3.1 Research sample

The research sample included 192 respondents who were grouped according to the following parameters P1- achieved qualification before entering 1st employment, P2 – company size, P3 – sector of the economy, the company belongs to, P4 – current position (Table 1).

Table 1. Research sample parameters

Parameter	Parameter Value	Number of Answers	% of Incidence
P1 – Achieved qualification before entering 1st employment	secondary with school-leaving examination	46	23.96 %
	tertiary	146	76.04 %
P2 – Company size	small (up to 49 employees)	38	19.79 %
	middle (50 to 249 employees)	57	29.69 %
	large (250+ employees)	97	50.52 %
P3 – Sector of economy	primary	36	18.75 %
	secondary	124	64.58 %
	tertiary	32	16.67 %
P1 – Current position	manager	111	57.81 %
	non-manager	81	42.19 %

Evaluation of the sample based on the parameters:
the sample consisted of 23.96 % respondents with secondary education completed with a school-leaving examination and 76.04 % respondents with tertiary education (P1).

As for the size of the companies (P2), the largest proportion belonged to large companies (50.52 %), then middle-sized companies (29.69 %), and lastly, small companies (including micro-sized ones) (19.79 %).

The third parameter (P3) divided the sample in the following way, 64.58 % of respondents worked in the secondary sector (industry, construction, …), 18.75 % in the primary sector (agriculture, extraction of raw materials and minerals, …) while the tertiary sector (various services) had the lowest representation (16.67 %).

The working position of the respondents (P4) was divided into 57.81 % employees working in managerial positions and 42.19 % occupying non-managerial positions.

We excluded from further processing all those values of the parameters which had <5 incidences.

3.2 Evaluation of the indicators and their results

The values of the indicators and their percentual incidence are presented in Table 2.

Table 2. Results of the evaluation of the studied indicators

Indicator	Indicator Value	Number of Answers	% of Incidence
VZ1 Placement according to qualification	yes	145	75.52 %
	no, one level above	4	2.08 %
	no, one level below	23	11.98 %
	a very different position	20	10.42 %

VZ2 Reasons for not being placed according to qualification	there was no free position matching the qualification	24	12.50 %
	lack of knowledge, skills, and experience	12	6.26 %
	placement based on inheritance, privileges etc.	4	2.08 %
	influence of acquaintances	4	2.08 %
	discrepancy between qualification and placement due to higher wage	3	1.56 %
	not commented	145	75.52 %
VZ3 Employee's interest in further training after entering employment	yes	150	78.13 %
	no	42	21.88 %
VZ4 Employee's target of further training	Development of knowledge and skills within the field of study	47	24.48 %
	Development of knowledge and skills in another field of study	37	19.27 %
	Development of knowledge and skills for a managerial career	18	9.38 %
	Additional training for a specific type of work	31	16.15 %
	Formal tertiary education – achieving qualification	25	13.02 %
	Acquiring key competences	8	4.17 %
	Self-study necessary for solving current problems	4	2.08 %
	No education	22	11.46 %

Based on the indicator values VZ1 (work placement according to qualification) we can state that the majority of respondents (56.52 %) were placed according to their qualification. Less than 24% of respondents were placed on a different level, out of which 11.98 % below their qualification and 10.42 % at a different post.

The main reason for not being placed according to their qualification was based on the answers, lack of free posts corresponding to their qualification (12.50 %). 6.26 % admitted a lack of knowledge, skills, and experience.

78.13 % of respondents consider further education and training after entering employment (VZ3) important and expressed their interest in it but a relatively large part of respondents (21.88 %) is not interested. Despite that, approximately 10% of them chose one of the alternatives of indicator VZ4.

In terms of indicator VZ4, the highest proportion (24.48 %) expressed their interest in further development of their qualification, 19.27 % is interested in other fields of study. The 10% threshold was passed by 16.15 % who would prefer additional training for a specific type of work and 13.02 % who would like to increase their qualification by formal tertiary education.

3.3 Results of statistical validation

The statistical validation of hypotheses was carried out based on the chi-square coefficient and Cramer's V. The results are summarised in Table 3, Table 4, Table 5, and Table 6.

Table 3. Results of statistical validation of differences according to qualification (P1)

Association	Chi square	df	Cramer´s V
VZ1/P1	1.41	3	0.09
VZ2/P1	2.84	4	0.25
VZ3/P1	0.74	1	0.06
VZ4/P1	3.46	7	0.13

The results in Table 3 do not show statistical significance in any association. Therefore, we accept the $1H_0$ hypothesis and reject the $1H_1$ that is, we were not able to demonstrate any differences in the answers of respondents with secondary or tertiary education neither concerning their work placement or attitude towards further education.

Table 4. Results of statistical validation of differences according to company size (P2)

Association	Chi square	df	Cramer´s V
VZ1/P2	14.79*	6	0.20
VZ2/P2	18.44*	8	0.44
VZ3/P2	3.98*	2	0.14
VZ4/P2	28.72*	14	0.27

*$p<0.05$

The data in Table 4 confirm a statistical significance in all indicators. Base on that, we reject the $2H_0$ hypothesis and accept the $2H_1$ hypothesis and state that both, the approach towards employee placement based on qualification and access to further training is different in micro, small, middle, and large companies.

Table 5. Results of statistical validation of differences according to the sector of the economy (P3)

Association	Chi square	df	Cramer´s V
VZ1/P3	10.35	6	0.16
VZ2/P3	13.32	8	0.38
VZ3/P3	1.98	2	0.10
VZ4/P3	28.41*	14	0.27

*$p<0.05$

Table 6. Results of statistical validation of differences according to work position (P4)

Association	Chi square	df	Cramer´s V
VZ1/P4	2.98	3	0.12
VZ2/P4	2.94	4	0.25
VZ3/P4	4.09*	1	0.15
VZ4/P4	6.44	7	0.18

*$p<0.05$

The results in Table 5 and Table 6 show statistical significance just in two cases (VZ4/P3 and VZ3/P4), therefore, in both cases the null hypotheses $3H_0$ and $4H_0$ are accepted and we rejected the alternative hypotheses $3H_1$ and $4H_1$. That means that we could not prove that either the sector of economy or the work post of the respondents has a statistically significant impact on their placement and interest in further education and training.

4 Conclusion

The main aim of the paper was to study the importance of the qualification achieved within the formal educational system on entering employment and the interest in and need for further education which would accompany the career development. Numerous individuals perceive career as equal to managerial post but interest in managerial training seems to be relatively low. Specialist training is much more preferred in companies operating in Slovakia. The most preferred is further training within the same field as the achieved qualification, and skills and knowledge specific for one's work position. We can state that interest in training in other fields is high due to the general requirement of interdisciplinarity. A great many attend different forms of formal education often accompanied by a change of qualification. We considered the start of a career essential for the placement of employees into positions according to their qualification and their interest in further professional training important for career development.

We discovered inadequacies in placing the employees according to their qualification; ¾ of the respondents were placed on posts according to their qualification but some 12% on lower positions. It is noteworthy that 11% were placed on posts requiring different qualification. It was caused by an inappropriate choice of field of study or by better remuneration system in other fields even in posts requiring a lower level of education.
The reasons for not placing the employees according to their qualification were lack of free posts and more than 6% mentioned inadequate knowledge, skills, and experience. Concerning the evaluation of graduates on entering their first employment by the employers, we can indicate that practical skills were considered more important than the acquired knowledge. Skill development is included in formal education only partially. Just a few fields of study at the secondary or tertiary level include practical training as part of the programme. The evaluations provided by employers pointed at further deficiencies in graduates' capacity of intuitive, creative, and strategic thinking. Meeting the requirements of employers is possible only if it is based on cooperation between educational institutions and companies.

80% of respondents showed interest in further education within their field of study or some other but 20% were not interested. Further education is the reflection of aspirations, interests, and needs in building one's career. Though, part of those who are not interested in further education eventually participates in education as it is necessary to preserve their employment.

The statistical validation of the hypotheses confirmed two of our assumption. One is, that the size of a company is an important factor influencing our research topic – placement of employees and their approach towards further education. The other confirmed fact is that the sector of the economy does not influence it. The remaining two assumptions concerning the influence of the level of qualification and work post (manager/non-manager) were in our sample refuted.

The topic of education and career is quite extensive, and the paper could cover only part of our empirical research. We intend to continue the analysis of various aspects entering the decision-making process concerning the choice of education and career development. We would like to engage in the optimisation of educational choices utilising the services of educational and career councillors. Our ambition is also to increase the size of our sample.

Acknowledgement

The paper was elaborated within VEGA No. 1/0328/21 Post-pandemic business management: identifying temporary and sustainable changes in sequential and parallel management functions in the context of the COVID-19 pandemic – proportion 100 %.

References

1. R. Bromber, The 60-Year Curriculum: New Models for Lifelong Learning in the Digital Economy. *Distance Education* **42,** 1, 173-176 (2021)
2. I. Biao, Benin: Governance, lifelong learning and development. *Journal of Adult and Continuing Education* https://doi.org/10.1177/1477971421999411 (2021)
3. S. Dorian, N. Mirea, Lifelong Learning in Context of Industry 4.0. *Acta Technica Napocensis - Series: Applied Mathematics, Mechanics, And Engineering* **64,** 1-S1, 55-62 (2021)
4. S. Cronholm, Lifelong Learning: Principles for Designing University Education. *Journal of Information Technology Education: Research* 35-60 (2021)
5. A. Tuckett, Europe's lifelong learning markets, governance and policy: Using an instruments approach. *International Review Of Education* **66**, 5-6, 893-896 (2020).
6. B. Beláňová, Human Resources Policy in Relation to Cybernetic Security in Slovak Medium and Small Companies. *In Current Problems of the Corporate Sector 2020.* Paris, Édition Diffusion Presse Sciences, 1-10 (2020)
7. H. Lauder, The roles of higher education, further education and lifelong learning in the future economy. Journal of Education and Work **33**, 7-8, 460-467 (2020)
8. M. Laal, P. Salamati, Lifelong learning; why do we need it? *Procedia-Social and Behavioral Sciences* **31**, 399-403 (2012)
9. D. N. Aspin, J. D. Chapman, Lifelong learning: concepts and conceptions. *International Journal of Lifelong Education* **11**, 19-38 (2007)
10. Ch. Wilding, *Emoční inteligence: Vliv emocí na osobní a profesní úspěch*. Praha, Grada Publishing (2010)
11. J. Suchý, P. Náhlovský, *Pozitivní emoce: Jak je posilovat a rozvíjet v osobním i pracovním životě.* Praha, Grada Publishing (2012)
12. M. Flešková, V. Dolinská, *Aplikovaná psychológia pre ekonómov a manažérov.* Bratislava, Iura Edition (2010)
13. A. Slezáčková, *Průvodce pozitivní psychologií*. Praha, Grada Publishing (2012)
14. Z. Macková, *Psychológia*. Bratislava, Vydavateľstvo Ekonóm (2012)
15. R. Kohoutek, *The history and present in career counselling.* http://rudolfkohoutek.blog.cz/1205/1 (2016)
16. M. Jigau, *Kariérne poradenstvo: Kompendium metód a techník.* http://web.saaic.sk (2011)
17. L. Beková, Š. Grajcár, D. Lepeňová, E. Uhríková, M. Zvariková. *Kariérové poradenstvo v Slovenskej republike.* Bratislava, Slovenská akademická asociácia pre medzinárodnú spoluprácu, centrum Euroguiance (2014)
18. M. Vaculík, *Assessment centrum: Psychologie ve výběru a rozvoji lidí.* Brno, NC Publishing (2010)
19. M. Daňková, *Koučování. Kdy, jak a proč*. Praha, Grada Publishing (2015)
20. Y. M. Kundi, S. Hollet-Haudebert, J. Peterson, Career adaptability, job crafting and subjective career success: the moderating roles of lone wolf personality and positive perfectionism. *Personnel Review* (2021)
21. R. F. Belchior, R. Lyons, Explaining entrepreneurial intentions, nascent entrepreneurial behavior and new business creation with social cognitive career theory–a 5-year

longitudinal analysis. *International Entrepreneurship and Management Journal* 1-28 (2021)

22. M. Szarková, Reasons of Employees' Fluctuations from the Perspective of Employers in the Slovak Republic. *In Hradec Economic Days* **9**, 404-413 (2019)
23. R. Hanák, *Dátová analýza pre sociálne vedy.* https://statistikapspp.sk/ucebnica/datova-analyza-pre-socialne-vedy (2016)

 https://doi.org/10.1051/shsconf/202111503011

CREATIVE INDUSTRIES: CHALLENGES AND OPPORTUNITIES IN XR TECHNOLOGIES

Hana NOVAKOVA[1*] , and *Peter* ŠTARCHOŇ[1]

[1]Tomas Bata University in Zlín, Faculty of Multimedia Communications, Univerzitní 2431, 760 01 Zlín, Czech Republic

Abstract

The paper aims to point out current challenges and opportunities in extended reality (XR) technologies that cover augmented, mixed and virtual reality (AR, MR, VR) within the frame of creative and related industries and current technological progress particularly based on the results of primary research. While the XR technologies struggle with limitations of the proper research on one side, multiple research projects including this one, have shown significant acceptance and potential of the XR technologies, especially in the ongoing digital era accelerated by the COVID-19 situation. The research presented in this paper was held with the small groups of artists in France who had opportunity to test the augmented reality mobile application on artworks from various artists. This and other cited research projects show that the acceptance rate of the XR technologies by the informed audience strongly predominates over its penetration in such population which can be interpreted as significant market opportunity. This milieu has strong impact on the democratization of the XR industry in the recent years with substantial investments by business sector. Meanwhile these technologies have become powerful enough to raise serious concern of their impact on ethical issues, social distancing and other related topics.

1 Introduction

From the low penetration rate of the extended reality technologies, it might seem that there is not big chance for the XR to become successful technology with all the current challenges it faces: especially technical, financial and ethical. While the economic impact of VR alone was supposed to achieve approx. 15bn USD, by the year of 2020 according to Adaola [1], some users feel repulsion to the digital worlds or heavy headsets, addiction issues or social distancing. However, during the research conducted in France with the university art students about their attitudes to augmented reality (AR) and another research of the perception of virtual reality (VR) by children and their parents, the obtained results show the opposite. The aim of this article is to highlight the current challenges and

* Corresponding author: h1_novakova@utb.cz

opportunities in XR and to put the results from the conducted research projects in broader context of the current situation on the XR market and technological progress.

Technological development plays crucial and positive role in the XR hardware and software democratization, both on the developers' and users' side. Not only the upcoming 5G that allows users to transfer big data instantly, but also each new successful XR content means a huge technological leap. According to Zhu [2] "in the future of 5G communication and IoT (Internet of Things) applications under human-machine interaction will drastically affect the lifestyle of human in the social networking aspect, and this kind of device can improve the intelligence of machines based on the big data acquired from AI techniques." For example, in VR gaming the high-end users already benefit from Hife-Life: Alyx, while the massive spread of Czech-made game Beat-Saber (bought by Facebook) has demonstrated there is a serious increase of the number of the new users which makes the industry more attractive for investments,and development. However, Lee [3] argues that large number of current VR application domains suffer from software issues since no single system supports satisfactorily all the aspects of creation of a virtual reality application.

Currently the market is awaiting the advanced holo-technologies by Microsoft for experts and developers, Apple Glasses for early-users and Facebook glasses for massive spread of the technology. Once the masses start using VR, that by then will be probably called extended realities (XR), the fancy light glasses combined with amazing and useful content become less expensive, even more users will be attracted. These new users will profit from better XR hardware and software which again would attract more investments, research and development. As the XR content and technologies become more accessible financially and technologically for wider groups of people the democratization of the XR becomes reality.
Another argument for the importance of XR are the new paradigms brought by the pandemic. More than year lasting pandemic with repetitive lockdowns has changed the way mass audiences consume education, entertainment and express themselves in arts, humanities, business, science etc. There is a significant decline from the massive public experiences (for example concerts or movie-theatres for hundreds of present visitors) in favor to the individually shared consumption. All this within the quote of Oculus evangelist Palmer Lucky:" Digital worlds were not made to separate people in their basements but to connect them through the new metaverses created by human imagination." According to Hain [4] the interdisciplinarity, synergistic effects, organized design process and educational strategies are the key principles for the protection of the local industrial heritage.

2 Theoretical background / Literature review

The digital worlds have experienced a long and dramatic progress from original cyberpunk ideas by William Gibson or Neal Stephenson to the most immersive experiences of nowadays. „Some recent studies by marketing scholars have proposed a framework for research in digital marketing where VR and AR, that is, XRs, are identified as critical digital technologies that will lead to new marketing opportunities," concludes in his study Alcañiz [5]. Gobiera [6] argues that the widely accepted model called „Milgram's Reality – virtuality continuum" is already an outdated concept as the XR technologies are in service of the reality itself and calls in his paper for more proper research of these technologies: „Due to a shortage of investment, we are not using its full potential."

According to Grizewski's [7] research conducted on VR marketing communication on 150 observations of respondents tested on VR presentation with Oculus Rift hardware, video and printed advertisements, the VR technology has significant and positive impact on all three communicated media: the acceptation of the offer, the VR technology and the presentation itself. Confronted with the Mine's [8] comparison of the historical virtual environments created in various US leading colleges and universities, the development of the interaction techniques has since the start compensated for the limitations of working in virtual world while simultaneously exploiting its power. This argument was later updated from the perspective of the narrative of the virtual worlds in the Ryan's [9] monography where she explains the trade-of among the interactivity and immersion in virtual environments. Javorik's [10] psychological analysis later demonstrated a strong link between immersion, affective responses, and positive user experience. Another study focused on the XR opportunities in tourism, states that tourism may not easily recover to pre-pandemic times. According to Lacka [11] and later confirmed by Kwok [12] "there are various ways how tourism enterprises can navigate the prohibitive travel climate, firstly for survival, and subsequently transform towards thriving in the post-pandemic new normal by exploiting XR as technological innovation."

3. Methodology

Opposite to Munoz-Leiva [14] who started the first-ever longitudinal review of the state of art in merchandising research spreading into 60-year period, that provided holistic approach, this research on augmented reality (AR) was held two weeks in October 2020 in Aix-en-Provence, France with small groups of university art students with potential to capture the current creative industries trends. Prior to the study and based on the Cawood [15] and Fiala findings on AR. Each group of students was asked to read and sign an informed consent in an electronic form. The methodology consisted in three parts: Part 1 consisted of a workshop with advanced AR technologies in artworks from various authors. Part 2 was based on observation of the researchers and Part 3 on a semi-structured interview consisting of grading, answers from selected options and open answers of respondents recorded by the art students themselves through the application. The limitation of the research is mainly the number of respondents during pandemic, which was in total 21. This research is, when it is meaningful, supported by the data from another VR research on perception of children and parents based among others on the previous findings of analysis of human gestures in the 3D spaces by De Amicis and Soave [16] and also Guillaume [17] VR technological concepts combined with Lucas [18] This second research was held in September 2020 in Zlín, Czech Republic with 39 respondents, out of which 18 were parents and 21 children. Both research projects showed further areas of exploration and the authors plan to collect more data. As shown by Naz [19] lower quantity of participants as limitations are quite common in VR experiments as each participant of the research usually requires separate time due to the individual VR experience.

4. Results

The source of all the figures in this section are from the own research.

Figure 1 and 2 show the attitude of the French artist respondents towards the XR mobile technologies and their will to repeat the experience in the future to find out the acceptation rate of the technology. Only 19% of answers were neutral in the case of XR mobile applications. All of the other answers were positive.

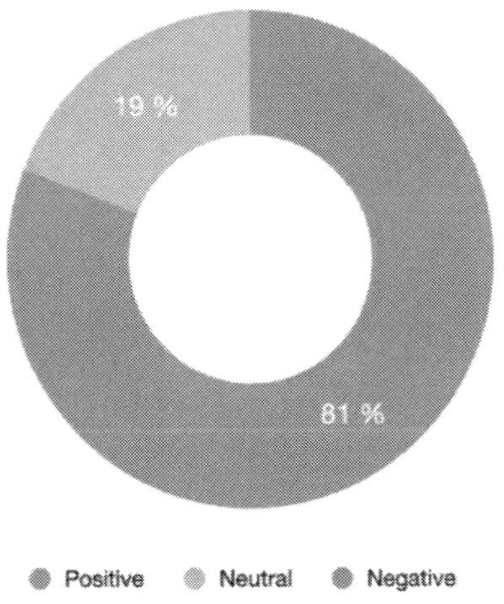

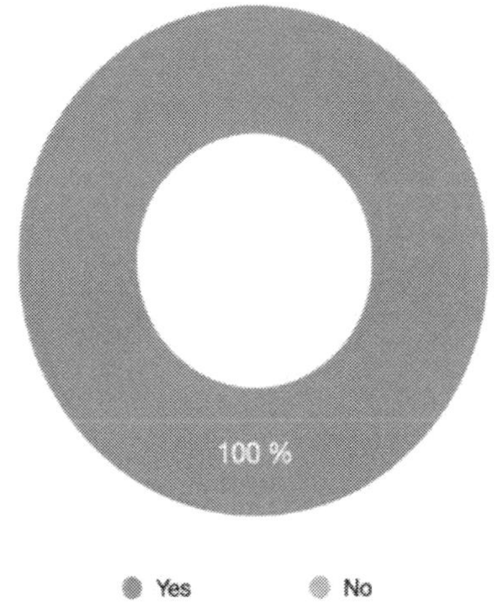

Fig. 1. Attitude towards XR mobile application. **Fig. 2.** Will to repeat the experience in the future.

Figure 3 and 4 show the ideal length of the AR experience according to the users: 43% of the experiences were supposed to last from 11 to 20 sec. and 33 % below that time. No respondent wanted to hold the phone over 1 minute for the experience. The willingness to create AR projects among the potential art students and teachers was positive in 81%.

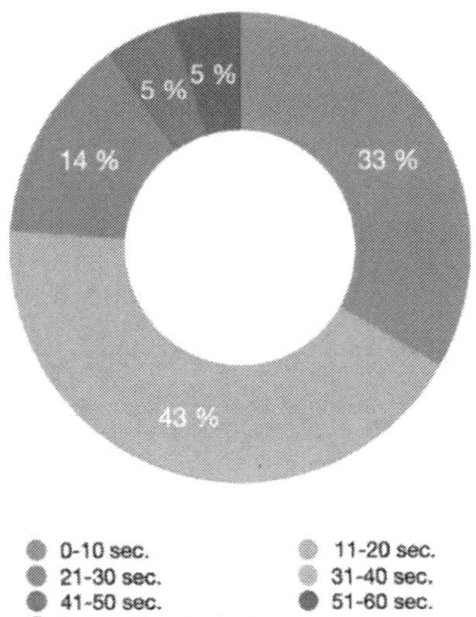

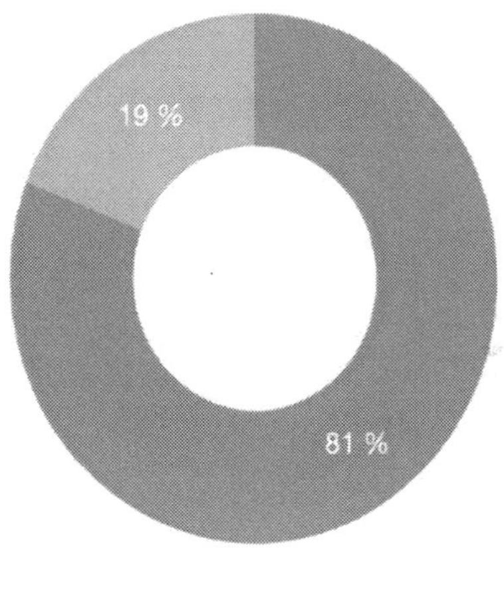

Fig. 3. Preferred length of AR mobile experience. **Fig. 4**. Would you use AR for your projects?

The following chart shows the best use for AR according to the respondents when 38% would use AR for 2D or 3D animation, almost one fifth for AR posters and one fifth of the respondents would bring new life for artworks. Next best use would be AR illustrations in the books (14%), architecture and design (5%) or additional information (5%)

 https://doi.org/10.1051/shsconf/202111503011

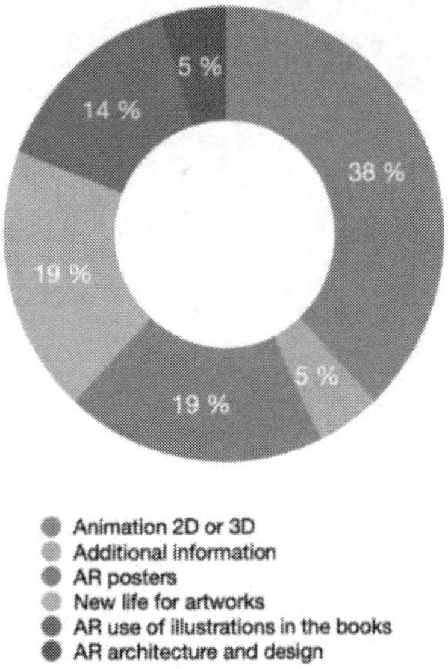

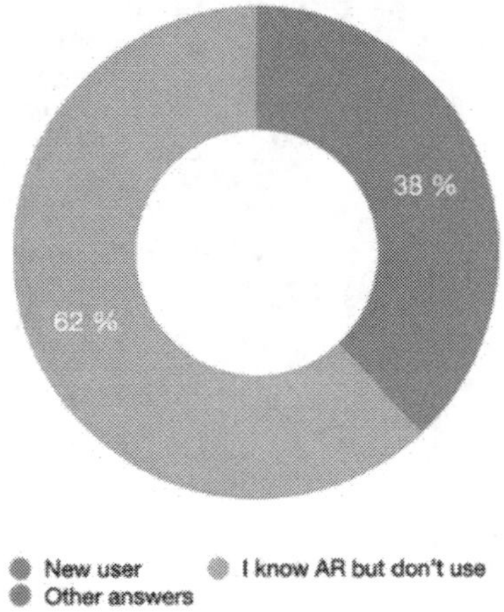

Fig. 5. The best use for AR in Arts. **Fig. 6**. New or experienced users?

The last four figures define 38% as new users and 62% respondents as not new to AR but not using it. 67% of the respondents as female and 33% as male.

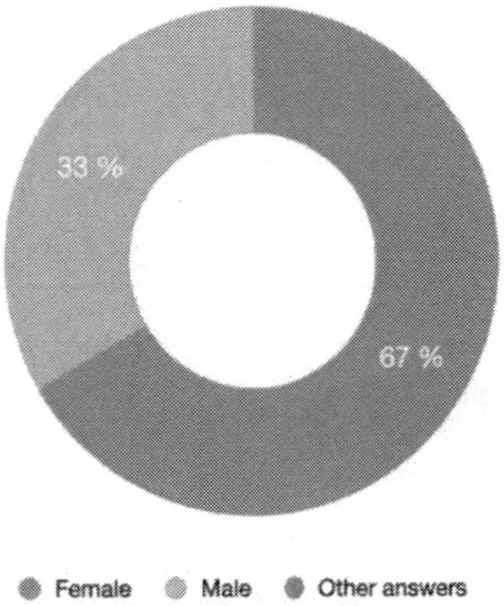

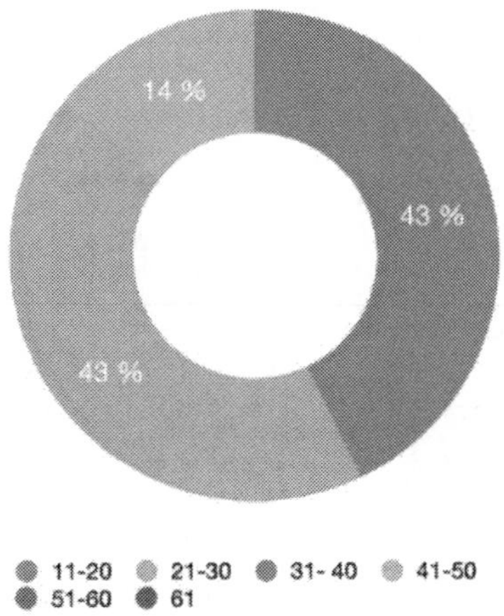

Fig. 7. Gender of the respondents.

Fig. 8. Age of the respondents.

43% of the respondents were 20 years old or younger, another 43% of the respondents were 21 to 30 years old and the time they have dedicated to arts vary from no answer to several decades.

The last Figure summarizes how many years did the respondents spend deicating their time to creation of varioius forms art from sculpture or painting to any form of modern digital or visual artsmand the answers show a great variety.

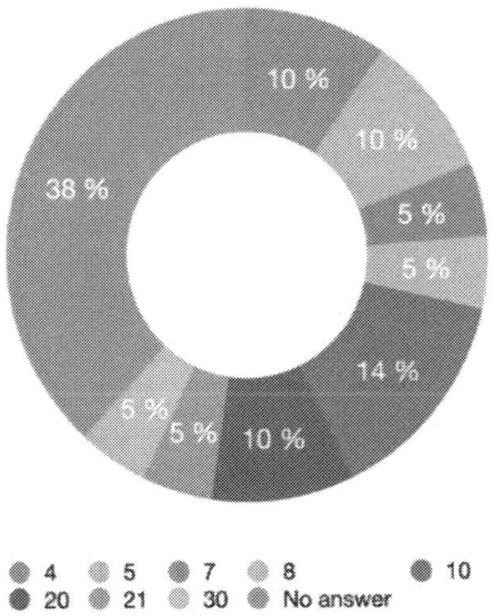

Fig. 9. How many years are you in Arts?

5. Discussion and interpretations

The attitude towards XR mobile applications during the research held in France among art students on AR (Fig. 1) was positive in 81% of the cases. There was no negative answer. About one fifth was neutral. The will of the respondents to repeat the XR experience (Fig. 2) in the future was 100%. The high acceptation rate was probably among other reasons caused by previous workshop where the technology was tested by the respondents and they had opportunity to experiment with it. It is possible that the results would differ in the audiences that had no previous chance to try the experience. The positive acceptation rate was also obtained in another research held for children (100% will to repeat the experience) and their parents (96% will to repeat the experience) in the Czech Republic where the VR software was tested on total of 39 respondents. Next questioning of the art students intended to find out how long mobile AR experience is the most popular (Fig. 3) and the results show that shorter is better. The interpretation of such results consists of the fact that people are usually not keen to hold their mobile devices in one stable and intact position for long time with one third of the respondents considered the time below 10 seconds is preferable. The figure 4 shows the 81% acceptance rate of the AR projects among the French artists from the point of view of creators. On the other hand, the current immersive VR experiences make sense in the length of about 10-12 min. depending on the immersive rate of the content and interactivity of the narrative. The best use of AR projects from the point of view of the artists is according to the art related respondents shown on chart on Figure 5. Moving objects or animation have higher acceptation rate than static pictures. The last four charts (Fig. 6, 7, 8 and 9) are about the respondents: There were more women than men among respondents, the artists were often new to the technology or haven't use it. Majority of the respondents (86%) were aged under 30 years of age and the number of years they spent in arts varies from no answer to 4 up to 30 years.

6. Conclusion

The key findings of the conducted research projects show that the acceptation rate by various groups of population who had chance to experiment with the XR technologies are significantly high: from 81% to 100% depending on (1) the respondents (French art students capturing the latest trends, children or parents in the Czech Republic) and (2) the question asked (Would you like to repeat AR or VR experience in the future? Would you use AR technology for your projects?). At the same time the penetration rate of these technologies is rather low as the research shows and often new for the respondents. The creative industries and related fields are though experiencing unprecedented period of opportunities, combined with the factors predicted by Lévy [20] such as digitalization and upcoming 5G, rapid technological progress combined with as Carayannis [21] points out industry 5.0, society 5.0, transformation, blockchain, and the role of AI via the internet of things architectures and significant investments by the ICT companies in XR. The positive effects or externalities of XR are visible both, for the users and the producers and touch a wide range of activities based on creativity, digital worlds, knowledge or information: from advertising, architecture, design to gaming, software and publishing. Such era can be interpreted as new potential of the XR technologies as the cross-industry discipline. The optimistic conclusion for the developers is expected to arrive with amazing potential also for the users who can benefit from "More for Less" factor deriving from the democratization of the industry, thus more content for lower prices, more immersive experience with amazing data transfer and cloud computing in seconds or less. At the same time these technologies have become powerful enough to raise serious ethical, legal and social questions e.g. how to cope with the digital social distancing and other related topics.

References:

1. T. Adaola, L. Páduaba, M. Foncesa, L. Agrellosc, Sousaa Luís Magalhãesd Emanuel Peresa, A rapid prototyping tool to produce 360° video-based immersive experiences enhanced with virtual/multimedia elements, *Procedia Computer Science*, **138** Pages: 441-45 , Published: (2018)
2. M. Zhu, Z. Sun, Z. Q. Shi, T. He Huicong Liu, T. Chen, Chengkuo Lee, Haptic-feedback smart glove as a creative human-machine interface (HMI) for virtual/augmented reality applications, *Science Advances – applied sciences and technologies*, **6**, DOI: 10.1126/sciadv. aaz8693, (May 2020)
3. S. H. Lee, S.H. Nam and J. Y. Kim, Analysis of a Quality Evaluation Model for VR Contents, *International Journal of Grid and Distributed Computing*, **11** pp.97-110 http://dx.doi.org/10.14257/ijgdc.2018.11.2.09, South Korea, (2018)
4. V. Hain, R. Löffler, V. Zajíček, Interdisciplinary Cooperation in the Virtual Presentation of Industrial Heritage Development, *Procedia Engineering*, **161**, p. 2030-2035, ISSN 1877-7058, https://doi.org/10.1016/j.proeng.2016.08.798 (2016)
5. Alcañiz M, Bigné E and Guixeres J (2019) Virtual Reality in Marketing: A Framework, Review, and Research Agenda. *Frontiers in Psychology* **10**, doi: 10.3389/fpsyg.2019.01530, (July 2019)
6. P. Gobiera; De Obiera Silva; About Reality: relations between museums and virtual reality, *Virtual Creativity*, **9**, numbers 1&2, Intellect Ltd. Article, DOI: https://doi.org/10.1386/vcr_00005_1 (2019)
7. F. Grudzewski, A. Marcin, G. Mazurek, K. Piotrowska, Virtual reality in marketing communication - the impact on the message, technology and offer perception - empirical study, *Economics and business review*, **4**, DOI: 10.18559/ebr.2018.3.4, (Jul 2018)
8. R, Mine, Mark, *ISAAC: a meta-CAD system for virtual environments, Computer-aided design*, **29**, pp 547-553, Elsevier Science Ltd. 0010-4485/97, PII: 0010-4485(96)00095-4
9. M. L. Ryan, *Narration as virtual reality: immersion and interactivity in literature and electronic media, Translation:* Eva Karasová. Prague: Academia, Možné světy. ISBN 978-80-200-2507-4. Retreived (2020).
10. A. Javorik, Marder, B; Pizzetti, M; Warlop, L,, Augmented self-The effects of virtual face augmentation on consumers' self-concept, *Journal of Business research*, **130**, pages 170-178, DOI: 10.1016/j.jbusres.2021.03.026, (2021)
11. A. O. J. Kwok & Sharon G. M. Koh (2020), COVID-19 and Extended Reality (XR), *Current Issues in Tourism*, **17**, DOI: 10.1080/13683500.2020.1798896 (July 2020)
12. Lacka, E., Assessing the impact of full-fledged location-based augmented reality games on tourism destination visits. *Current Issues in Tourism*, **23**, 345–357. https://doi.org/10.1080/13683500.2018.1514370 , Google Scholar, (2020)
13. L. Buchholtz; Fine, Gary Alan; Wohl, Hannah Art markets in crisis: how personal bonds and market subcultures mediate the effects of COVID-19, *American Journal of Cultural Sociology* 8:462–476 https://doi.org/10.1057/s41290-020-00119-6 , (2020)

14. F. Munoz-Leiva, Mer Lopez, F. Liebana-Cabanillas, S. Moro, Past, present, and future research on self-service merchandising: a co-word and text mining approach. *European Journal of Marketing*, DOI: 10.1108/EJM-02-2019-0179, (2021)
15. S. Cawood; Fiala, Mark; *Augmented Reality: A Practical Guide*; Paperback: 328 pages; Publisher: Pragmatic Bookshelf; **1** ISBN-13: 978-1934356036, (Jan, 2008).
16. R. De Amicis; Soave, Marco; *Analysis of Human Gestures in the 3D Space to Control Multimedia Interfaces*, Fondazione Graphitech, Via alla Cascata 56/c, 38123 Povo (TN), Italy (2020)
17. P.M. Guillaume; Pascal Guitton, ed. *Virtual reality: concepts and technologies*. **2**. Boca Raton, FL: CRC Press. ISBN 04-156-8471-4. (2011)
18. S. Lucas, Erin Pangilinan; Vasanth Mohan; (23 April 2019). *Creating Augmented and Virtual Realities: Theory & Practice for Next-Generation Spatial Computing*, 300p., O'Reilly, ISBN-10: 1492044199. Retrieved in (2021)
19. A. Naz, R. Kopper, P. Ryan, *Emotional Qualities of VR Space* IEEE Virtual Reality (VR), March 18-22, Los Angeles, CA, USA 978-1-5090-6647-6/17 (2017)
20. P. Lévy; Cyberculture, *Rapport au Conseil de l'Europe,* Odile Jacob, ISBN 978-2738105127, Retrieved in (2020)
21. E. Carayannis; K. Christodoulou; P. Christodoulou, S. Chatzichristofis; Z. Zinonos, Known Unknowns in an Era of Technological and Viral Disruptions-Implications for Theory, Policy, and Practice, *Journal of the knowledge economy*, DOI: 10.1007/s13132-020-00719-0, (Feb, 2021)

What organizational steps are necessary, to successfully establish a culture of open innovation in the company?

Marco Pister *

SEEBURGER AG, Edisonstr. 1, 75015 Bretten, Germany

Abstract. This paper investigates in open innovation, organizational culture, internal and external collaboration and the steps which are necessary to establish a culture of open innovation in the company. Open innovation is one of the most relevant topic in research and development. The research results concentrate on, how leaders can shape a corporate culture and environment in which open innovation processes are embrace and can implemented successfully. Innovation processes have a high risk factor. This is precisely why it is crucial to create a corporate culture that is open to failure and also accepts less successful attempts. The attitude of the management is one of many key factors.

1 Introduction

1.1 Topicality and Relevance

The contemporary business environment is increasingly shaped by the need for innovative products and services, which is accentuated by Edelbroek, Peters, and Blomme [7], who argue that typical life cycles of products become shorter, thus amplifying the need for companies to innovate more frequently. Innovation itself is therein characterized as a complex process: "Innovation can be viewed as an organizational process, often initiated by organizations' internal Research and Development departments, which comprises three main elements: idea generation, (…) idea promotion, (…) and idea realization" [7].
This trend towards an increased necessity for innovation is – among other reasons – being driven by fast developments in the technology sector, where new possibilities in regard to technological solutions increasingly offer new chances for companies [21]. Kratzer, Meissner, and Roud [12] argue in these regards that while the need for stronger innovation is evident and a business imperative, it should not be forgotten that innovation and the underlying processes always bear the threat of failure. This seems especially true for disruptive technologies and associated products and services – while such approaches might be deemed drivers of the market, they always bear the risk for companies to experience failures with accordingly negative consequences that have to be viewed throughout the lens of corporate risk management. However, the authors further argue that the solution to this

* marco.pister@web.de

challenge does not lie in avoiding innovation, but rather in finding strategies to minimize the associated risk. One of these strategies lies in what Müller and Thoring [13] describe as failing early: As typically costs, and thus the risk, of innovation processes scale drastically with time, the authors argue that it is imperative for innovators to achieve the critical failure stage at an early point in the development while the investment is still comparatively low.
Failures in innovation processes typically arise from a lack of acceptance on the market [13]. As new products always bear the risk of not meeting market demands and thus failing financially, the authors argue that it is crucial to not await the fully finished and presentable product before approaching potential customers, but to approach the potential market already at early stages of development, in order to make sure that the market demands are understood correctly and are thus met. This early evaluation, the authors argue, allows companies to minimize the chances of failure while still at a low-investment stage. This strategy, Müller and Thoring [13] showcase, can be described as one aspect of open innovation processes, which are considered to be a contemporary and successful way to foster corporate innovation. This paper seeks to assess in how far such strategies are dependent on company internal factors and how leadership processes can influence a company's ability to utilize open innovation.

1.2 Research Statement and Research Question

The present work, as described, builds on the state of research regarding open innovation processes and their relevance for successful product and service development within the contemporary business environment. While the general feasibility of such an approach to innovation seems well established within the existing scientific literature on the topic, it seems unclear how organizations can achieve a working environment in which open innovation processes can easily be implemented. This work acknowledges that leaders play an important role in creating such an environment, while arguing that the connection between leadership, corporate culture, and success in implementing open innovation processes seems less well established. While leadership can also be discussed throughout the lens of thought leadership, the focus within this publication lies on the application of leadership to the field of internal communication, where leaders create the necessary conditions in which open innovation strategies and subsequently innovation success can prosper. The leading research question of this publication on leadership, culture, and innovation, thus, is as follows: How can leaders shape a corporate culture and environment in which open innovation processes are embraced and can be implemented successfully?

2 Open Innovation – Approaches, Benefits, and Challenges

2.1 Early Developments and Comparison to Other Innovation Approaches

Open innovation is one of the most relevant topics in research and development. The idea behind the concept of open, and thus collaborative, innovation is that it is no longer possible for a company to innovate in isolation from its environment [13 & 5]. In order to remain competitive, it is necessary to draw on different partners and resources and thus to open up one's own boundaries, both in relation to potential competitors and to representatives of the market in which one operates. Following these reflections, Chesbrough [5] formulated an attempt to define what Open Innovation is: it is a paradigm

according to which companies should draw on both external and internal ideas and approaches and pave the way to the market, including in cooperation with partners.
While much of the scientific literature on Open Innovation focuses on large, multinational companies, findings such as those of van de Vrande, de Jong, Vanhaverbeke, and de Rochemont [19] show that the topic is also of increasing relevance to small and medium-sized companies: According to the Dutch study, medium-sized companies in particular were very active in the area of Open Innovation, primarily following the motive of being competitive in this way. The authors argue that for younger, less established companies in particular, open innovation can also pose a threat in terms of corporate culture. Opening up to cooperation with other companies and even competitors could lead to drastic changes within the corporate and communication culture and even to an exodus of employees.
This in turn ties in with the assumptions of Jabbour and Santos [11] and makes it clear why HR management (HRM) is so important for innovation projects. HRM can, on the one hand, serve as a safeguard against such potentially harmful influences and, on the other hand, lead to an open corporate culture that welcomes innovation. In terms of open innovation, the aim is to create a corporate culture that is open to collaborative working and fosters a positive culture of error.

2.2 Leadership and Open Innovation – Chances and Challenges

Open innovation communities, as Fleming and Waguespack [8] describe them, in many cases lack the structure of more classical corporate settings and rely heavily on implicit types of collaboration. This seems especially true for the more open formats which only partially are driven by classical leadership, as the participation is in many cases not described as mandatory: "leadership in such communities depends more on the trust and mobilization of peers than on approval of superiors. To wit, members cannot be fired or forced to participate in any activity, nor can they be compelled to pay attention to any other member" [8].However, as the authors further argue, this implicit setting with a less direct approach to leadership does not negate the necessity of strategic leadership, rather, it elevates the demands for leaders. As they can typically not rely on hierarchical and organizational structures only, Fleming and Waguespack [8] argue, leaders have to increasingly rely on their persuasive capabilities or, as Bass and Avolio [3] explain, the transformative power of transformational leadership. Such a leadership approach is characterized by the authors as one that relies less on the transactional nature – thus, on rewarding behavior throughout corporate structures and processes –, and rather on enabling and motivating employees. Transformational leaders, therefore, typically aim to shape an environment and a relationship between the organization and its employees that leads to employees being intrinsically motivated to work for the company and to follow its goals.
This is also acknowledged for the field of open innovation, where authors such as Edelbroek, Peters, and Blomme [7] showcase the influence that transformational leadership can have on the quality of the outcome of open innovation processes. The authors argue that transformational leadership is a feasible tool for leaders to foster the organizational engagement of their employees. As transformational leadership aims to foster such commitment and – as studies cited by the authors indicate – does so successfully, it can be shown that transformational leaders can increase pro-active and more innovative behavior of their employees. Also, by developing a higher level of commitment and experiencing emotional support through transformational leaders, employees tend to experience stronger self-efficacy, which in turn can influence their creativity- and innovation-related behavior positively. The authors were also able to showcase this by utilizing an empirical study where they assessed how increased organizational commitment can actually influence the quality of open innovation processes. In a similar vein, Naqshbandi, Tabche, and

Choudhary [14] established a significant connection between the usage of transformational and enabling leadership styles and open innovation.

3 Corporate Culture

3.1 Corporate Culture and Leadership

Corporate culture is described by Appelbaum et al. [1] as a collection of characteristics of an organization that contribute to the organization's overall mood or environment, implying that it is a highly contextual concept.

In this regard, the scholars equate an organization's organizational culture to a person's personality, in which personality psychologists such as Goodman, Disabato, and Kashdan [9] summarize a collection of beliefs, interaction and communication patterns, and fundamental personality characteristics for organizations and their culture, which Denison [6] describes similarly.

Corporate culture is increasingly playing a supporting role in organizations, as summarized by Ramdhani, Rahmdhani, and Ainisyifa [15] "Corporate culture is one of a determinant factor in enhancing to achievement of organizational goals and objectives. Corporate culture as an important aspect of not only in organizational behavior but also as a tool in understanding how organizations function".

This aspect of contemporary entrepreneurship is thus critical not just for practice but also for scientific observation, which is increasingly concerned with the question of how culture itself has a formative effect on various aspects of corporate performance on the one hand, and how culture can be shaped in the context of the corporate vision on the other (as will be shown in the following sections).

Guiso, Sapienza, and Zingales [10] use Goldman Sachs and a quote by its Vice President, Greg Smith, as a well-known example of how the public perceives and values business cultures. According to the writers, the latter said in an interview that he has always believed that the company's culture is a major factor in its growth. He explains the dominant culture in this regard as "revolved around teamwork, integrity, a spirit of humility, and always doing right by our clients" [10].

However, the writers assert that this corporate culture, which is defined by sustainability and social responsibility, cannot be sustained in the long run, demonstrating that corporate cultures need appropriate treatment.

Corporate culture may be shaped by a variety of factors, but writers such as Bolton, Brunnermeier, and Veldkamp [4]) emphasize the managers' actions and describe how managers may significantly contribute to defining an organization's culture through their engagement and value exchange.

Corporate culture, in any case, is critical for managers, as shown by Syafii, Thoyib, and Nimran [4], who demonstrate that corporate culture is a critical variable that both represents and affects manager behavior. Transformational leadership can be described as leading in this way [2] This leadership approach seeks to achieve the best possible fit between organizational ideals, visions, and priorities and the desires and motivations of workers. The establishment and preservation of a complementary organizational culture is represented in this context as a mediating force between leadership and entrepreneurial results, as shown by the aspect of creative power [18].

3.2 Openness and Innovation

The importance of corporate culture on the topic of (open) innovation is also highlighted by Kratzer, Meissner, and Roud [12]. Especially given the inherently risky nature of innovation processes, the authors explain that it is crucial to create a corporate culture open to failure and accepting of less successful trials: "For all of the features, the management's attitude towards risky innovation projects is one of many key factors. However, the attitude towards risk is especially important as it correlates strongly with the central features of the innovation concept, namely the newness of the solutions and the willingness to change" [12]. Thus, the authors explain both on theoretical frameworks and their own empirical data that an open corporate culture and collaboration style does not only influence employees' willingness to participate in such endeavors but also favorably shapes the success of such projects. This is confirmed by a recent study by Yun, Zhao, Jung, & Yigitcanclar [22], who argue that corporate culture can in general be considered to be among the driving forces behind innovation: "Innovation culture, which is a kind of static and open innovation culture, is a transdisciplinary culture with the aim to pragmatically integrate anything desirable, necessary, useful, feasible, and appropriate; innovation culture is mainly considered to be an aspect of organizational culture, where the societal innovation culture always sets the context“ [22].

4. Discussion

4.1 Critical Discussion and Limitations

While the present work aimed to establish the connection between corporate leadership, corporate culture, and success in the implementation of open innovation strategies, the approach can only be discussed as exploratory in nature. Building on the contemporary state of research – as described throughout the previous sections of this publication –, the author managed to showcase in how far leaders can use their communication approaches to shape an open-minded corporate culture which, subsequently, should allow for the usage of open innovation. At the same time, it was highlighted that such strategies do not only offer chances on a competitive market [16] but can also be seen as challenging in regard to their implementation: one of the core issues named in this context is the changing role of employees within companies. Especially the networking-aspect arising from the intensified collaboration between clients and internal employees is described here as a challenge for many: While more classical ways of collaboration between companies and their clients saw interaction points mostly in sales teams or key account management, the contemporary approach sees increasing collaboration also between teams of researchers. While this challenge is discussed within this work, it also needs to be concluded that open innovation does not only have a communication or collaboration component but also one that focuses on the way mistakes and failures are addressed within the company. As the introduction of this paper with regard to Edelbroek, Peters, and Blomme [7] stated, in the business environment described throughout this publication, it seems necessary to acknowledge that innovation without failure seems impossible, and indeed, a majority of innovation processes do not turn out to be successful. Thus, authors such as Yun, Zhao, Jung, and Yigitcanlar [22] argue that it needs to be one of the key characteristics of the corporate culture to allow and embrace mistakes and successless developments as a natural aspect of innovation, especially in highly disruptive fields. Thus, leaders are recommended to shape their communication and attitude in a forgiving and embracing way. While the present work was able to establish this general framework, it remains unclear with which tools and strategies these goals can be achieved. This leads back to the explorative nature of the work: Derived from the state of research, it can be summarized that a relationship between the relevant

variables seems to exist. However, the complexity of the topic and the sheer amount of potentially influencing factors and variables highlights the necessity for further empirical work on the topic. This is perceived as critical in light of the existing literature which mostly points a theoretical perspective on the topic. Over the analysis of the recent scientific work on the topic, a lack of systematic reviews or meta-analyses was identified, which might be a feasible way to foster understanding on the topic. As it was argued within the introduction of this topic, it seems to be clear that open innovation processes are a legitimate tool to foster innovation success, and leaders can create conditions in which such processes prosper. However, further clarification seems necessary on the exact nature of the influence of leadership, subsequent organizational culture, and open innovation success.
Concludingly, however, it can be summarized that leaders have a twofold influence: First, in an optimal scenario, they enable and motivate their employees to go above and beyond the necessary requirements and to believe in themselves. Second, they shape a corporate culture in which innovation can grow.

4.2 Outlook

As Yun, Won, and Park [21] argue, open innovation processes might gain further relevance in an increasingly competitive market environment shaped by faster product life cycles. Thus, it can be assumed that it will remain a core challenge for leaders to create the right conditions for further implementation of suitable approaches. At the same time, however, leaders worldwide are combating a disruptive change in organizational collaboration brought on by the recent COVID-19 pandemic: As the crisis continues to influence societal and business life, the typical mode of collaboration in many companies changed in a significant way, as Waizenegger, McKenna, Cai, and Bendz [20] summarize with regard to the increasing usage of tele-work tools and virtual meetings. While open innovation is a type of R&D focusing heavily on interactive formats and collaboration between individuals, this pandemic-induced change might also affect the usage of such innovation strategies and will require leaders to find ways to lead the corporate cultures necessary for successful innovation into an increasingly digitalized work environment. While the work environment shifts, the challenges may continue to grow. This also is highlighted by Fleming and Waguespack [8], who concludingly highlight that findings from classical, i.e., physical, interactions in the field cannot necessarily be applied to digitalized communication: “Caution should also be exercised before generalizing the results to all open innovation communities, particularly those that lack physical interaction” [8].

References

1. Appelbaum, S., Bartolomucci, N., Beaumier, E., Boulanger, J., Corrigan, R., Dore, I., ... & Serroni, C., Organizational citizenship behavior: a case study of culture, leadership and trust. *Management decision*, **42**(1), 13-40 (2004)
2. Bass, B. M., & Avolio, B. J. (Eds.), *Improving organizational effectiveness through transformational leadership*. sage. (1994)
3. Bass, B. M., & Riggio, R. E., The transformational model of leadership. Leading organizations: *Perspectives for a new era*, **2**, 76-86 (2010)
4. Bolton, P., Brunnermeier, M. K., & Veldkamp, L., Leadership, coordination, and corporate culture. *Review of Economic Studies*, **80**(2), 512-537 (2013)
5. Chesbrough, H. W., *Open innovation: The new imperative for creating and profiting from technology*. Harvard Business Press (2003)
6. Denison, D. R., Bringing corporate culture to the bottom line. *Organizational dynamics*, **13**(2), 5-22 (1984)

7. Edelbroek, R., Peters, P., & Blomme, R. J., Engaging in open innovation: The mediating role of work engagement in the relationship between transformational and transactional leadership and the quality of the open innovation process as perceived by employees. *Journal of General Management*, **45**(1), 5-17 (2019)
8. Fleming, L., & Waguespack, D. M. (2007). Brokerage, boundary spanning, and leadership in open innovation communities. *Organization science*, **18**(2), 165-180.
9. Goodman, F. R., Disabato, D. J., & Kashdan, T. B., Integrating psychological strengths under the umbrella of personality science: Rethinking the definition, measurement, and modification of strengths. *The Journal of Positive Psychology*, **14**(1), 61-67 (2019)
10. Guiso, L., Sapienza, P., & Zingales, L., The value of corporate culture. *Journal of Financial Economics*, **117**(1), 60-76 (2015)
11. Jabbour, C. J. C., & Santos, F. C. A., The central role of human resource management in the search for sustainable organizations. *The International Journal of Human Resource Management*, **19**(12), 2133-2154 (2008)
12. Kratzer, J., Meissner, D., & Roud, V., Open innovation and company culture: Internal openness makes the difference. *Technological Forecasting and Social Change*, 119, 128–138. doi:10.1016/j.techfore.2017.03.022 (2017)
13. Müller, R. M., & Thoring, K., Design thinking vs. lean startup: A comparison of two user-driven innovation strategies. *Leading through design*, **151**, 91-106 (2012)
14. Naqshbandi, M. M., Tabche, I., & Choudhary, N., Managing open innovation. *Management Decision*, **57**(3), 703-723 (2019)
15. Ramdhani, A., Ramdhani, M. A., & Ainissyifa, H., Conceptual framework of corporate culture influenced on employees' commitment to organization. *International Business Management*, **11**(3), 826-830 (2017)
16. Rauter, R., Globocnik, D., Perl-Vorbach, E., & Baumgartner, R. J., Open innovation and its effects on economic and sustainability innovation performance. *Journal of Innovation & Knowledge*, **4**(4), 226-233 (2019)
17. Shafie, S. B., Siti-Nabiha, A. K., & Tan, C. L., ORGANIZATIONAL CULTURE, TRANSFORMATIONAL LEADERSHIP AND PRODUCT INNOVATION: A CONCEPTUAL REVIEW. *International Journal of Organizational Innovation*, **7**, 30-43 (2014)
18. Syafii, L. I., Thoyib, A., & Nimran, U., The role of corporate culture and employee motivation as a mediating variable of leadership style related with the employee performance (studies in Perum Perhutani). *Procedia-Social and Behavioral Sciences*, **211**, 1142-1147 (2015)
19. Van de Vrande, V., De Jong, J. P., Vanhaverbeke, W., & De Rochemont, M., *Open innovation in SMEs: Trends, motives and management challenges. Technovation*, **29**(6-7), 423-437 (2009)
20. Waizenegger, L., McKenna, B., Cai, W., & Bendz, T., An affordance perspective of team collaboration and enforced working from home during COVID-19. *European Journal of Information Systems*, **29**(4), 429-442 (2020)
21. Yun, J. J., Won, D., & Park, K., Entrepreneurial cyclical dynamics of open innovation. *Journal of Evolutionary Economics*, **28**(5), 1151-1174 (2018)
22. Yun, J. J., Zhao, X., Jung, K., & Yigitcanlar, T., The culture for open innovation dynamics. *Sustainability*, **12**, 1-21 (2020)

SHS Web of Conferences **115**, 03013 (2021) https://doi.org/10.1051/shsconf/202111503013
Current Problems of the Corporate Sector 2021

Leadership in Conditions of Industry 4.0

Diana Puhovichová[1,*], *Nadežda* Jankelová[1]

[1] University of Economics in Bratislava, Faculty of Business Management, Management Department, Dolnozemská cesta 1, Bratislava, Slovakia

Abstract. Nowadays, technological development and innovation play an important role in every enterprise. This is particularly reflected in the increase in competitiveness of any enterprise. It is Industry 4.0 that wants to lead to possible profound changes in several areas that go beyond the industrial sector. Leadership is essential for the successful promotion of a culture of innovation. Therefore, leaders and leadership style play a key role in the paradigm shift towards Industry 4.0. The aim of the theoretical research is to present the key characteristics of Leadership 4.0, the necessary skills of Leaders 4.0 and to introduce appropriate leadership styles in the context of Industry 4.0. Through the available domestic and foreign literature, we have provided a theoretical reality of what leadership in Industry 4.0 conditions should look like. According to results, that we have found out during theoretical research, we have reached conclusions that the core attributes of Leadership 4.0 include communication, knowledge and understanding, well-set standards and methods, coaching, expectations, openness and transparency, trust, employee orientation, and a culture of mistakes. We also pointed out that within the presented leadership matrix, the digital leader represents the most appropriate style within Industry 4.0, and among other things, this leader should dispose of cognitive skills, business skills, interpersonal skills, and strategic skills.

1 Introduction

Increasing global competition is forcing businesses to expand and personalize their product and service offerings to meet the needs of their customers through mass customization. Due to the volatile and highly competitive market, companies are required to continuously increase the flexibility of their business processes. In fact, technological advances are at the root of the Industry 4.0 concept, which first emerged at Hannover Messe in 2011, and was initiated by a German government to improve the performance of the industrial sector, and to cope with the increased complexity of manufacturing systems by improving systems integration. In research, Industry 4.0 refers to a set of concepts and technologies that enable decentralized decisions based on real-time connectivity and communication between systems and machines to develop and improve processes, products, and services [16]. This change is driven by various ongoing global megatrends such as globalization, urbanization, individualization, and demographic changes that will

* Corresponding author: diana.puhovichova@euba.sk

significantly affect the entire business environment in the future. On one hand, the increase in globally interconnected business activities will increase complexity in production networks. On the other hand, volatile demand and customized products will additionally influence production and planning processes. These demanding requirements will force businesses to adapt their entire operations, including structures, processes, and products [2].

The process of transforming an enterprise to Industry 4.0 conditions depends on the adaptation of leadership to the Industry 4.0 era - a phenomenon often characterized as leadership 4.0. For a successful transformation to Leadership 4.0, managers need to be qualified and promoted to meet and be prepared for such challenges [10]. It is therefore evident that a culture of openness to change in a business flows from the right leadership.

In the following theoretical research, we will discuss the impact of Industry 4.0 on leadership in enterprises. We will provide a comprehensive concept of what leadership in Industry 4.0 should look like so that companies are able to adapt to the conditions of the new phenomenon. Through professional literature and articles, we will determine the basic concepts that will give us a better understanding of the mentioned issue.

2 Theoretical Background

2.1 Industry 4.0

The development of digitalization over the last few decades has led to the so-called Fourth Industrial Revolution. This revolution is also known as Industry 4.0. Engineer and economist Klaus Schwab, founder and president of the World Economic Forum has declared that Industry 4.0 will fundamentally change the way we live. The new phenomenon aims to combine the strengths of traditional industries with state-of-the-art Internet technologies. According to experts, this industrial revolution is based on the Internet and enables communication over large networks between humans and cyber-physical systems [5,7]. Industry 4.0 is thus defined as the fourth industrial revolution, a new level of organization and control of entire value chains throughout the product lifecycle. This cycle includes the fulfillment of individualized customer requirements and extends from idea, actual order, development and production, delivery to the customer and recycling. The key is the availability of all necessary information in real time through the interconnection of all instances involved in value creation, as well as the ability to derive the best possible value stream based on the resulting data. By linking people, objects and systems, dynamic, real-time optimized, self-organizing, inter-enterprise value networks are developed that can be optimized based on various criteria such as cost, availability, and efficiency of resources [2].

PwC's study, "Industry 4.0: Building Digital Enterprise", shows that the technology needed to implement Industry 4.0 is ready, the problem is the lack of digital culture, vision, training, and a shortage of experts [3]. From the above, it is evident that these problems can be solved by the right leadership.

For the successful implementation of the enterprise transformation to Industry 4.0 conditions, nine core technologies are required, which must be part of the entire enterprise system. The core enabling technologies that make up the Industry 4.0 concept include: big data and analytics, autonomous robots, simulations, horizontal and vertical integration of systems, the Internet of Things, cybersecurity, cloud computing, additive manufacturing, and augmented reality. The following table shows all the above technologies that form the basis for the successful operation of an enterprise in the conditions of Industry 4.0.

Table 1 Pillars of Industry 4.0

Nine pillars of Industry 4.0	
Big Data and Analytics	The term Big Data refers to large, diverse, and complex data sets that influence organizational decision-making in relation to business strategy. Therefore, there is a need to increase the levels of data and improve technological capabilities that accelerate the competitiveness of enterprises in the form of increased productivity, innovation. The big data framework could be described as: data as a tool (solving traditional value chain problems using existing capabilities), data as an industry (new ventures and the development of software systems to process big data) and data as a strategy (building data resources by developing new innovative business models). The process of big data analytics involves the analysis of large data sets that provide information on customer preferences, algorithms for correlations, trends, and other information. We can predict failures through analytics and thus reduce the probability of errors [8].
Autonomous Robots	Robots are used in the manufacturing industry to solve complex tasks that are not easily solvable by humans. Current improvements in industries would make the use of robots less complicated and easy to use. Different human-robot interfaces create a close collaboration of robot usage and human brain. The use of more industrial robots in factories is accelerating with Industry 4.0. They would have applications in multiple fields such as manufacturing, logistics, distribution activities and could be controlled remotely by humans through mutual collaboration [1].
Simulation	Industry 4.0 simulation refers to the 3D simulation of products, materials and manufacturing processes that use real-time data to reproduce the physical world in a virtual model. Operators use simulation to identify weaknesses in their products, test and optimize machine settings and processes in the virtual world [10].
Horizontal and vertical integration	Vertical integration refers to flexible and reconfigurable systems inside the manufactory to the extent that they are fully integrated with each other to achieve agility. Horizontal integration refers to the integration of partners within supply chains. An industrial network collects large amounts of data to optimize system performance and send it to the cloud. This coordination mechanism forms the framework of the smart factory. Therefore, manufacturing systems are designed as a self-organized structure that integrates each physical object through smart networks. In addition, cloud-based systems enable vertical partners to integrate with each other through common platforms. Product and process flows would be visualized and tracked by members of the supply chain [8].
Internet of Things	The concept of the Internet of Things represents structures that allow objects to communicate with each other. The IoT is expected to open many economic opportunities and is considered one of the most promising technologies with great disruptive potential. The idea was formulated in 1999 to describe a system in which the material world communicates with computers (data exchange) using sensors that are everywhere. In this approach, not only objects but also processes, data, people and even animals or atmospheric phenomena are treated as variables - a system of all things [17].
Cyber Security	Cybersecurity is another important issue that could have a destructive impact on the business environment due to the malicious intent of terrorist attacks. For this reason, it is necessary to find preventable solutions and build defensive systems against the negative effects of terrorist incidents. There are solutions that destroy cyber terrorist attacks, in the form of analyzing previous terrorist attacks through "radiation control" before future attacks occur. In addition, it is important to build a national defense system, and train employees against cyber-attacks. Although cyber warfare solutions will cost businesses a significant amount of money, the expected total will ultimately not be high due to the potential negative effects of attacks [11].
Cloud Computing	Cloud computing is a new parallel and distributed system that consists of interconnected and virtualized computers used based on service level agreements between the service provider and customers. It provides significant benefits to businesses by freeing them from the need to build an underlying hardware and software infrastructure and the associated investment. Cloud computing and the Internet of Things together improve intelligent perception and M2M-based communication and ensure efficient use of resources in production systems [18].
Additive	Additive manufacturing, also defined as 3D printing, deals with the production of

Manufacturing	customized goods based on customer requirements. The most common method is the prototype and 3D printing method to produce small batches, thereby gaining the advantage of smaller inventories and overproduction [8].
Augmented reality	Augmented reality is defined as an interactive technology that enables a correspondence between a virtual world and its users, with the virtual world being used as part of the real world [8]. The technology enables the delivery of information in real time. A typical example is virtual reality glasses, which can be used, for example, to train employees in an enterprise [10]. Augmented reality provides control of users' movements using sensor technology to control certain tasks [8].

We share the view of Črešnar and Nedelko [6] that the above-mentioned technologies will bring significant changes in the enterprise, but they are not only focused on the hard, technological aspects, but the organizational environment is expected to become more multicultural, multidisciplinary, agile, collaborative, and open, which will also affect the soft aspects related to people. As a result of these changes, the question of what leadership in Industry 4.0 should look like: "What values, qualities and attributes leaders should dispose of?"; or "How to get to the forefront in the Industry 4.0 era?"

2.2 Leadership 4.0

Leadership focuses on the behavior of senior executives, which sets the general framework in the business and influences the corporate culture, without focusing on the individuals of the business. These aspects are applicable to every employee in the company and have an impact on customers, suppliers, and partners. However, the question is how the leadership of enterprises can be transformed to introduce new business models and new ways of working brought about by digitalization. The solution lies in the implementation of so-called digital leadership, also known as leadership 4.0 [4].

Digital leadership (leadership 4.0) is fast, cross-cutting, team-based, and collaborative leadership with a strong focus on innovation. The leader's personal competencies, his or her mindset, as well as his or her ability to apply new methods and tools, such as design thinking, are crucial dimensions for leaders. Design thinking is a methodology used by designers to solve complex problems and find desirable solutions. Design thinking is solution-focused and action-oriented thinking aimed at creating a preferred future. In business management, it could be used as a step-by-step process to find and implement solutions that increase profits, efficiency, and customer satisfaction; and to keep the business ahead of the innovation curve. Industry 4.0 requires businesses to think in new ways [14].

This phase of digital transformation is very demanding and full of challenges that require non-traditional leadership skills and approach to compete in this situation for the survival of companies. For example, Apple Inc. is the most innovative company; this is not due to the technical skills of CEO Steve Jobs, but to his leadership style. Similarly, Microsoft Corporation owes its success to Bill Gates' different leadership style. Amazon.com is one of the early adopters of 4.0 technologies with a net worth of $150 billion. In addition, the second largest company, Adidas is increasing its use of artificial intelligence (AI) technologies in products and is using 3D printing. Company executives have expressed their belief that Change 4.0 is mandatory for those who understand this digital massive wave. In the United States, the non-profit Smart Manufacturing Leadership Coalition (SMLC) is pushing for technological transformation as it has found that leaders understand global competitiveness and challenges. In addition, a PWC report analyzed from 235 European companies that only 3.3% of leaders are investing in Industry 4.0 because they assume that digitalization is mandatory for the sustainability of their business [19].

CEO Rajeev Peshawaria of ICLIF (Leadership & Governance Centre, Malaysia), after analysing 255 definitions and real case studies, developed a leadership formula [15], as shown below.

Leadership = (charisma + personality) x (power and authority) + (skills and competencies + best practices) x (cognitive intelligence + emotional intelligence)

In addition, the book Open Source Leadership [15] provides a definition of leadership from the perspective of the Fourth Industrial Revolution, confirming that in the wave of digital disruption, leadership is the art of harnessing human energy to create a better future.

In this world of constant change in technology, mergers, increasing global competitiveness, diversity of ideas, along with a mix of different opinions, experiences and digitalization, the forms of business must be based on leadership that is in everyone's interest. In the fourth industrial revolution, leadership is not about power and status, but about a fundamental change of mindset. We need leadership that will prevent the world from becoming a victim of Industry 4.0. Indian Prime Minister Modi said: "The administrative mind-set of the last century will not help us move forward into the new century". Former Malaysian minister Datuk Seri Idris Jusoh said: "Leaders should seize the potential opportunities that Industry 4.0 brings with great responsibility and wisdom by providing digital leadership to their institutions [19].

We share Zakaria and Nasir's view that forward-looking leaders have the potential to anticipate future changes and, as a result, can prepare businesses to successfully fight against Industry 4.0. The Fourth Industrial Revolution undoubtedly has the power to change many things from the physical world to the digital world. However, one thing is for sure, leaders need to fully prepare because the old ways are no longer enough for businesses to operate in Industry 4.0 conditions.

3 Research design

The aim of the theoretical research is to point out the importance of the modern phenomenon Industry 4.0, which blurs the boundaries between physical, biological, or digital spheres, and to show its impact in leadership through literature and scientific articles. The object of the theoretical research is the leadership of enterprises that are affected by the growing impact of Industry 4.0 We focus on the issue of values of digital leadership in conditions of Industry 4.0, the basic characteristics of Leader 4.0 and the leadership styles that should be applied in enterprises in conditions of Industry 4.0.

To analyze the topic, we will use accessible domestic and foreign sources that describe in detail the issues of Industry 4.0, the basic technologies that form its concept, the essence of leadership and leadership 4.0, and the emerging changes in the field of leadership, which are caused by the impact of Industry 4.0. Through the information gathered, we will complete a theoretical elaboration of the issue.

To elaborate the theory, we used the basic scientific methods: analysis, synthesis, generalization, as well as induction and deduction. From the acquired information, we have created a theoretical knowledge base on leadership that reflects the needs of Industry 4.0.

By elaborating the mentioned issues, we developed the following research questions:

- What core values should business leadership dispose of in conditions of Industry 4.0?
- What should a leader in Industry 4.0 be like? What qualities and skills should leader 4.0 dispose of?
- What business leadership styles are suitable to Industry 4.0?

The research questions were answered in the results and discussion of the theoretical research.

4 Research results and discussion

For businesses, leadership plays an important role in Industry 4.0, as changing the way a business is run requires far-reaching strategic decisions at all levels of the enterprise. In this context, future leaders will need to be more responsive to the perceived patterns and signals that Industry 4.0 points to in a constant field of growth, change and opportunity. Therefore, in the following section of the theoretical research, we will offer answers to the research questions, providing the necessary characteristics of leadership in Industry 4.0 terms:

What core values should business leadership dispose of in conditions of Industry 4.0?

Based on the characteristics of Industry 4.0 and Industry 4.0-compliant leadership outlined in Chapter 2, the most influential attributes of successful leadership of the company and employee are:

1. **Communication:** intensive communication between management and employees
2. **Knowledge and understanding:** knowledge of the significance of the changes initiated by digitization
3. **Methods and standards:** appropriate methods and standards implemented within the corporate culture and identity
4. **Coaching:** Managers with coaching skills who become good role models for the company's employees
5. **Expectations and orientation:** guiding the level of management and meeting the expectations of the company
6. **Openness and transparency:** an open and transparent atmosphere between employees and management
7. **Trust:** trust and freedom in the workplace regarding the time, place, and content of work
8. **Employee focus:** Focus on employee growth and well-being
9. **Bug culture:** living an open bug culture in the enterprise that enables the innovation needed in an Industry 4.0 environment [4].

What should a leader in Industry 4.0 be like? What qualities and skills should dispose of?

Digital technologies have an impact not only on the information technology field, but also on business management. Leadership 4.0 is leadership in the age of Industry 4.0, which is handled by Leaders 4.0 These leaders are called digital leaders. Not every leader in an enterprise is also a digital leader. Germany, as a showcase country where Industry 4.0 originated, is thriving when it comes to Industry 4.0 and the Internet of Things. However, this does not mean that the managers of these (German) enterprises are automatically digital leaders. What makes a digital leader a leader is not the area in which the enterprise operates. It's a matter of leadership style and the ability of these leaders to inspire their employees to innovate and to sustain and develop their ideas. We consider the following elements to define digital leaders: **organizational goals** (ability to move from fixed cycles of employee performance appraisal to the ability to understand that situations determine the need for employee appraisal, together with the ability to establish routine feedback procedures), **people** (ability to allocate tasks based on the situation and the competence of the team; linking the skills of managers and employees to create the competence of network intelligence), **change** (high level of willingness and capability for change; fostering a high level of agility between market, customer, partners and employees), **output** (the ability to manage processes, evaluate tasks and results with teams, use resources in accordance with competencies - cross-hierarchically and cross-functionally - instead of controlling contracts, resource plans and evaluating results within project boundaries, as traditional leaders do), **mistakes and conflicts** (creating an open atmosphere with the effect of learning from mistakes, and an atmosphere of cooperation in conflict resolution), **communication** (the ability to create a transparent framework for information distribution), as well as **innovation** (the knowledge that innovation can be learned; the ability to transform old structures through the use of multidisciplinary teams and creative processes in a flexible working environment) [14].

The acquisition and refinement of leadership skills is influenced by individual differences in cognitive capacity, personality traits, temperament, abilities to control emotions, identities and values resulting from both factors, cultural context, and personal experiences [12].

According to Mumford [13], leaders in Industry 4.0 conditions should be equipped with the following leadership skills, which can be divided into 4 groups: 1) **cognitive skills** that executive leaders need to understand the required complex behavioral patterns, including creative thinking, decision making, and strategic problem solving; 2) **interpersonal skills**, which are defined as goal-oriented behaviors used in personal interactions to achieve a desired state of affairs; 3) **business skills** include organizing, negotiating, and managing personal, financial, and material resources; 4) **strategic skills**, which are associated with conceptualizing the mission and vision of the company. These four groups of leadership skills consist of other relevant skills, which are listed in the table below.

Table 2 Groups of Leadership Skills

COGNITIVE SKILL	INTERPERSONAL SKILL	BUSINESS SKILL	STRATEGIC SKILL
Speaking	Social perceptiveness	Operations analysis	Visioning
Active listening	Coordination	Management of personnel resources	Systems perception
Writing	Negotiation	Management of financial resources	System evaluation
Reading comprehension	Persuasion	Management of material resources	Identification of downstream consequences
Active learning			Identification of key causes
Critical thinking			Problem identification
			Solution appraisal

What business leadership styles are suitable to Industry 4.0?

Oberer and Erkollar's study [14] developed a two-dimensional Leadership 4.0 matrix that depicts the types of styles that can be applied in companies in Industry 4.0 conditions. It will be based on the orientation towards innovation and technology and towards caring for people. The matrix is divided into four quadrants. In each quadrant, one leadership style that is best suited the particular leader. The X axis describes the capabilities of the individual and focuses on the orientation towards technology and innovation. The Y axis defines the leader's people focus. For example, a person with a high level of education in a technology-related field is likely to want a technology-oriented leadership style, such as TL (Technology Leader 4.0) or DL (Digital Leader 4.0, if they also have a high employee orientation). A person who generally seeks employee engagement in decision making is more likely to prefer an employee-centric leadership style such as SL (Social Leader 4.0). A leader who is without a specific orientation, neither to people's interests nor to technology/innovation, with a low orientation to people as well as a low orientation to technology/innovation, then we talk about the so-called FL (Freshmen leader 4.0). The following table shows the different styles that are part of the two-dimensional leadership 4.0 matrix.

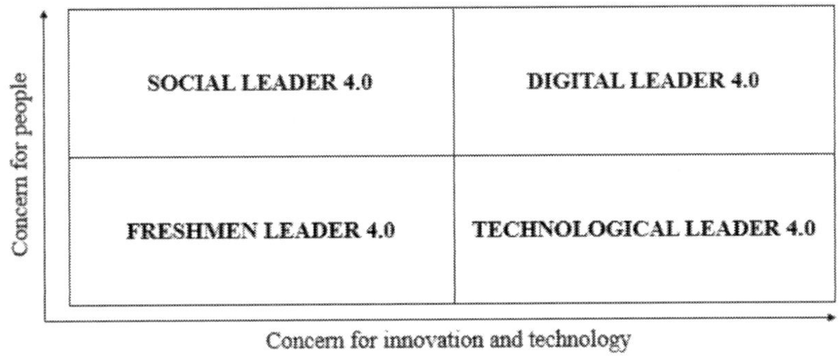

Fig. 1. Leadership matrix

FL focuses on traditional structures with a primary focus on the end product; it has no focus on employees; customer needs are of negligible importance, as well as for new technologies. SL disposes of the ability to create a friendly atmosphere for employees regardless of innovation and technology. Employee support is important. TL focuses on the ability to determine how new technologies can be leveraged to provide a higher value. Innovation orientation and low employee focus prevails. DL focuses on the ability to understand how technology affects people, and how the organizational model aligns with human nature. These are equally essential elements of digital leadership. The digital leader "Leader 4.0" is the most productive style of Leadership 4.0.

5 Conclusion

After explaining the importance of leadership in the digital age, this article reveals that digitized enterprises need a new type of leadership, called leadership 4.0. This should foster an agile environment that is essential for driving innovation forward quickly and creating higher customer satisfaction. Through leadership 4.0, on one hand, the culture of enterprises is being influenced and, on the other hand, the work of enterprises itself is being influenced. It is up to the companies whether the leadership is able and willing to adapt to the new phenomenon of Industry 4.0 and thus sustain and develop their competitiveness.

Through the literature and professional scientific articles, we have highlighted the essence of leadership in Industry 4.0 conditions, thus fulfilling the main objective of the theoretical research. We have also provided answers to the above research questions, in which we have noted what basic attributes leadership should disposes of, we have listed the basic and necessary characteristics of a leader 4.0, and we have provided an overview of appropriate leadership styles in the form of a leadership 4.0 matrix.

Acknowledgement

This paper is a partial output of the VEGA ME SR project No.1/0017/20 Changes in the application of managerial functions in the context of the fourth industrial revolution and adaptation processes of enterprises in Slovakia. And a partial output of the Project for young teachers, researchers, and PhD students No. I–21-111-00 The impact of the fourth industrial revolution on managerial functions.

References

1. M. Aiman, et al. Industry 4.0 : A review on industrial automation and robotic. Jurnal Teknologi. *Sciences & Engineering* **78**, pp. 137-143 (2016)
2. C. J. Bartodziej, The concept industry 4.0. *The concept industry 4.0*, pp. 27-50 (2017)
3. A. Bencsik, Challenges of Management in the Digital Economy. *International Journal of Technology* **11**, pp. 1275-1285 (2020)
4. S. Bolte, J. Dehmer, J. Niemann, Digital Leadership 4.0. *Acta technica napocensis – Series : Applied mathematics, mechanics, and engineering* **61**, pp. 637-646 (2018)
5. M. Brettel, N. Friederichsen, M. Keller, M. Rosenberg, How Virtualization, Decentralization and Network. Building Change the Manufacturing Landscape: An Industry 4.0 Perspective. *International Journal of Information and Communication Engineering* **8**, pp. 37–44 (2014)
6. R. Črešnar, Z. Nedelko, Understanding Future Leaders : How are Personal Values of Generations Y and Z Tailored to Leadership in Industry 4.0 ? *Sustainability* **12**, pp. 4417 (2020)
7. N. Daneshjo, M. Majerník, E. Danishjoo, E, More Exact Approaches to Modernization and Renewal of the Manufacturing Base. *TEM Journal* **6**, pp. 445–449 (2017)
8. G. Erboz, How to Define Industry 4.0 : The Main Pillars of Industry 4.0. *Managerial trends in the development of enterprises in globalization era*, pp. 761-767, 2017.
9. S. Helming, F. Ungermann, N. Hierath, N. Stricker, G. Lanza, Development of a training concept for leadership 4.0 in production environments. *Procedia Manufacturing* **31**, pp. 38-44 (2019)
10. M. Chiarvesio, R. Romanello, Industry 4.0 Technologies and Internationalization: Insights from Italian Companies. *International Business in the Information and Digital Age* ***13***, pp. 357-378 (2018)
11. H.S. Cho, T.H. Woo, Cyber security in nuclear industry – Analytic study from the terror incident in nuclear power plants (NPPs). *Annals of Nuclear Energy* **99**, pp. 47–53 (2017)
12. R. H. Lord, Identity, deep structure and the development of leadership skill. *The leadership Quarterly* **16**, pp. 591-615 (2005)
13. T. Mumford, M. Campion, F. Morgeson, The leadership skills strataplex: Leadership skill requirements across organizational levels. *The Leadership Quarterly* **18**, pp. 154-166 (2007)
14. B. Oberer, A. Erkollar, Leadership 4.0 : Digital Leaders in the Age of Industry 4.0. *International journal of organizational leadership* **7,** pp. 404-412 (2018)
15. R. Peshawaria, Minimum supervision, maximum performance. *Leader to Leader*, **87**, pp. 42-47 (2018)

16. S. Richard, R. Pellerin, J. Bellemare, N. Perrier, A business process and portfolio management approach for Industry 4.0. *Business Process Management Journal* **27**, pp. 505-528 (2021)
17. M. F. Skender, I. Ali, Digitalization and Industry 4.0 (2019)
18. A. Ustundag, E. Cevikcan, Industry 4.0 : Managing the Digital Transformation. *Industry 4.0 : Managing the digital transformation*, pp. 1-286 (2018)
19. N. Zakaria, A. Nasir, A. Akhtar, Are the Leaders Ready to Embrance Industry 4.0 ? *Journal of Advanced Research in Dynamical and Control Systems* **11**, pp. 543-549 (2019)

Improvement of OSH Management through the holistic approach of incident controlling

Kristina Schreyer[1,*], *Arman* Dehghani[2], and *Eleonora* Fendeková [2]

[1]IU, Faculty of Business Management, Berg-am-Laim-Str. 47, 81672 MÜNCHEN, Germany
[2]EUBA, Faculty of Business Management, Dolnozemská cesta 1/b, 852 35 Bratislava, Slovak Republic

Abstract. This paper deals with the performance optimization of occupational safety and health management (OSHM) with the help of the integration of near miss management into existing KPIs of the company. In this context, two independent processes are carried out: risk assessment and near miss management. In this work, the holistic view of these two processes is investigated as an optimization potential OSHM. Based on a literature review, the framework of the qualitative method applied here is established. After definitions and general information concerning OSHM, the processes of risk assessment and near miss management in different organisations (23 international plants for risk assessment and 6 German plants for near miss management) with established processes according to BPM are examined. The chosen approach is combined with the method of the PDCA cycle to relate the results of the investigation to the CIP existing in the OSHM. The result was elaborated in the form of a recommendation for action: optimization of existing processes, integrated hazard and near miss management as a more sustainable solution and proposal for the implementation of a KPI for better control of incidents based on predictive risk assessment.

1 Introduction

In management and steering of organization the performance indicators are important as key figures for measuring team performance in a process which demonstrate employees' ability by comparing it to their previous performance or the average level of industry and the growth potentials. The optimization of key figures in the field of human resource management is one of the companies' main objectives. Despite the significant role played by occupational safety and health management (OSHM) in this field, it is almost impossible to measure OSH using quantitative and especially prospective key figures. OSH is mostly considered an extra expense and thus a source of costs as well. This is mostly because it is scarcely possible to observe the practical advantage of workplace safety policies directly and instantly. Decreasing the resultant simplified vision is an important task in OSHM. [1] In a study, the German Social Accident Insurance (DGUV) proves the benefit of OSH for a company based on Return on Prevention (ROP).[2]

* Corresponding author: elenora.fendekova@euba.sk

The legal and normative framework obligates the enterprises to determine and control the hazards and potential dangers at the workplace at an early stage. [3,4] Over the past decades, these regulations have enabled the EU to achieve a significant reduction in accidents and an increase in preventive OSH, which is to be further improved.[5] According to a survey conducted in 2015, 54 percent of German companies carry out a risk assessment.[6] This component of OSH is stated to be a competitive factor, as the costs of possible occupational accidents are avoided through appropriate OSH measures.[7] In addition, near misses could also contribute to improving the global effectiveness of an OSHM. A near miss can be defined as a dangerous situation and/or an unsafe act in which the chain of events could have resulted in an accident if it had not been broken. [8]

Accidents and near misses are usually characterized by common causes. In addition, near misses are a more relevant source of information for increasing the level of safety in an organization because they indicate deficiencies in safety systems, the so-called "weak signals" without causing high consequences. [9] Despite the similarity of accidents and near misses, they are recorded and monitored independently of each other. One of the reasons could be the lack of a legal requirement to record and control near misses. This work aims to develop a recommended approach for optimizing the OSHMS based on the integration of near misses into the risk assessment process, which is to be monitored by a suitable KPI.

2 Theoretical Background / Literature review

2.1 OSHMS and modern risk assessment

The positive effects of implementing occupational safety and health management systems (OSHMS) at organizational level, both in the reduction of hazards and risks and in productivity, is now widely recognized by governments, employers, and workers. [10] The entrepreneurial duties which are derived from the legal conditions are correspondingly complex and relate to both the normative, strategic and operative management level.[11] Employers are thus obliged, to carry out risk assessments of workplaces, define responsibilities in occupational safety, inform and train employees, monitor compliance with occupational safety regulations, etc. In order to be able to carry out these tasks in a systematic and formalized manner, the implementation of an operational OSHM is essential.[10] At the same time, the introduction of an OSHMS means that support from the top management and the consideration of OSH requirements in all processes of an organization secured. [12]

OSHM is the targeted and systematic planning, implementation, and monitoring of OSH as a management task. OSHM is above all a strategy and methodology for the systematic planning, organization, operation, and monitoring of OSH. It does not contain any claims beyond the valid public law obligations.[12] The model of modern OSH is based on a holistic understanding of safety and health at work, emphasizes the personal responsibility of an organization, its managers and employees, emphasizes the management task of "occupational safety and health", involves managers more closely in OSH and relies on regular evaluation with the aim of continuous improvement, integrates occupational safety into the operational processes and regards it as a process.[1] In detail, modern OSH comprises the following elements shown in Figure1:[1]

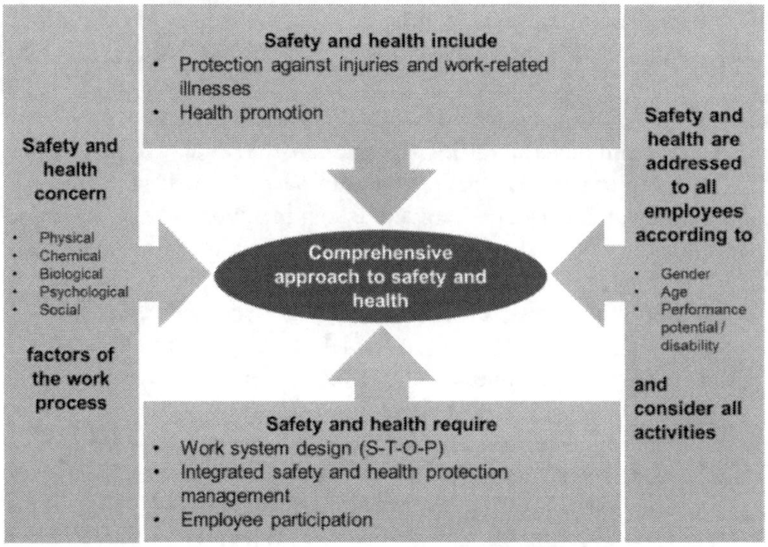

Fig. 1. Main elements of the OSH management system, own Illustration based on [1]

2.2 Risk assessment

Risk assessment is one of the most important steps in ensuring safety and health at work. [13] It is defined as the "systematic identification and evaluation of hazards". [1] This can be carried out both retrospectively and prospectively. Schlick et al. emphasize the prioritization of preventive OSH because of its predictive risk evaluation for the avoidance of accidents or potential hazards. [14] Risk assessment in accordance with the legal requirements (e.g. German OSH Act) is a concept for action to improve safety and health.

The central operational preventive instrument for this is the general overview of the working conditions to be created for the overall operation and action to be derived from it. So, the improvement of working conditions will be steered focused on the key priorities steered. The target management for overall company action can be justified in this way and effectively controlled. [1] The methodical approach for the assessment of working conditions according to §§ 5 and 6 German Occupational Safety and Health Act is shown in Figure 2.

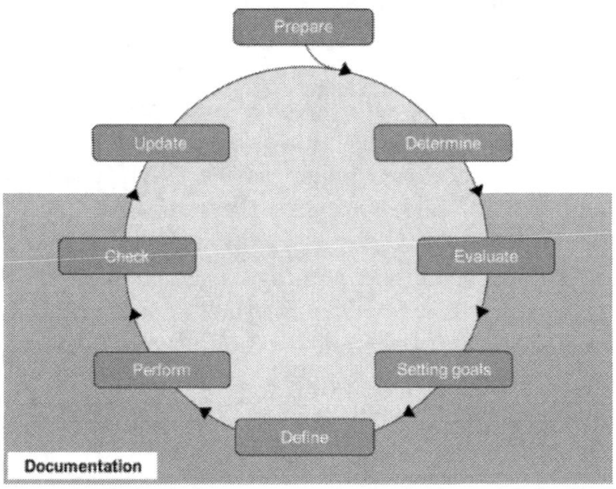

Figure Chyba! Dokument neobsahuje žiadny text so zadaným štýlom.. Methodical approach for risk assessment, own illustration based on [1]

2.3 Near miss

A near miss is the result of a dangerous event at work that does not result in an injury to the employee, i.e., it is not an accident. [15] Nesmith, Keating, and Zacharias clarify that this presents an opportunity to gain valuable information to prevent future accidents.[16] This is visualized by the so-called accident pyramid and shown in Figure 3. Near misses form the basis of the accident pyramid. The time of a lost workday is considered a criterion of accident severity and thus an event occurs less frequently the more severe the accident is. [17]

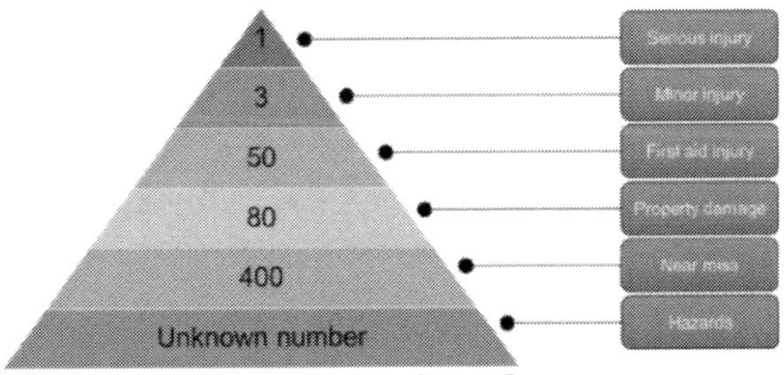

Figure 3. Accident Pyramid own illustration based on [17,18]

3 Methodology

The aim of the work is to develop a recommendation for optimizing the OSHM by means of integrating near misses into the risk assessment process. The basis for the recommended approach is the following research questions (Q):

Q1) How can the prevention process of OSHM be optimized through near miss management?

Q2) Which key figure(s) can make this optimized process sustained?

The following theses (T) result from the research questions:

T1) By taking a holistic view of hazards and integrating near misses into the hazard assessment process, the OSHM prevention process will be optimized.

T2) The numerical risk assessment available in OSHM serve as a suitable key figure for this new holistic approach.

To achieve answers to the set objectives and research questions, the framework of the work is first established based on a literature review. The basis of the proposed recommendation is the process analysis of the risk assessment and the near miss reporting of selected sites. Within the framework of this investigation it comes to the application of the qualitative method of BPM in combination with PDCA-Cycle. The method selection was based on criteria of familiarity and applicability in relation to similar procedures of working in OSHM. The contents of the cyclical methodology correspond to both the PDCA cycle and the intended CIP. The proposed approach requires a qualitative analysis of the existing processes as well as the inclusion of quantitative measurability for the follow-up of the targeted process optimization. The existing sub-processes in OSHM are similar to this approach. This also speaks for the use of the methods mentioned.

The first phase is the process survey and includes the exact delineation of already existing processes from one another. The findings from this are summarized in the process documentation to form an as-is process model. Based on the as-is documentation, existing weaknesses can be identified as part of the process analysis. The causes determined from this serve as input for a renewed process conception, which results in a desired process model. This is followed by the process implementation of the desired model. Here, the necessary changes are implemented in the new as-is process. Continuous monitoring of the processes should be a continuous task, even if process controlling is mapped as the last step in this cycle. To analyse the presented current process, a qualitative method is used, as mentioned earlier. To uncover possible sources of waste of underutilized resources, a non-value-added analysis is applied, with the aim of making the process content more effective, as recommended by Dumas et al. [19] As noted, the chosen approach is combined with the quantitative method of the PDCA cycle. The data collection was carried out in 23 international sites to analyse the risk assessment process and 6 German sites for near miss management, all with the same production and similar structure and working culture.

4 Results

The analysis of risk assessment process shows that active updating is insufficiently integrated into the process and must therefore be regarded as a process gap. The result of the investigation on risk assessment process in 23 entities can be presented as Figure 4. The weakness thus lies in the fact that prospective findings are not proactively considered, and no consequent implementation of countermeasures are taken to improve safety based on the investigation of near misses. The first two steps of the PDCA cycle, planning and execution, are considered, but not all the events required by the risk assessment guidelines are included in the third step of review. Thus, the cyclical approach is incomplete. This type of waste shows that the existing potential of the relevant safety instructions is not sufficiently exhausted and is disregarded as a link in the process chain. The possible adversary to the found process weakness is consequently the inclusion of near misses or recognized potential hazards in the risk assessment process. More specifically, the output of a near miss must serve as a renewed input to the CIP of the risk assessment. This change is visualized as a target model in Figure 5.

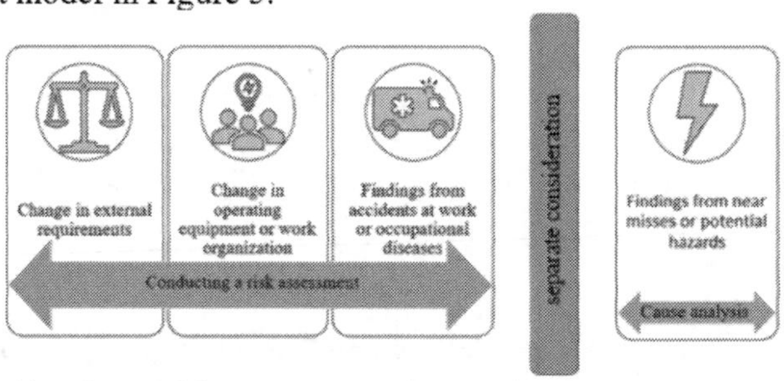

Figure 4. separate consideration of risk assessment and near miss management, own illustration

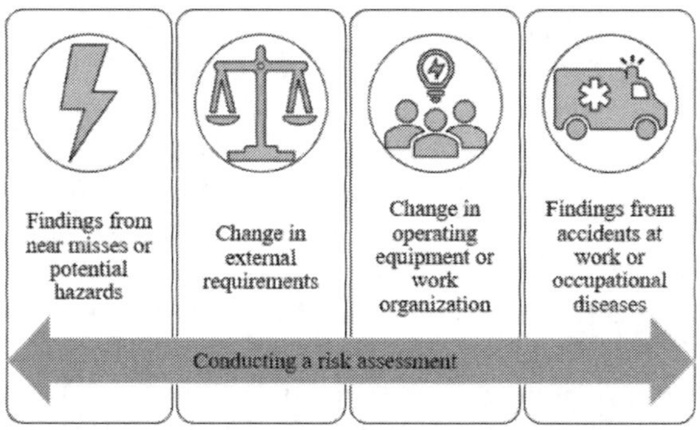

Figure 5. holistic risk assessment taking considering near miss management, own illustration

Proactive analysis and handling of near misses enables the inclusion of potential safety information and thus promotes safety and health protection of employees. The analysis of near miss management of 6 plants shows two common process weaknesses in all cases. First, the lack of reference to the overall accident analysis. The comprehensive analysis of all accidents enables direct reference to accident causes and insufficient measures to be taken in relation to each other. The second weakness can be defined as a missing indicator for the evaluation of occupational safety regarding incidents.

The integrated linking of incidents to each other supports the occupational safety of employees and promotes the prospective prevention of near misses. When similar incidents are repeated, it is an indicator that a more thorough investigation and analysis is needed to eliminate the cause of a potential accident. [17] Accordingly, it enables the assessment of the effectiveness and quality of the safety level and the countermeasures that have already been implemented. The additional use of key performance indicators, besides the reasons of performance evaluation, can also serve as an instrument for early warning. In the context of the present work, this is a very great benefit for the company.

It is recommended to include the risk matrix according to "Nohl", which is already used for risk assessment, in the processing of near misses. In most cases, the recognized hazard potential is classified as having a low impact and an unlikely frequency of occurrence, but a subsequent event can lead to an increase in these criteria. [20] Figure 6 visualize this relation.

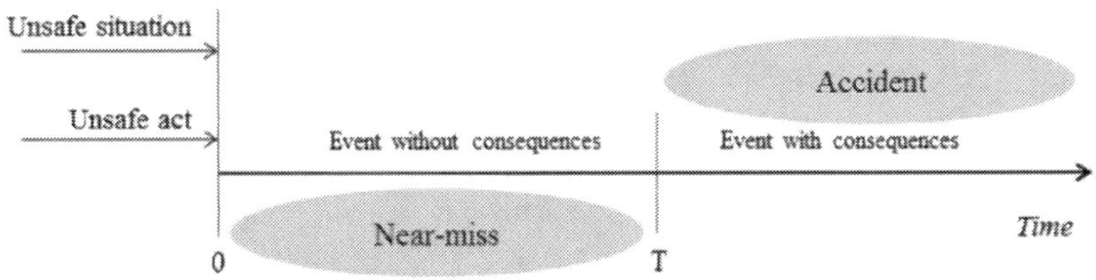

Figure 6. relation between near miss and accident.[9]

An early warning instrument of this kind can measurably highlight changes and anomalies. Thus, early forecasts are possible and negative developments can be interrupted by appropriate countermeasures. [20] The integration of an early warning indicator can be based on the Predictive Risk Index (PRI) according to Chen & Yang to predict an accident. [15] With the help of the PRI, the occurrence of an event can be predicted in that, for example, an increasing trend indicates a greater probability of a future accident. The existing system can be extended to include the PRI with little effort.

5 Discussion and Conclusion

The main component of the investigation result is the existing process gap and the missing reference to key figures. These factors are significantly decisive for the continuous improvement and effectiveness of the OSHM. The system-based implementation of a dependent consideration of all incidents is associated with little effort using IT tools. The integration of the PRI represents a further interesting optimization. It has been explained why an early warning indicator is an important instrument in the context of the reference to key figures. The core of the advantage over other possible indicators is the explicit promoting of preventive occupational safety, which makes such a PRI indispensable. The integration of a PRI into the organizations' KPI system is also rated as low cost.

Furthermore, the essential criteria for optimizing the OSHM were identified and executed during the process analyses. A number of recommendations for action emerge from the comparison of these results with the practice of existing systems:

- Integration of near misses into the event-related review of risk assessments.
- Comprehensive analysis of causes and measures for all reportable incidents.

- Concept development and integration of key figures for optimized measurability of accident occurrence, exemplified by a PRI.

Since this is a study of comparable companies with similar structures and production, it must be considered that a scattering of the company structure and/or culture can partially influence the results.

In OSH, the control of complete process chains by means of key figures offers great potential for reducing company costs and for preventive support of employee safety and health. The digital transformation taking place in the context of Industry 4.0 enables the collection of ever greater volumes of data and thus the opportunity to improve the performance of existing control instruments. Preventive action on potential hazards is indispensable for the continuous improvement of the OSHM.

Acknowledgment: The authors would like to thank the participating contact persons of the respective plants and for the information provided.

References

1. Bundesanstalt für Arbeitsschutz und Arbeitsmedizin (BAuA). (Hrsg). Indikatoren und Parameter zur Bewertung der Qualität des Arbeitsschutzes im Hinblick auf Arbeitsschutzmanagementsysteme. German federal institute for OSH, **959**, 177 (2002). View. 21.01.2021.

   https://www.baua.de/DE/Angebote/Publikationen/Schriftenreihe/Forschungsberichte/2220/Fb959.pdf?__blob=publicationFile
2. D. Bräunig, T. Kohlstall, Berechnung des internationalen "Return on Prevention" für Unternehmen: Kosten und Nutzen von Investitionen in den betrieblichen Arbeits- und Gesundheitsschutz. Deutsche Gesetzliche Unfallversicherung *DGUV Report* **1** (2013)
3. German OSH Act, Gesetz über die Durchführung von Maßnahmen des Arbeitsschutzes zur Verbesserung der Sicherheit und des Gesundheitsschutzes der Beschäftigten bei der Arbeit (Arbeitsschutzgesetz - ArbSchG). Arbeitsschutzgesetz vom 7. August 1996 (BGBl. I S. 1246), das zuletzt durch Artikel 427 der Verordnung vom 31. August 2015 (BGBl. I S. 1474) geändert worden ist. [Online]. https://www.gesetze-im-internet.de/arbschg/ArbSchG.pdf
4. DIN ISO 45001, Occupational health and safety management systems - Requirements with guidance for use (ISO 45001 :2018), 2018
5. EU-Kommission, Mitteilung der Kommission an das Europäische Parlament, den Rat, den Europäischen Wirtschafts- und Sozialausschuss und den Ausschuss der Regionen über einen strategischen Rahmen der EU für Gesundheit und Sicherheit am Arbeitsplatz, www.ec.europa.eu/social/main.jsp?catId=151&langId=de, 4-5, 2014
6. Bundesministerium für Arbeit und Soziales (BMAS), Sicherheit und Gesundheit bei der Arbeit - Berichtsjahr 2017 : Unfallverhütungsbericht Arbeit, view : 3.03.2021 von www.baua.de/dok/8809570, 28, 2018
7. M., Kittelmann, L., et al., Handbuch Gefährdungsbeurteilung : Teil 1 Grundlagen und Prozessschritte der Gefährdungsbeurteilung, BAUA,6, 2021 https://www.baua.de/DE/Angebote/Publikationen/Fachbuecher/Gefaehrdungsbeurteilung.pdf?__blob=publicationFile&v=37 doi: 10.21934/baua:fachbuch20210127
8. S. Jones, C., Kirchsteiger, W., Bjerke, The importance of near-miss reporting to further improve safety performance. *Journal of Loss Prevention in the Process Industries* **12**, 59-67(1999).

9. M.G. Gnoni, G. Lettera, Near-miss management systems: A methodological comparison, *Journal of Loss Prevention in the Process Industries* 1, doi:10.1016/j.jlp.2012.01.005 (2012),
10. International Labour Organisation : Guidelines on occupational safety and health management systems - ILO-OSH 2001. Geneva : International Labour Office. (2009)
11. J. Brauweiler, A. Zenker-Hoffmann, Arbeitsschutzmanagementsysteme nach OHSAS 18001. Grundwissen für Praktiker. Wiesbaden : Springer Gabler, 1 (2014)
12. A., Ritter, Arbeitsschutzmanagementsysteme. In : Wörterbusch Arbeitssicherheit und Gesundheitsschutz, 9 Aufl. Wiesbaden : Universum Verlagsanstalt, 1 (2009)
13. K., Knöll, P. Lugbauer, Arbeitsschutz, Arbeitsmedizin und Gefährdungsbeurteilung – Zukunftsorientierte Ausrichtung im Unternehmen. In Simmel, M. & Graßl, W. (Hrsg.), Betriebliches Gesundheitsmanagement mit System, Wiesbaden : Springer Fachmedien Wiesbaden. doi : 10.1007/978-3-658-26956-2_6, 44 (2020)
14. C., Schlick, et al., Arbeitswissenschaft (4. Auflage). Berlin : Springer. Abgerufen von http://www.springer.com/, 394 (2018).
15. J., R., Chen, Y., T., Yang, A predictive risk index for safety performance in process industries, doi: 10.1016/j.jlp.2004.03.001, *Journal of Loss Prevention in the Process Industries*, **17**, 233–242, (2004).
16. G., Nesmith, J., T., Keating, L., A., Zacharias, Investigating process safety near misses to improve performance. Process Safety Progress, doi: 10.1002/ prs. 11563 Ggg, **32**, 170–171(2013)
17. F., W., Nerdinger, G., Blickle, N., Schaper, Arbeits- und Organisationspsychologie , Springer-Lehrbuch, 4., vollständig überarbeitete Auflage. Berlin, Germany : Springer. Abgerufen von http://www.springer.com/, 544 (2019)
18. J., Wallace, Developing Effective Safety Systems; IChemE, 1995; ISBN 0852953585
19. M. Dumas, et al., Fundamentals of business process management (Computer science, Second edition). Berlin, Germany : Springer. Abgerufen von http://www.springer.com/, 23 (2018)
20. H., U., Küpper, et al., Controlling: Konzeption, Aufgaben, Instrumente (6. Aufl.). s.l. : Schäffer-Poeschel Verlag. Abgerufen von http://site.ebrary.com/lib/alltitles/docDetail.action?docID=10773156, 478 (2013).

SHS Web of Conferences **115**, 03015 (2021) https://doi.org/10.1051/shsconf/202111503015
Current Problems of the Corporate Sector 2021

Debunking the Myths of Vision and Mission Statements: A Content Analysis of FT Ranked Top 100 Business Schools

Dr Kamran Ahmed Siddiqui[1,*]

[1]College of Business Administration, Imam Abdulrahman Bin Faisal University, Saudi Arabia.

Abstract. The purpose of this paper is to answer seven famous myths about vision and mission statements by analysing the vision and mission statements of FT Ranked Top 100 Business Schools in the world. First myth states that all business schools have both vision and mission statements. Second myth says both mission and vision statements are equal in length. Third myth states that all nine components must be present in all mission statements. Fourth myth assumes faculty is the hallmark of business school's mission. Fifth myth states technology is at the centre stage of business school's mission. Sixth myth assumes research is not part of business school's mission. Finally, seventh myth says leadership is no more part of business school's mission. FT ranking for Top 100 Business Schools was selected for this study and business school's vision and mission statements were collected from their official websites. Content analysis was used as the major data analysis technique. Data analyses using descriptive statistics proved all seven myths wrong.

1 Introduction

World's top business schools [1] have similarities and differences in their focuses, roles and functions [2]. These differentiations are reflected in their strategic planning especially their crucial elements of strategic planning e.g., vision and mission statements [3][4]. Vision statements normally explains the long-term goals of the university and these statements control and guide the future practices [4]. On the other hand, mission statements reflect the reason for being an organization. These statements eventually guide the functions of the organization [4].

2 Methods

Top-100 business school were selected for this study based on FT ranking [1]. Business school's vision and mission statements were collected from their official websites. This technique of collecting vision/mission statements through official websites is quite common in earlier studies on vision/mission statements of universities [2][3][4][5], medical schools [6], law schools [7], non-profit education institutions [8] or even in business schools [9].

* Corresponding author: KASiddiqui@iau.edu.sa

This study employs Content Analysis, as it is commonly used technique in contemporary business studies. It provides opportunity to collect verbal, print and electronic data by both ways qualitatively and quantitatively [10]. Content analysis or textual analysis is "any technique of making inferences by objectively and systematically identifying specified characteristics of messages" [11]. In other words, it is "summarizing, quantitative analysis of messages that relies on scientific method (including attention to objectivity, inter subjectivity, priori design, reliability, and hypothesis testing) and is not limited as to the types of variables that may be measured or the context in which the messages are created or presented" [12]. In the literature, content analysis has been extensively used in analysing mission statements for example in different industries in USA [13]; in cross-country analyses [14]; in public libraries [15]. Even in academia, many studies on mission statement have used content analysis for example in top 100 global brands [16]; in social work school [17]; or even in engineering schools [18]. Content analysis has also been used in many studies focused on analyzing university's vision/mission statements [19][20] [21]; in FT ranked European Business Schools [22] and even in AACSB-accredited schools [23]. The framework used in this study to analyse mission statements is based on the 9 possible components and 10 characteristics of mission statements. It also involves 10 components of vision statements. This study used the components and characteristics of mission statements and components of vision statements and their operational definitions; possible explanations keywords as coding schemes taken from earlier studies [24]. This framework of nine components has extensively been used in earlier to analyse mission statements [22][16].

3 Results and Discussion

This study provides four sets of analyses on mission and vision statement including (1) Word count and Vocabulary density; (2) Frequently used keywords; (3) Major themes in mission statements; (4) Examination of components and characteristics of mission statements.

Table 1 presents a comparison of vision and mission statements based on the word and vocabulary density of mission and vision statements. This table suggest that vision statements are shorter in length (average number of words = 47) as compared to corresponding mission statement (average number of words = 63). Table 4 also suggest that vision statements have higher vocabulary density (0.32) as compared to corresponding mission statements (0.226). Vision statements having relatively higher vocabulary density indicates simpler text with words reused.

Table 1 Word count and Vocabulary density: Comparisons of Vision and Mission statements

Word Count & Vocabulary Density	Vision Statement	Mission Statement
Total Number of Words	2539	5751
Average Number of Words	47	63
Minimum Number of Words	3	9
Maximum Number of Words	200	304
Unique Number of Words	812	1299
Vocabulary Density	0.320	0.226

Table 2 provides a list of top five frequently keywords in vision and missions statements. Four keywords are common in both vision and mission statements including 'business', 'research', 'leader', and 'world'. Two keywords are distinct; 'students' appeared among vision statement and 'knowledge' appeared more in mission statements.

Table 2 Most frequently used keywords in vision and mission statements

Vision Statement	Mission Statement
Business (50)	Business (92)
Research (28)	Leaders (62)
World (26)	Research (60)
Leaders (21)	World (52)
Students (19)	Knowledge (44)

Table 3 presents components of mission statements with frequency of occurrence. It suggest that not all nine components of mission statements are fully replicated in FT Ranked Top 100 Business school's mission statements. Highest frequency is for 'Concern for public image'.

Table 3 Components of Mission Statements with Frequency

Components of Mission Statements	Frequency
Concern for Public Image	53
Products or Services	39
Markets	39
Customers	27
Concern for Survival, Growth, Profitability	15
Concern for Employees	13
Self-Concept	10
Philosophy	7
Technology	4

Table 4 presents selected characteristics of mission statements i.e., Social responsibility and Environmental responsibility. Rest of characteristics of mission statements and all components of vision statements does not show any variance. It shows a significant number of business schools are committed towards social responsibility. Interestingly 16 business schools have shown social responsibility in their vision statements. Ironically, environmental responsibility has not been reflected by any top 100 business school in their mission or vision statements.

Table 4 Selected Characteristics of Mission Statements with Frequency

Components of Mission Statements	Frequency
Socially Responsible 16	45
Environmentally responsible 0	0

https://doi.org/10.1051/shsconf/202111503015

Based on the presence of vision and/or mission statement FT ranked top 100 Business Schools classified into three clusters (Fig. 1). Cluster 1 include business schools having Mission Statements only. 47 business schools fall in this category while average rank is 43 (Table 5). Cluster 2 include business schools having both mission and vision statements. 44 business schools fall into this category and their average rank is 56 (Table 6). Cluster 3 include business schools having vision statements only. This seems to be smallest cluster as compared to other two having only 9 business schools listed in this category with an average rank is 58 (Table 7). It means business schools having only mission statement relatively higher rank as compared to other two clusters.

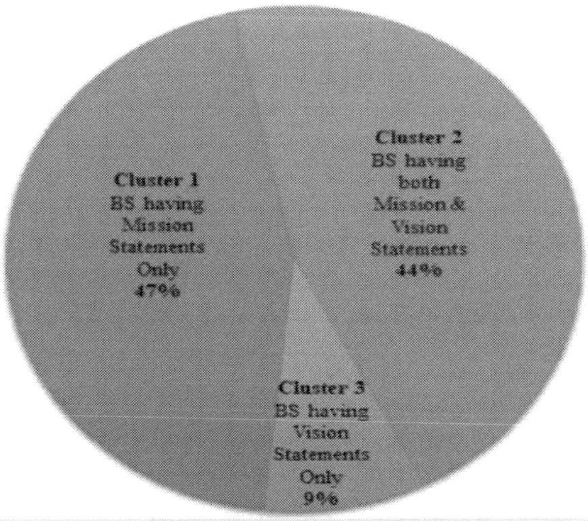

Fig. 1. Clusters of Business Schools based on Vision and Mission Statements

https://doi.org/10.1051/shsconf/202111503015

Table 5 Cluster 1 - Business Schools having Mission Statement Only

FT Rank 2020	Business School & University	Country
1	Harvard Business School	US
2	University of Pennsylvania: Wharton	US
3	Stanford Graduate School of Business	US
4	Insead	France
6	MIT: Sloan	US
7	London Business School	UK
8	Columbia Business School	US
10	University of Chicago: Booth	US
11	Northwestern University: Kellogg	US
12	University of California at Berkeley: Haas	US
14	Yale School of Management	US
16	Duke University: Fuqua	US
18	University of Virginia: Darden	US
19	University of Cambridge: Judge	UK
19	HKUST Business School	China
21	University of Oxford: Saïd	UK
22	New York University: Stern	US
24	Esade Business School	Spain
25	UCLA Anderson School of Management	US
29	SDA Bocconi School of Management	Italy
30	University of Michigan: Ross	US
31	Georgetown University: McDonough	US
34	University of Florida: Warrington	US
35	Nanyang Business School, NTU Singapore	Singapore
36	University of Southern California: Marshall	US
45	Vanderbilt University: Owen	US
47	Emory University: Goizueta	US
47	University of Washington: Foster	US
50	City Business School (Cass), University of London	UK
54	Sungkyunkwan University GSB	S Korea
57	Rice University: Jones	US
63	Singapore Management University: Lee Kong Chian	Singapore
66	Rotterdam School of Management, Erasmus University	Netherlands
70	George Washington University	US
72	Ohio State University: Fisher	US
73	University of Maryland: Smith	US
74	SMU: Cox	US
79	EMLyon Business School	France
81	ESMT Berlin	Germany
81	University of Texas at Dallas: Jindal	US
83	Brigham Young University: Marriott	US
87	Western University: Ivey	Canada
88	AGSM at UNSW Business School	Australia
88	Essec Business School	France
94	University of Toronto: Rotman	Canada
96	University of California at San Diego: Rady	US
100	University of Georgia: Terry	US

Table 6 Cluster 2 Business Schools having both Vision and Missions Statements

FT Rank 2020	Business School & University	Country
5	CEIBS	China
9	HEC Paris	France
13	IESE Business School	Spain
15	National University of Singapore Business School	Singapore
16	Dartmouth College: Tuck	US
23	Cornell University: Johnson	US
25	IMD Business School	Switzerland
27	Indian Institute of Management Bangalore	India
28	Indian School of Business	India
37	Shanghai Jiao Tong University: Antai	China
38	Renmin University of China Business School RMBS	China
39	University of North Carolina: Kenan-Flagler	US
40	Indiana University: Kelley	US
42	Indian Institute of Management Calcutta	India
43	Warwick Business School	UK
44	Washington University: Olin	US
45	Alliance Manchester Business School	UK
47	Shanghai University of Finance and Economics	China
50	CUHK Business School	China
52	IE Business School	Spain
55	Imperial College Business School	UK
56	HKU Business School	China
57	University of Notre Dame: Mendoza	US
59	Pennsylvania State University: Smeal	US
60	Babson College: Olin	US
62	Durham University Business School	UK
64	WHU – Otto Beisheim School of Management	Germany
65	University of California at Irvine: Merage	US
67	Boston University: Questrom	US
71	Mannheim Business School	Germany
74	University of Pittsburgh: Katz	US
76	University of Rochester: Simon	US
77	University of Connecticut School of Business	US
78	University of Minnesota: Carlson	US
84	The Lisbon MBA Católica | Nova	Portugal
85	Purdue University: Krannert	US
86	Texas A & M University: Mays	US
91	McGill University: Desautels	Canada
92	Wisconsin School of Business	US
93	Miami Herbert Business School	US
95	University of Edinburgh Business School	UK
97	Macquarie Business School	Australia
98	City University of Hong Kong	China
99	University College Dublin: Smurfit	Ireland

Table 7 Cluster 3 - List of Business Schools having Vision Only

FT Rank 2020	Business School & University	Country
31	Carnegie Mellon: Tepper	US
33	Fudan University School of Management	China
40	University of Texas at Austin: McCombs	US
53	Georgia Institute of Technology: Scheller	US
61	Indian Institute of Management Ahmedabad	India
68	University of St Gallen	Switzerland
69	Northeastern University: D'Amore-McKim	US
80	Melbourne Business School	Australia
88	Edhec Business School	France

Table 8 presents cluster wise information of most frequently used words. Different cluster presents almost similar patterns of most frequent words total information mentioned Table 3. Four keywords are common in both vision and mission statements including 'business', 'research', 'leader', and 'world'. Two keywords are distinct; 'students' appeared among vision statement and 'knowledge' appeared more in mission statements.

Table 8 Comparisons of Business School Clusters based on most frequent keywords

Cluster Information	Five most frequently used words	
	Vision	Mission
Cluster 1 Business Schools having Mission Statements Only Number of Business Schools = 47 Average Rank = 43		Business
		World
		Leaders
		Research
		Knowledge
Cluster 2 Business Schools having both Mission & Vision Statements Number of Business Schools = 44 Average Rank = 56	Business	Business
	Research	Research
	World	Leaders
	Leaders	World
	Students	Knowledge
Cluster 3 Business Schools having Vision Statements Only Number of Business Schools = 9 Average Rank = 58	Leaders	
	Business	
	Research	
	Students	
	World	

4 Conclusions

This paper attempts to answer seven myths about vision and mission statements by analysing the vision and mission statements of Top 100 Business Schools in the world.

Myth 1: All business schools have both vision and mission statements.

False. Based on the analysis presented earlier (Fig 1; Table 5,6 & 7) 47% business schools have only mission statements; 44% business schools have both mission and vision statements; and only 9% business schools have only vision statements.

Myth 2: Both mission and vision statements are equal in length.

False. Vision statements are shorter in length as compared to corresponding mission statement (Table 4). Vision statements have relatively simpler text with multiple words reused.

Myth 3: All nine components must be present in all mission statements.

False. Not all nine components are replicated in all FT Ranked Top 100 Business school's mission statements (Table 3).

Myth 4 Faculty is the hallmark of business school's mission.

False. Concern for faculty and staff has only 13% presence in FT Ranked Top 100 Business school's mission statements (Table 3).

Myth 5 Technology is at the centre stage of business school's mission.

False. Technology is the least mentioned item among nine components in FT Ranked Top 100 Business school's mission statements (Table 3).

Myth 6 Research is not part of business school's mission.

False. Research is one of the four most important keywords used in FT Ranked Top 100 Business school's mission statements (Table 3).

Myth 7 Leadership is no more part of business school's mission.

False. Leadership is one of the four most important keywords used in FT Ranked Top 100 Business school's mission statements (Table 3).

5 Recommendations

This paper supports recommendations provided by earlier research on the subject area [25]. Firstly, it supports the development of separate criteria to evaluate mission and vision statements for academic institutions.

It also supports the recommendation of need for uniqueness by Business Schools. Under the prevailing situation and competitive market, it is a very difficult situation for business schools to differentiate among other business schools. Now it is time for Business schools decision makers including deans, directors and other senior members to create uniqueness in their business schools, starting with uniqueness mission statements.

References

1. FT Business School Ranking, available at: http://rankings.ft.com/businessschoolrankings/rankings (2020)
2. T. Bayrak. A content analysis of top-ranked universities' mission statements from five global regions. *International Journal of Educational Development*, **72**, 102-130 (2020).
3. M. Breznik, K. M. Law. What do mission statements reveal about the values of top universities in the world?. *International Journal of Organizational Analysis*. 27, 1362-1375 (2019).

4. Taş, M. A., Çiçek, H., & Yastioğlu, S. The Future In The Context Of Mission, Vision And Goals: Case Of Mehmet Akif Ersoy University Strategic Plan Studies (2020).
5. J. Jungblut, M. Jungblut. All different? All equal? differentiation of universities' mission statements and excellence initiatives in Germany. *Science and Public Policy*, **44**, 535-545 (2017).
6. R. M. Lewkonia. The missions of medical schools: The pursuit of health in the service of society. *BMC Medical Education*, **1**, 1-7 (2012).
7. J. Organ. Missing missions: Further reflections on institutional pluralism (or its absence). *Journal of Legal Education*, **60**, 157–167 (2010).
8. E. A. Wilkerson, C. A. Evans. A content analysis of mission statements in nonprofit education. *The Journal of Nonprofit Education and Leadership*, 8 (2018).
9. E. Dedousis. An Analysis of Mission Statements of Tertiary Institutions: Business Colleges in UAE. *e-Journal of Business Education and Scholarship of Teaching*, **12**, 31-51 (2018).
10. P. Mburu, T. Matenge, D. Amanze, R. Makgosa. Corporate branding in Botswana: A content analysis of visual brand elements. *Journal of Business Theory and Practice*, **1**, 262-284 (2013).
11. David, M. E., David, F. R., & David, F. R. Mission Statement Theory And Practice: A Content Analysis And New Direction. *International Journal of Business, Marketing, & Decision Science*, **7,** 2014.
12. K. Neuendorf. The content analysis guidebook Sage Publications, Inc. In *Library of Congress. CA: United States* (2002).
13. Peyrefitte, J., David, F. R. A content analysis of the mission statements of United States firms in four industries. *International Journal of Management*, **23**, 296 (2006).
14. Bartkus, B. R., Glassman, M., McAfee, R. B. A comparison of the quality of European, Japanese and US mission statements:: A content analysis. *European Management Journal*, **22**, 393-401 (2004).
15. Crawford Barniskis, S. Deconstructing the mission: A critical content analysis of public library mission statements. *The Library Quarterly*, 86(2), 135-152 (2016).
16. Cortés Sánchez, J. D. Mission statements of Top 100 Global Brands worldwide: Text mining and visualization. *Intangible Capital*, **14**, 584-603 (2018).
17. Holosko, M. J., Winkel, M., Crandall, C., & Briggs, H. A content analysis of mission statements of our top 50 schools of social work. *Journal of Social Work Education*, 51, 222-236 (2015).
18. Creamer, E. G., & Ghoston, M. Using a mixed methods content analysis to analyze mission statements from colleges of engineering. *Journal of Mixed Methods Research*, **7**, 110-120 (2013).
19. M. Hladchenko. The organizational identity of Ukrainian universities as claimed through their mission statements. *Tertiary Education and Management*, 22(4), 376-389 (2016).
20. Deus, R. M., Battistelle, R. A. G., & da Silva, G. H. R.. Sustainability insights from the mission statements of leading Brazilian Universities. *International Journal of Educational Management,* (2016).
21. J. R. Warchal, A. I. Ruiz, D. You. Do Psychology Department Mission Statements Reflect the American Psychological Association Undergraduate Learning Goals?. *Psychology Learning & Teaching*, 16, 369-374 (2017).
22. G. Mazurek, P. Korzynski, A. Gorska, A. Palyga. Mission statements in FT ranked European Top 100 Global Brands-a content analysis. *European Research Studies Journal*, XXIII (1), 639-649 (2020).
23. B. Orwig, R. Z. Finney. Analysis of the mission statements of AACSB-accredited schools. *Competitiveness review: An International Business Journal*, (2007).

24. K. Falah, K. Siddiqui, Analyzing Mission Statements of Universities in Saudi Arabia, Academy of Strategic Management Journal, **20**, 1-11 (2021)
25. F. David, & F.R. David. *Strategic management: A competitive advantage approach, concepts and cases*. Florence: Pearson–Prentice Hall (2016).

Perspectives, implementability and measurability of ethical principles in management

Georg Sievers M.Sc.[1*]
[1] GMS Wangen, Säntisweg 10, 88239 Wangen, Germany

Abstract. Ethical behaviour in companies is built on principles, rules, values and resources. As integrative ethics, the macro level ensures a long-term and sustainable orientation of the company in the context of business ethics. The macro level ensures that the company can change its role from "expectant" to "creator" with regard to corporate ethics. The mezzo level prevents unethical behaviour in companies by offering them the opportunity to reflect on their own behaviour in the context of (government) guidelines. Finally, the micro level is considered essential. It creates the systemic preconditions for companies to be able to fulfil their ethical duties and take responsibility for their actions. Partnerships and networks play a central role in this context. Against this background, the existence of a regulatory framework, the measurement of ethical corporate behaviour with the help of standards and reporting, the existence of a sustainable corporate mission statement as well as the availability of necessary resources and the documentation of sustainable behaviour in the company can be derived from the literature as central success factors for the implementation of ethical standards in companies. These factors are compared by means of empirical research with a focus on the packaging industry in Germany, Switzerland and Austria so that the existence of a guiding regulatory framework in society and a sustainable corporate mission statement emerge as the final and particularly relevant characteristics for an ethically-sustainable company, the referencing of one's own behaviour to sustainability indices, the use of the same (in particular the Dow Jones Sustainability (DJSI) and the Financial Times Stock Exchange (FTSE4 Good) indices) in reporting, the reference to the Sustainable Development Goals (SDGs) in the formulation of the corporate objective and the use of voluntary standards in sustainability reporting.

1 Introduction

Attributes of ethical behaviour in business are built on principles, rules, values and resources. The concept of principles becomes relevant when ethical thinking takes hold in such a way that morality becomes contentious, i.e. when discussions about "right" and "wrong" as well as "good" and "evil" arise. In this context, a "morally right action" is understood to be a qualitatively "right" action, which can then also be described as "ethically correct" [1, 2].

Principles then ensure that it is possible to classify what is "right" or "wrong" or "good" or "bad". These principles are codified within the company within the framework of (corporate) mission statements and thus develop into rules or corporate values. The common support of cultural and ethical values should lead to increased integration of the individual, improved coordination of the most diverse areas of the company, increased motivation of all those involved and thus ideally also to an increase and improvement of the company's success [1, 3]. Finally, resources are necessary for implementation, as only if these are available is it possible to maintain competitiveness and competitiveness in the company. Resources thus present themselves as a prerequisite and a basis for integrative ethics which can be understood as a union between ethics and business at the micro, mezzo and macro levels. The "macro level" comprises the state, the government and the legislator, the "mezzo level" comprises companies, organisations and institutions, and the "micro level" comprises the acting individuals in organisations. With regard to the "macro level", the institutions then show themselves either in the form of the

* Corresponding author: g.sievers@web.de

separation of powers (legislature, executive, judiciary and controlling) as well as in political coinages (fiscal, tax, social, monetary and foreign trade policy). This separation into micro, mezzo and macro levels is also referred to in the literature as the "subject matter of business ethics approach" [4]. Business ethics is thus based on the three levels mentioned above, which can serve as a basis for establishing ethical structures in companies and thus aim to reduce gaps between corporate behaviour and the documentation of sustainable behaviour in companies [5]. In addition to taking into account the three levels in companies, this can also be done in particular through cooperation between companies and the state [6]. The "level approach" - as it is abbreviated - assumes that business ethics is reflected in three social dimensions in the company [4], all of which are equally relevant for the implementation of business ethics, whereby there is a high degree of freedom in the design of the three levels and interdependencies do not necessarily exist [6]. Accordingly, the three levels or dimensions provide for the offer of legal framework conditions, create the basis for private ownership, entrepreneurial freedom and fair competition at the macro level [7, 8] and at the same time prevent unethical behaviour through incentive and sanction structures [4].

Against this background, it is important to question how it can be possible to operationally develop the "level approach" in the corporate context in order to exist and be perceived as an ethically-sustainable company.

2 The state of the problem in the literature

The macro level ensures a long-term and sustainable orientation of the company in the context of business ethics. The macro level ensures that the company can change its role from "expectant" to "creator" with regard to corporate ethics [2].

The mezzo level prevents companies from behaving unethically by offering them the opportunity to reflect on their own behaviour in the context of (state) requirements. Finally, the micro level is considered essential. It creates the systemic preconditions for companies to be able to fulfil their ethical duties and take responsibility for their actions. Partnerships and networks play a central role in this context [4].

The literature is based on the following assumption: If companies take ethical considerations into account at the micro level, as measured by relevant sustainability indices and reporting standards for sustainability, they can have an "ethics-improving" effect at the mezzo level in cooperation with other companies, organisations and institutions, and positive effects of their behaviour are ultimately reflected at the macro level in the form of macroeconomic factors [4].

At the macro level, macroeconomic factors can be useful in recognising whether states, governments or legislative institutions are implementing an "integrative ethic" [9].

Thus, on the macro level, the company and the state are addressed as two of three institutions that can be described as "places of morality". Through virtues and the adherence to values and norms, moral behaviour can be demonstrated and ultimately result in ethical behaviour [10,11] which can be seen in concrete terms in the national economy in the offer of relevant products and services that meet sustainable criteria. At the mezzo level, relevant requirements in the organisational process of the company ensure that a company behaves "correctly" - in the sense of "ethically correct". The Sustainable Development Goals (SDGs) serve as guidelines in setting goals for companies and can be operationalised in (sustainability) reporting through relevant standards such as those of the Global Reporting Initiative (GRI), the OECD or the UN Global Compact [12]. Finally, at the micro level, the focus is on the question of which microeconomic factors should be taken into account for the integration of issues into the corporate context or by management. The "Environmental Social Governance" (ESG) standards or the requirements of the "German Sustainability Code" (DNK) for reporting can be mentioned here as examples. The "Environmental Social Governance" (ESG) concept is intended to ensure the protection of the environment, compliance with social standards and good and sustainable corporate governance [13]. Through corporate principles, behavioural guidelines and a so-called "Code of Conduct" as well as orientation towards a vision (goal), a mission (self-image) and an orientation towards action (values) in the company, the standards can then be guidelines for the actions of each individual and ensure a positive appreciation of the actions of those concerned which can be achieved through orientation towards or the use of relevant market indices. the use of relevant market indices such as the German Environmental Social Governance Index (DAX 50 ESG), the Dow Jones Sustainability Index (DJSI), the Financial Times Stock Exchange (FTSE4Good) and the STOXX SI Index [11].

Taking these levels into account and considering the interrelationships, it can ultimately be possible to positively accompany a change that is permanently present in the company and to address the needs of the various stakeholders equally [14]. Business ethics examines whether and to what extent moral standards can be integrated in corporate policy and in the behaviour of companies in order to finally achieve the application of a moral code of conduct to all economic levels within the framework of economic ethics [10]. The challenge for companies in the context of (alternative) ethical behaviour is to integrate the (necessary) self-interest or the pursuit of profit and the pursuit of ethical values in corporate policy which can be done by taking into account the "Golden Rule" - i.e. the reciprocity of human action - as well as the "Practical Syllogism" in the form of a concept of responsibility that can be understood as "ethical economics".[15] defines "ethical economics" as a "theory of the integration of economics and ethics".

From this, the literature can define the "ethical minimum" in questions of accounting and in questions of corporate organisation from the perspective of the stakeholders as framework conditions that can be understood as the consideration of ethical requirements in the company to ensure economic and legal capacity to act [16]. Divided into the three levels of analysis, activities in fiscal, tax, social, monetary and foreign trade policy can be identified for the macro view, following [16, 17, 18]. For the mezzo view, requirements within the framework of a corporate culture or through laws, regulations and guidelines can be mentioned and, for the micro view, it applies that the individual or the company perceives different roles in the real economic world (e.g. as entrepreneur, politician, customer or owner) and shows ethically "correct" (sustainable) behaviour here. From the literature, the following aspects can be named as the first relevant success factors for the formulation, operationalisation and implementation of ethical standards: existence of a sustainable corporate mission statement, existence of entrepreneurial resources, existence of a guiding regulatory framework, documentation of sustainable behaviour in the company as well as behaviour analogous to and measurement of entrepreneurial behaviour on indices and with the help of reporting standards [19, 20].

3 Research design

On the basis of comprehensive literature research, the status quo can first be ascertained within the frame-work of research on the topic under investigation. This literature research and the literature analysis based on it are oriented towards the key words sustainability, corporate ethics, ethical minimum, reporting, corporate structure as well as reporting/standards and the corresponding English counterparts. Based on this, an empirical vali-dation of the findings from the literature research is carried out. An online survey with the aim of ascertaining success factors for the implementation of ethical standards is used to ascertain the mezzo level, an expert sur-vey in the form of guided expert interviews with the aim of supplementing, modifying and excluding success factors from the online survey and literature research is used to ascertain the micro level [21], and an analysis of business and sustainability reports with the aim of ascertaining relevant macroeconomic indicators in the context of ethics and sustainability is used to ascertain the macro level. As a result, an (overall) index is created from these three empirical approaches, i.e. success factors that are relevant for ethically-sustainable behaviour of a company are derived in comparison with the results from the literature research. The result is a model for an ethically responsible company and recommendations for action are derived for companies - depending on the actual situation of the company. The validation of the model or the derivation of the weighting of the factors is carried out with the help of a logistic regression of the categories (variables) and a "receiver operating characteristics curve" (ROC) analysis is used to determine the cut-off value in order to derive degrees of fulfilment between companies with the highest possible selectivity.

4 Results of the research

4.1 Results of the online survey

A total of 136 participants took part in the online survey and completed it, 258 participants were approached - this represents a response rate of 52.71 percent. 4 participants were contacted via LinkedIn, 51 via Facebook, 7 participants via XING and 1 participant via Twitter. 67.5 percent of the respondents said that the company they represent comes from Germany, 21.7 percent from Austria, 7.2 percent from Switzerland and 2.4 percent from France (indicating the company's headquarters). The majority of respondents (29.7 percent) are located at the level of "Chief Executive Officers" (CEOs). With regard to the Sustainable Development Goals (SDGs), there is a "high" use/awareness of the SDGs with a focus on "Clean Water and Sanitation" (46.34 percent), "Industry, Innovation and Infrastructure" (42.86 percent) and "Peace, Justice and Strong Institutions" (48.15 percent). In terms of the effectiveness of the standards with regard to the public (stakeholders, shareholders) as well as with regard to the internal impact (optimisation of sustainable processes and structures, offer of sustainable products and services), the respondents rate the standards differently with a maximum value of five. The best values are given to voluntary standards, the "Environmental Social Governance" (ESG) indicators and the standards of the "Organisation for Economic Co-Operation and Development" (OECD standards). With regard to the implementation of sustainable or ethical orientation in companies, these focus in their implementation on the areas of processes, products and management. The focus is on clearly defined and standardised processes with a focus on sustainable products and a commitment to sustainability on the part of the management [19].

4.2 Results of the expert survey

Within the framework of the expert interviews, 8 experts - including one group interview with 3 experts - were interviewed. The interview guide focused on the question of how integrative ethics can be realised in the company and how implementation can be positively accompanied in the company. Here, the focus was on indica-tors, reporting standards and macroeconomic indicators, e.g. in the form of key business figures. The experts come from the packaging/sustainability context and are active there in company management or as company advisors. As a result, the experts formulate guiding principles in the company which include a statement on the working conditions, the health burdens at the workplace, the working hours and questions about the heaviness of the work as relevant factors for an ethical or sustainable orientation in the company. In second place is the existence of a so-called "Code of Conduct" in the company, which contains corporate values (especially integrity, fairness, networking) and is the basis for a corporate philosophy. Finally, the use of standards in sustainability reporting (especially standards of the International Standards Organisation (ISO 26000) and the Global Re-porting Initiative (GRI)) is mentioned in third place, CO2 emissions) and focus on business parameters (social commitment expenditures, equity ratio, investment ratio, return on sales, fluctuation rate) as well as employee satisfaction), thus enabling the use of indices ("Deutscher Aktienindex" (DAX) and "Dow Jones Sustainability Index" (DJSI)) in reporting.

4.3 Results of the analysis of the annual and sustainability reports

The annual and sustainability reports as well as - if necessary - supplemented publications of the companies were analysed with regard to sustainability indices used/addressed, formulated goals for sustainable development in the company based on the "Sustainable Development Goals" (SDGs) and addressing sustainable standards in reporting. The reports of Covestro AG, BASF, Siemens AG, Krones AG, the Tetra Laval Group, Syntegon Technology (formerly Bosch Packaging), KHS GmbH - which belongs to the Salzgitter Group -, KUKA AG, Gerresheimer AG, Gurit AG, Lanxess AG, OMV AG and Semperit AG were examined. A "best-in-class approach" was chosen here with regard to the "top 10 companies" in the packaging sector (wood, paper, glass, metal, other). In a first step, the analysis shows that the majority of the companies refer to the Global Reporting Initiative (GRI) or the United Nations (UN) Global Compact in their reports. Sustainability indices are referenced comparatively little in the reports. A second step shows that only about half of the companies explicitly refer to the Sustainable Development Goals (SDGs) in their reports. These are then also reported with a view to the key figures offered. In some cases (e.g. Siemens AG), the companies do refer to these indicators, but there is no concrete statement on fulfilment. This is only described verbally. Only two of the researched companies (Gerresheimer AG and Gurit AG) do not refer to the targets.

4.4 Consolidation of the results: basis for the overall index formation

Building on the success factors identified up to this point, the following success factors can serve as a basis - expanded by the findings from the analysis of the annual/sustainability reports (cf. Chapter 4.3.3) - for the formation of the overall index in the stated order of priority: existence of a guiding regulatory framework in society, existence of a sustainable corporate mission statement ("Code of Conduct") as a manifestation of a corporate philosophy, referencing of one's own behaviour to sustainability indices ("Dow Jones Sustainability Index" (DJSI), "Financial Times Stock Exchange"(FTSE4 Good)-Index), reference to selected "Sustainable Development Goals" (SGDs) ("Clean Water and Sanitation", "Industry, Innovation and Infrastructure", "Peace, Justice and Strong Institutions") in the formulation of corporate goals, use of voluntary standards in reporting, existence of corporate resources, focus in implementation on the areas of processes, products and management as well as internal documentation of sustainable behaviour in the company.

4.5 Logistic regression of variables and ROC analysis

Finally, the weighting of the individual categories is derived with the help of a logistic regression of the variables. The aim is to make a distinction between companies that fulfil the requirements and those that do not. Based on this, a receiver operating characteristics curve (ROC) analysis is used to determine the cut-off value. The result is a model that shows how strongly the variables are expressed in the company. In other words: the higher the value of the validated values, the more the company fulfils the requirements for ethical-sustainable behaviour in the company. In this way, the probabilities can be read whether the company will show ethical-sustainable behaviour or not on the basis of the recorded values of the variables. The following variables were given a final weighting: existence of a guiding regulatory framework, existence of a sustainable corporate mission statement, referencing of own behaviour to sustainability indications, reference to the Sustainable Development Goals (SDGs) as a corporate goal, use of voluntary standards for (sustainability) reporting, existence of corporate resources, focus in implementation on the areas of processes/products/management and internal documentation of sustainable behaviour in the company.

5 Research recommendations and discussion

5.1 Implications for theory

It became clear in the course of the investigation that there is comprehensive research regarding ethical issues, ethical issues within companies and also issues relating to sustainability. There is however a lack of a consistent connection of these deliberations which goes beyond merely applying corporate ethics. Connection means here that companies are offered concrete recommendations for action - based on the starting point - in a manner that implementation can take place. Here this thesis can close a gap by (a) providing, through the link between for methodological approaches (literature research, online questionnaire, expert interviews, analysis of business/sustainability reports) a holistic approach to identify the relevant factors in a corporate context and (b) subjecting these variables to an assessment in practice/making this possible. In this regard the following applies:

- a consideration of the ethical and sustainable objectives in the company with regard to the company reputation appears sensible. However, it is not a matter of orienting measurement of fulfilment to the company indicators but to the relevant factors for success.
- ensuring a division of these factors for success as much as possible which requires a validation of the variables with the help of empirical research methods.
- the opportunities to transfer in a beneficial manner ethical values in the company are to be found fundamentally in taking on board these factors for success, in the conduct of management (top management) (role model function of top management), the integration of ethics management into the strategic management of a company and transparent actions on the part of staff, shareholders and stakeholders.

The key question in the course of this thesis was: How can, based on key ethical values, implementation of ethical principles be possible in commercial management? This question can be answered as follows: the existence of a company model which contains company objectives and company values provides the initial basis for the company activity but also for the successful application of ethical principles in management. Based on this the following variables can apply as factors for success in order to make possible the objective of implementation of ethical principles in management:

- the existence of a guiding framework of order
- conduct analogue and measurement of corporate behaviour using indices and with the help of reporting standards
- the existence of corporate resources
- documentation of sustainable conduct in the company

Regarding the guiding framework what is relevant is that the actions of organisations/companies or individuals in companies/organisations are frequently regarded as an end in itself and not as a necessary part of a comprehensive code of conduct. In this regard one has to ascertain that first of all each individual is responsible in terms of his abilities for the shaping and securing of his life and for the shaping of his immediate environment and therefore also for his conduct in an organisation. Ultimately one cannot, however, dispense with a code of conduct as this provides the framework for conduct which corresponds to company guidelines. The existence of this code of conduct is the precondition that the invisible hand of the market leads to maximum well-being for all those concerned and balance in markets. Only through values, norms and conventions which are formed, according to Smith, in markets/society, then manifesting themselves in a code of conduct will a framework of orientation be formed – which should be understood as a framework of order – for all actions of companies, organisations and individuals. Thus, an unjust distribution of goods is prevented and stable imbalances are excluded in the long term as expressed in the moral philosophy of Smith whose morality is based on values such as sympathy and the existence of a moral society. However, it is harder to go down this path if the number of those individuals affected is higher and society becomes an anonymous large society without neighbourhood warmth and this leads to a loss of subjective conduct security. As a consequence, the influences differ for each individual increasingly and societal conduct control fails.

In this manner a society can develop increasingly into a route towards slavery. Only through close communication with society can societal conflicts be resolved which brings with it again the necessity for values to exist which society despite all the anonymity which has arisen can still constitute a basis. Whether this has to be crowned with success however may be quite questionable because then economic policy would fundamentally be laying down the law and the division of power on the part of the state has to be questioned as a consequence and the extent to which large, anonymous societies make possible an influencing of individuals at all is debatable.

However, in the age of societal change and increasing intercultural opening of society these attempts to create a framework of values appear now appear insufficient. Aspects of alienisation which attach themselves to globalisation, digitalisation, individualisation etc. make for a framework of order which contains sanctions which then determines again the actions of the individual.

For the usage/assessment of conduct in companies using indices and the application of reporting standards one can state that these are to be understood as a means to an end/as a toolkit in order to make possible the operational implementation of standards in companies and ensure their realisation.

The availability of resources is part of this as without these realisation is not possible. The same applies to the documentation of implementation as the corresponding monitoring structures can only take place through such documentation.

Building upon this the following hypothesis was devised: Fundamentally ethics and economy are a contradiction but a world without ethics is inconceivable and the pursuit of economic goals of efficiency and effectiveness will only be possible together with the pursuit of ethical objectives. As a justification for this the following can be listed:

- In the course of the literature research, it became clear that through the existence of a guiding framework of order in society and the corresponding implementation in companies through mission statements, visions, missions and values a culture can be created which not only makes possible ethically oriented company conduct but which even promotes it.
- The then implemented ethical – sustainable direction in the company promotes the corporate image and reputation and makes for a higher level of trust on the part of consumers.
- Thus, not only societal added value is to be recognised through ethically sustainable conduct in the company but consumers may also prefer such companies in their selection choices.

5.2 Implications for corporate practice

For corporate practice what applies to companies is that they should orient their actions to the factors of success [22, 23]:

- the existence of a sustainable company mission statement,
- referencing their own conduct to sustainability indices (Dow Jones Sustainability Index, FTSE4GOOD Index, other index),
- reference to the SDGs as corporate objectives (Clean Water and Sanitation, Industry/Innovation and Infra-structure, Peace/Justice and Strong Institutions, other),
- the usage of voluntary standards for (sustainability) reporting (GRI, ISO26000, UN Global Compact),
- the existence of corporate resources,
- recognising which criteria are relevant for measuring sustainability and can thus be included in the assessment of rating agencies, for example, focusing in implementation on the areas processes, products and management and
- internal documentation of sustainable conduct in the company.

The more this occurs the more ethically sustainable conduct in the company is both perceived and realisable and the chain of actions assumed in the hypothesis will take place.

Based on a comprehensive literature research on ethical-sustainable behaviour in companies, a set of variables and indicators was derived by validating these findings with the help of empirical methods (online survey, expert survey, analysis of business and sustainability reports) which serve as a basis for corporate management. Signals are set which, when perceived and aligned with business policy, will enable the demand for ethics and sustainability to be realised in the company in a targeted manner. It has become apparent in the course of the study that this possibility cannot be regarded as given in the research presented in the literature to date. Although there are extensive considerations in the literature on ethics and sustainability in the corporate context and the combination of these into an integrative ethics is also being discussed there have been no approaches to date. However, there has been a lack of approaches that have derived concrete variables in order to submit maxims for action for companies. The formulation of recommendations for action is still open. In a next research step, the identified and weighted variables will be separated into the sectors "red", "yellow" and "green", which should represent the need for action for the companies. Differentiated recommendations for action are then derived.

6 Conclusion

Ethical conduct in companies is based on principles, rules, values and resources. The macro- level ensures in the form of integrative ethics long-term, sustainable direction for the company in the context of corporate ethics. In this manner the macro-level ensures that the company can change its role from expectant to creator with respect to corporate ethics. The mezzo-level prevents unethical conduct on the part of companies by providing the opportunity to reflect upon self-conduct in the context of (state) instructions. The micro-level applies ultimately as an essential level. It creates the systemic preconditions such that companies can meet their ethical obligations and take responsibility for their actions. Partnerships and networks play in this context a key role. Given this background from literature central success factors can be identified for the implementation of ethical standards in companies: the existence of a framework of order, measuring ethical – corporate conduct with the help of standards and through reporting, the existence of a sustainable company mission statement and the existence of the necessary resources and documentation of sustainable conduct in companies. These factors are replicated through empirical research with a focus on the packaging industry in Germany, Switzerland and Austria such that ultimately and in particular relevant features for sustainable, ethical companies the following are shown: the existing of a guiding framework in society, a sustainable company mission statement, referencing self-conduct to sustainability indices, the usage of the same in reporting, reference to the SDGs in the formulation of the objective of the company and the usage of voluntary standards in sustainability reporting. If then a company presents itself as relatively well-developed in its implementation, then it is a matter of communicating the successes in order to achieve further positive effects in terms of reputation and image and, in addition, to intensify the application of methods not yet exploited sufficiently intensively. However, should a company demonstrate a backlog in the implementation of ethical – sustainable standards then it is a matter of identifying through a comprehensive analysis of the current state the fundamental needs for action, to prioritise these and then to implement them.

References

1. E. Göbel. Unternehmenspolitik – Grundlagen und praktische Umsetzung, 3rd edition, Konstanz: UVK (2003).
2. D. Smrekova. Morals, ethics and applied ethics. *Filozofia*, **52**, 149-160 (1997).
3. S. Ozen. Ethic, the Theory of Ethics, the Ethics of Science, and Work, 8131-148 (2015).
4. A. Remišová, A. Lašáková. Unethical practices in the Slovak business environment – Entrepreneurs vs. State? *Ethics & Bioethics in Central Europe*, **10**, 78-95 (2020).
5. J.L. Fernández, J. Camacho. Effective elements to establish an ethical infrastructure – An exploratory study of SMEs in Madrid region. Journal of Business Ethics, 138, 113-131 (2016).
6. K.E. Goodpaster. Business Ethics. In: Becker, L./Becker, C. (Ed.), Encyclopedia of Ethics. New York: Garland, 111-115 (1992).
7. P. Ulrich. Integrative Wirtschaftsethik – Grundlagen einer lebensdienlichen Ökonomie. Berlin/Stuttgart/Wien: Haupt (1997).
8. E. Beauchemin, L.P. Cote, M-J. Drolet. Conceptualising Issues in the Conduct of Research – Results form a Critical and Systemic Literature Review. *Journal of Academic Ethics*, **4**, 20201 (2021).
9. J.A. Marceta. Respect the Author – a Research Ethical Principle for Readers, Journal of Academic Ethics, 19, 175-185 (2021).
10. M.G. Velasquez. Business Ethics – Concept and Cases, 6th edition, New York: Prentice Hall (2005).
11. C. Nietsch-Hach. Ethisches Verhalten in der modernen Wirtschaftswelt, 2nd edition, Konstanz: UVK (2016).
12. D. Brockhoff, G. Engelhardt, H. Yabroudi, L. Karg, A. Aschenbrenner, C. Felber. Publizitätspflicht zur Nachhaltigkeit – Entwicklung eines Anforderungskataloges für einen universellen Standard (PuNa-Studie), Potsdam: Institute für Advanced Sustainability Studies (IASS) (2020).
13. F. Knecht, S. Reich. Wertschöpfungsketten – ESG als kritischer Erfolgsfaktoren für das Management des gesamten Lebenszyklus. In: S. Bergius/T. Schulz (Hrsg.), CSR und Finance – Beitrag und Rolle des CFO für eine Nachhaltige Unternehmensführung. Berlin/Heidelberg: Springer, 373-389 (2014).
14. U.P. Kanning, T. Staufenbiel. Organisationspsychologie. Göttingen: Hogrefe (2012)
15. P. Koslowski. Ethik der Banken. München: Fink, (2009).
16. L.K. Trevino, K.A. Nelson. Managing Business Ethics – Straight Talk About How To Do It Right, 5th edition, London: John Wiley & Sons (2010).
17. S.O. Hansson. How to Perform an Ethical Risk Analysis (eRA), *Risk Analysis – An International Journal*, **38**, 1820-1829 (2018).
18. R.R. Kline. Engineering Case Studies – Bridging Micro and Macro Ethics, *IEEE Technology and Society Magazine*, **29**, 16-19 (2010).
19. . Van der Wal, M.A. Demircioglu. More ethical, more innovative? The effects of ethical culture and ethical leadership on realized innovation. *Australian Journal of Public Administration*, **79**, 386-404 (2020).
20. S.O. Hansson. The Ethics of Doing Ethics. *Science and Engineering Ethics*, **23**, 105-120 (2017).
21. S. Thompson. The Micro-Ethics of Everyday Life Ethics. *Cultural Studies*, **26**, 895-921 (2012).
22. C. Mackenzie. Ethical auditing and ethical knowledge. *Journal of Business Ethics*, **17**, 1395-1402 (1998).
23. M. Ashby. Ethical Regulators and Super-Ethical Systems, Preprint, 8 (2020).

SELECTED APPROACHES OF THE BEHAVIORAL ECONOMICS IN POST PANDEMIC MANAGEMENT

Norbert Súkeník[1*], *Nadežda* Jankelová[1]

[1] University of Economics in Bratislava, Faculty of Business Management, Management Department, Dolnozemská cesta 1, Bratislava, Slovakia

Abstract. Changes in organizational behavior, decision-making processes, human thinking and action are the subject of an exploration of the increasingly popular behavioral economy. We assume that her knowledge gained from various economic or psychological experiments in recent decades can help managers understand the specifics of human behavior and action. The Covid crisis and the pitfalls it brings pose new challenges for managers. Knowledge of behavioral economics and descriptive approaches to decision making allows us to understand how people act in real conditions. This knowledge can help managers streamline management and become better leaders. The paper deals with the benefits of behavioral economics for managers in the process of "reopening" the economy and its main goal is to highlight the knowledge and solutions of behavioral economics, usable for postpandemic management. To meet the goal, it is necessary to describe the changes and new specifics of the environment affected by the pandemic crisis in the first, theoretical part of the work. After analyzing these changes and evaluating them, we look for the answers offered by behavioral economics in the final part of the paper. Based on the empirically obtained knowledge of mainly foreign authors, we present several examples of their practical application in the newly created management environment.

1 Introduction

The COVID-19 pandemic has brought many changes. The personal and professional lives of most people on Earth have begun to transform dramatically. The oftenly involuntary transition to "homeoffice" brought about by the pandemic was the biggest shock to many organizations. If they did not want to forcibly suspend or terminate their activities, they had to reconsider their organizational habits and adapt them to the new conditions. Until then, teleworking was the domain of multinational companies and "freelancers". The advent of the pandemic and its associated constraints have created pressure that has led to a global change in the organization of work. Companies that have so far avoided digitalization have had to start building a technological infrastructure adapted to operate "online". Only the data that is still being collected will show how effectively

* Corresponding author: norbert.sukenik@euba.sk

companies have dealt with this obstacle. What is certain, however, is that almost all organizations have made significant changes. We are not yet able to predict which strategies and tactics will prove to be the most effective. Some changes will be permanent and some will gradually disappear. It is how to set up processes for the future, how to learn from a pandemic crisis and how to reopen and start a business is becoming the most pressing question today. Company managers must think strategically, with a view to the future.

The problems that come with "reopening" will require quality solutions. We do not know for sure to what extent it was possible to predict the beginning of the Covid crisis, but it is very likely that it will end once. Due to the high frequency of vaccinations, some countries are rapidly reaching or overcoming the inflection point, and the issue of reopening is becoming more and more relevant every day. We cannot forget that every problem is both a challenge and often a new opportunity to succeed.

The organization is based on the people who make it up. The more companies relied on technology to move to distance forms of work, the more they would have to rely on their people to return to some of the original patterns of work organization. The same employees may be physically returning to workplaces, but each of them has experienced the consequences of the pandemic spread of a dangerous virus in the past. In terms of their behavior, significant changes could have occurred. It is the manager's job to be able to respond to these changes.

2 Theoretical Background

2.1 The impact of a pandemic on the social climate and the behavior of individuals

The COVID-19 pandemic has so far infected more than 140,000,000 people, resulting in more than 3,000,000 deaths [1]. The relatively high infectivity and mortality rates for COVID-19 have forced countries to adopt new, unique approaches to preventing, suppressing and reducing the spread of the virus. The strategies used to the greatest extent are to wear protective face masks [2], monitor contacts [3] and shut down the economy [4,5].

One of the most controversial of these measures is economic shutdowns. These aim to reduce the risk of the virus spreading, especially through the air in workplaces where employees are in direct contact with each other or with the customer [4,6]. During an economic downtime, some employees may still work from home, especially in office positions. However, it is not possible to provide work from home for a large group of workers, especially for manual positions. While wearing a protective mask or monitoring contacts causes minimal economic damage [3], economic downtime causes major losses and leads to the greatest global recession since the Great Depression [7]. These losses cause the greatest harm to people working in the labor professions, as they are people from low-income groups and are therefore the least likely to have sufficient financial reserves [8-10]. This can put pressure on people and lead more often to fault behaviour.

Many scientists attribute the success of some countries in fighting the corona virus to a strategy of shutting down the economy, and it is undeniable that economic downtime is actually saving lives. However, they also result in economic losses which, through the ripple effect, cause other harmful consequences, such as a deterioration in the mental health of the population [5,6]. For this reason, during the COVID-19 pandemic, people developed strong attitudes towards economic shutdowns, so in the press, on social media and various professional or less professional discussion forums we can see strong arguments for and against economic shutdowns [5,11,12]. These attitudes collectively advance political decisions, but also encourage individual action. Lockdowns are most effective when all are governed by government regulations, but differing attitudes to the subject have led many individuals and organizations to disobey the order to suspend economic activity [13,14]. Disobedience ranges from disrespecting work from home to widespread public protests. Regardless of the belief in the appropriateness of economic shutdowns, they are exponentially less effective when people refuse to follow guidelines. Disobedience is only one consequence of the pandemic crisis, but it is also an illustrative example of how COVID-19 leads to behavioral changes that managers will have to cope with during reopening. It is the attitudes and behavior of employees that are studied by behavioral economics.

Undesirable behavioral abnormalities are likely to occur much more frequently after reopening in the workplace than before a pandemic broke out. It is the manager's job to monitor these deviations and to be prepared to take concrete steps to ensure the required behavior and safe return of the business to full operation.

2.2 Managerial challenges in the workplace after reopening

"From time to time, organizational environments undergo cataclysmic upheavals - changes that are so sudden and far-reaching that they change the trajectories of entire industries, overcome the adaptive capacities of resilient organizations, and transcend the understanding of seasoned managers." [15]

More efficient management and use of business resources in the post-pandemic era may require organizational hybridity, which involves the integration of different strategies, logics, and structural forms (eg, structural flexibility) to handle complex problems. In addition, companies need flexibility to effectively design and implement strategic changes, which requires fluidity, speed and setup that fosters innovative thinking and resilience [16]. This effort depends significantly on the dynamic capabilities of company managers [17]. And although senior executives have similar levels of cognitive ability [18], the use of new dynamic capabilities also requires the use of the cognitive and emotional skills of the senior management team and other critical stakeholder executives inside and outside the company [19]. The first challenge for managers will be to make effective use of dynamic managerial skills.

A changing workplace is a natural response to a pandemic crisis. Post-pandemic growth and changes in the labor market will inevitably lead to a change in some work processes. Technological innovations, such as increased informatization and automation, have reduced the need for manual work. This leads to a greater interest in artificial intelligence with a transformative effect on business [20]. There are currently a number of self-employed and globally mobile micro-enterprises that can move rapidly on demand [21]. Location-independent digital nomads are emerging who can work from anywhere. It also changes traditional employment structures, and creates pressure to be more flexible and based on the needs of their workers. The scale of this challenge has been influenced by the need to stay in one place due to COVID-19 measures, however, the lack of innovative action to adapt the work environment in post-pandemic growth can break employers. The second challenge is to respond to the demands of workers, not only in the context of a changing workplace, but the organization of work as such.

Changing the business environment requires an appropriate strategy for recognizing opportunities. Opportunities can be described as *"a stream of ever-evolving ideas driven and shaped by human social interactions, creative knowledge, and action at all times"* [22]. In order for managers to succeed in the newly created dynamic environment, they must seize opportunities. This means focusing on employees' new social needs and using individual creativity to help employees struggle to cope with the COVID-19 crisis. According to Dimov [22], individual creativity consists of: (1) personality, (2) inner motivation, (3) knowledge and cognitive abilities. Personality refers to the general dispositions of a person. Mostly, more creative people have a personality associated with being open to new ideas. This allows them to focus on new ways of thinking [20]. Creative managers have a natural curiosity in nature that allows them to delve deeper into new problems. They are prone to innovation and risk. Internal motivation is one of the reasons why some individuals may recognize certain opportunities better than others. The need for success can be a motivating factor that pushes individuals into creative activities. Knowledge and cognitive abilities relate to the information available to individuals and their ability to use it [20]. We see the third challenge in creative thinking and the willingness to bear the tolerable level of risk that innovative changes bring.

The reopening of workplaces thus brings challenges that will lead to "physical" and technological changes such as ensuring safety in the workplace, the introduction of new digital technologies, test stations and so. From our point of view, however, the second group of changes, which we can call behavioral, appears to be more important. "Physical" changes can only go as far as the behavior of all stakeholders allows. Behavior is the root of pandemic management and will continue to be root after it has passed. Therefore, we consider the most important challenge to be motivating and maintaining behavior, which will ensure the company not only a safe resumption of operations, but will support the management of all the above challenges and contribute to increasing the competitiveness of the organization.

The challenge seems even more important in the context of the increasingly common term "behavioral fatigue" by experts [23]. Behavioral fatigue can cause people exposed to long-term regulatory behavior to gradually disregard regulations. This leads to undesirable patterns of behavior [24]. Behavioral fatigue can also be transferred to the organizational behavior of employees. One possibility is that behavioral fatigue results from "ego depletion." According to this theory, self-control is similar to a muscle that can be tired [25,26]. Therefore, if people need self-control to comply with government regulations, they may be violated over time because they run out of psychic power. However, extensive attempts to replicate the findings on which the theory of ego depletion is based have failed [27], but this does not disprove the existence of behavioral fatigue. On the contrary, several studies empirically prove this [28–30]. Whether this phenomenon is real or not has not yet been resolved. But we think that individual "disobedience" can endanger and motivate all the workers involved, causing fear and stress. The key challenge, therefore, is to understand the behavior of members of the organization and to be able to successfully motivate them. Only in this way can the manager effectively manage the organization for a successful reopening.

3 Research design

The aim of the theoretical research is to point out the possible effects of the pandemic crisis on the social environment through professional literature and scientific articles. The object of the theoretical research is the changes in human behavior caused by the current situation. We focus mainly on deviations from what normative theories consider rational behavior. We explain how a change in conditions affects people's decisions and actions and how they change their organizational behavior. We extract from the literature those changes and challenges in the field of organizational behavior of employees, to which the knowledge of behavioral economics allows us to respond more flexibly.

To analyze the topic, we used available domestic and foreign sources that describe the organizational behavior of people, changes in the social environment due to the pandemic and list the challenges that these changes will bring to managerial practice. By processing the information obtained, we create the clearest possible picture of the reality that managers will have to face in the near future.

To process the theoretical part of the work, we used the basic methods of research: analysis, synthesis, generalization, induction and deduction. Based on theoretical knowledge, we solved the following research questions in the practical part:

- What deviations from rational behavior can managers expect in a post-pandemic work environment?
- How does behavioral economics explain these differences?
- How can managers use the knowledge of behavioral economics to ensure the required behavior of employees and thus more effectively face new management challenges?

We answered the answers to the research questions as part of the results and discussion of our work.

4 Research results and discussion

4.1 The contribution of knowledge of behavioral economics in reopening

Behavioral economics develops an ever-widening range of phenomena in human behavior. In the English literature on decision-making and behavior, we encounter the word "bias", whose literal meaning is inclination, deviation, prejudice, and the like. To explain the meaning, it seems more appropriate to start from the publication by V. Bačová, which refers to the work of M. Šinský, where he states that "prone to error is a free translation of the English word" bias "and expresses the difference between actual action (description) and prediction of certain normative models'. In his understanding, the propensity for error includes a tendency/inclination to make wrong decisions on a systematic and predictable basis [31].

In our work, we will address 6 selected tendencies to errors that, in our opinion, are related to behavior affected by COVID-19. We believe that this knowledge can help managers respond properly to changes in the behavior of members of the organization where they work. We focused on the following propensity for mistakes: bias by presence, bias in the status quo, framing effect, exaggerated optimism and self-confidence, affective heuristics and herd behavior. We will try to explain each of these preconceptions on the basis of the available literature. It should be noted that the nomenclature of these terms may differ from author to author in domestic literature. We have decided to use the terms that, in our opinion, best describe the meaning of the terms. We will describe the error tendencies in the context of the COVID-19 pandemic. We believe that they will help the reader to understand why people do not act in accordance with normative theories, that is, rationally. Understanding behavior will be key to its positive management in a post-pandemic economic environment.

4.1.1 Bias by presence

In the context that time as a parameter enters into every human decision, the costs and benefits of our choices take place at different points in time, meaning that many day-to-day decisions represent a compromise between immediate results (costs and benefits) and expected future results [32]. Presence bias or hyperbolic discounting is a nonlinear and inconsistent tendency for many individuals to prefer less benefit sooner over greater benefit later [32–34]. The current bias can lead to time-inconsistent preferences. An individual will make a plan for tomorrow, but once he arrives tomorrow, he may experience a change in preferences and revise his plan. For example, smoking has current benefits (temporary stress relief) as well as future costs (increased risk of lung cancer). In the case of COVID-19, for example, non-compliance with the rules may compromise the pleasure of breathing more freely without a protective mask (certain yield in the present) and an increased risk of infecting colleagues in the future (uncertain future costs). Uncertain future costs mean that not every non-wearing of the mask would lead to COVID-19 infection. So the so-called Myopic people (note myopia - myopia), who place more emphasis on the "here and now", are more likely to disregard preventive regulations to ensure workplace protection. Present bias is an explanation of why people do not behave in their best interests and why they have difficulty following beneficial preventive habits, such as safe distance in the workplace, washing their hands or wearing a protective mask [35].

Although present bias can lead to suboptimal patterns of behavior, this knowledge can also be used to provide the desired behavior. Such interventions, which represent immediate "cheap" rewards used as a way to increase immediate benefits from, for example, weight loss [36] or smoking cessation [37], have been shown to be effective in desired behavior change. In the short term, it is possible to increase workplace safety, for example by benefiting from a package of protective equipment, introducing free internet for employees who can work from home, or immediate reward for employees who are willing to continue working at home in case of a serious illness. It is up to the manager to evaluate the situation in his organization. Properly chosen immediate small and inexpensive rewards can lead to the desired change in behavior and bring a much greater benefit to the company in the future.

4.1.2 Bias in the status quo

The status quo bias is a disproportionate preference for the current situation among the options of the decision-maker and a reluctance to change this situation [38]. One reason is that people interpret the potential disadvantages of changing the current situation more strongly than the potential benefits of changing. However, thanks to the studies of behavioral economists, this reluctance can be turned into an advantage through the so-called "nudge". The concept of "nudges" was introduced in the behavioral economy to persuade individuals to behave rationally and make better decisions. Thaler and Sunstein [39] defined "nudging" as "any aspect of the architecture of choice that predictably influences individual decision-making without forbidding any options or changing economic incentives.", what they call the "architecture of choice," individuals may be influenced to make more reasonable decisions. The selection architecture can be used to create an environment in which it is easier to make optimal choices and more difficult to select suboptimal ones. The most appropriate option is to poke with a strong impact on guiding people's behavior in ways that suit their own long-term interests [39]. One of the most significant examples of "nuging" in practice is organ donation. Countries with an opt-out system (organ donation is assumed and organ donation is the default), have significantly higher rates of organ donation than opt-in countries (organ donation is not the default) [40]. Workplace COVID-19 prevention measures can also encourage people to engage in preventive behaviors, such as hand disinfection, by arranging default settings in the environment in which they work. An example could be the installation of disinfectant dispensers in suitable places in the workplace, e.g. at the entrances to buildings.

4.1.3 Framing effect

The framing effect relates to the fact that individuals' decisions often depend on the way in which the options are presented, and that decisions are often influenced by whether possible outcomes are designed in terms of profits or losses [41]. This concept is closely linked to the aversion to losses, which implies that the subjective perception of a reduction in utility from a particular amount of loss is about twice that of profit from the same amount. For example, the statements "the probability of surviving after 1 month of surgery is 90%" and "the probability of death within 1 month of surgery is 10%" elicit different reactions. Both statements provide the same information, but many people respond differently to the risk of surgery, with a 90% chance of survival versus a 10% chance of death [42].

This knowledge provides a perspective for formulating reports on workplace measures. It seems that messages designed to encourage people to take preventive behavior during COVID-19 (eg social distancing) should be worded in terms of losses, such as "If you do not wash your hands properly / If you do not follow the principles of social distance / If you do not wear a veil on workplace, we will be forced to suspend operations and reduce year-end remuneration. "However, this is not entirely true. This may be related to the demotivation of employees, who may perceive the reports formulated in this way as threatening. Unwanted pressure may be exerted on them, and therefore in this case it is more suitable e.g. "nudging" technique. However, the situation is different for health-related reports. Here, studies have shown that most people respond better to positive information in the sense of "If you follow anti-pandemic measures, you will protect you and your family and have a happy and long life [43]." It is up to the creativity and discretion of the manager to use this knowledge in specific situations in the workplace.

4.1.4 Exaggerated optimism and self-confidence

N.D. Weinstein found that people are unrealistically optimistic about the wide range of possible negative consequences of their actions and feel less than on average at risk of negative outcomes in their own lives [44]. Exaggerated optimism is the tendency of people to estimate the probability of positive future results as above average and the probability of negative future results as below average [45]. This can lead people to inadvertently take more risks than they would if they were aware of objective risk [46]. Exaggerated optimism helps explain a wide range of risky behaviors. For example, one study found that smokers significantly underestimated the risk of developing lung cancer compared to the risk of lung cancer in other smokers and even non-smokers [47]. Another similar study found that people at a subjective risk lower than the objective risk are more likely to believe that there is no risk of lung cancer if you smoke for only a few years, and believe that a large number of lung cancer patients will be cured [48].

People are aware of the risk of getting COVID-19 in suboptimal behaviors, such as not washing their hands or keeping their distance, but they probably believe that they are less likely than other people. In addressing such behavior, it is essential to provide quality feedback with accurate risk communication. It is good for people to illustrate possible risks and their effects with examples. Examples should be strong, which can be achieved by presenting the consequences of negative behavior on people like them (similar social group, age, status) who did not behave optimally [49]. Patterns are also a good example. We can achieve preventive behavior, for example, by pointing out that the same problem also applies to popular celebrities, and negative consequences have not been avoided even for people whose employees consider them to be of better quality than themselves.

4.1.5 Affective heuristics

Affective heuristics examine the commonly observed inverse relationship between perceived risk and perceived benefit. People rely on "affect" to assess the risk and benefit of a particular hazard [50]. Affective heuristics is a type of mental acronym in which people make decisions that are strongly influenced by their current emotions [51]. Emotions affect all types of decisions, big and small. A person is more likely to take risks or try new things when they are happy and less likely to reach a goal when they are troubled. Managing "inner feeling" is what behavioral economists call affective heuristics [52]. People's feelings about the relative "goodness" or "evil" of a particular person, object, or activity influence the decisions they ultimately make. Researchers have found that in a positive emotional state, a person is more likely to perceive their actions as an activity that brings high benefits and low risks [50]. When the emotional state is negative, people are inclined to believe that the decision will bring them low benefits associated with high risk. The study by Fischhoff et al. from 1978 plays an important role in the study of affective heuristics. Researchers have found that the perception of benefits and risks has been negatively correlated - the greater the perceived utility, the lower the perceived risk [53]. At the same time, it seems that the riskier the behavior, the less perceived the benefits. Certain behaviors, such as drinking alcohol and smoking, are considered high-risk and low-benefit, while other things, such as antibiotics and vaccines, are considered high-benefit and low-risk. Researchers have also found that emotions can also affect people's judgments about statistical information. In one study, patient recurrence rates were presented to clinicians, which were presented as either a probability (e.g., 30%) or a frequency (e.g., 30 out of 100). The study showed that physicians made more extreme judgments about the risk of recurrence when presented as a frequency [54]. This is because they see an absolute number behind the frequency, which for them means 30 people who have returned to disease behavior. This is also related to the already mentioned tendency to framing. Information in different forms can evoke different emotions and thus influence decisions.

According to the above findings, managers can purposefully evoke emotions associated with the behavior of employees in the workplace. The meaning of heuristics is that they make it possible to reduce the time needed to make a decision, by making this decision based on immediate emotion. If the manager is able to create positive emotions in the employees associated with the desired behavior, the employee will repeat these patterns of behavior naturally, without a deeper analytical assessment of the impact of his actions and will believe that he is doing the right thing.

4.1.6 Herd behavior

Social norms and the behavior of social groups, such as friends, family members and colleagues, influence the behavior of individuals. Herdiness occurs when people consider certain behaviors to be good or bad and mimic the observed behavior [55]. This characteristic of human behavior is observable in many areas, especially in economics and finance [56]. An experiment from Minnesota is well known, which proves this phenomenon empirically. Minnesota residents reported that more than 90% of Minnesotans paid taxes. This information had a significant impact on tax compliance. The effect of herd behavior was more pronounced than the other interventions used [39,57].

If a manager wants to encourage people to follow the required regulations, he should draw their attention to the group of people who do so. For example, informing people that everyone in the workplace is following the rules in an exemplary way can increase the motivation to follow the measures even for people who do not fully identify with their content.

4.2 Discussion

The COVID-19 pandemic with the onset of global vaccination may appear to be slowly fading, but its effects are beginning to show very slowly. Turbulent technological transformation, often forced innovation, new challenges and obstacles require original solutions. The economic, legal, social, technological and environmental environment is significantly transformed as a result of the pandemic. The pandemic has caught the world unprepared. All the more important is the ability to adapt quickly to new conditions. In addition to the problems, the pandemic will bring many opportunities. How managers respond to the challenges will be a determining factor in the success of companies in the post-pandemic era. Their skills and knowledge will determine the success of the organizations they represent. Every organization is based on the people who make it up. The importance of human capital continues to grow even in a period of raging pandemic, and we can assume that it will be even more pronounced after it erupts.

We can expect attempts to reopen workplaces in the near future. The same employees may return to their jobs, but the change in the country's social climate has meant that they can take different positions, behave according to new patterns of behavior, experience some emotions more intensely and act differently than we were used to before the crisis. How do managers deal with changes in employee behavior? Can they effectively manage people in new specific situations? A strategic approach to management requires leaders to respond to change with change. It is likely that the management approaches that worked in pre-crisis organizations will begin to appear ineffective and ineffective. Achieving the desired employee behavior will be one of the key challenges for managers when reopening.

Behavioral economics examines changes in organizational behavior, decision-making processes, human thinking, and action. We think that her knowledge can help managers manage new situations in connection with the behavior and actions of employees very effectively. Behavioral economics develops human behavior in specific situations and uses descriptive and prescriptive examination of decision-making in combination with the knowledge of psychology to be able to justify it. It is the behavioral economy that gives us the tools to understand human action. Understanding is the first step for a manager to be able to manage this behavior effectively.

In the final chapters, we present to managers selected findings of behavioral science, which we consider useful in their response to possible changes in organizational behavior that COVID-19 may have induced. We would be happy if we provoke a discussion on the topic of managing deviations in behavior in the post-pandemic era in Slovakia. The basis of behavioral economics is to investigate the issue by experiment. It is the scientific experiments, many of which we have mentioned, that help us to understand the issue of behavior. In Slovakia, too, we can find many experts who deal with the topic on a theoretical level. However, there is no empirical research that would examine the hypotheses of behavioral economics in the Slovak business environment. We believe that this gap will gradually begin to fill and the key importance of behavior in the post-pandemic world will encourage many Slovak experts to contribute to scientific work in the field of behavioral economics.

At the end of the discussion, we consider it necessary to defend a large number of sources of used literature. The knowledge of behavioral economics is mainly based on knowledge gained epmirically through economic and social experiments. The topic and the nature of the work imply the need to refer to several empirical researches to the individual knowledge, which empirically confirm this knowledge. Reducing the number of used sources of literature would reduce the value of the work and the empirically obtained knowledge would become unsubstantiated claims. We believe that in order to preserve the added value of the contribution, the reader will forgive us.

5 Conclusion

Based on our examination we can state a few conclusions. Recent research and available data confirm that we will see a number of changes in management practice as a result of the pandemic. The technological, social, economic, political and environmental environment is changing. These changes are coming more or less expected. It is the task of managers to respond to them and be able to eliminate threats on the one hand and seize opportunities on the other. One of the changes we can expect is a change in organizational behavior. Recent research suggests that the psychological pressure created by current crisis conditions will lead to a variety of undesirable variations in the behavior of individuals and groups. Every change means a challenge for the manager. Behavioral economics examines these deviations and its findings allow managers to prepare and respond more flexibly.

Managers will need to start using their dynamic skills. Creativity in management will be one of the key competencies in managing the new demands of employees. Only in this way can the manager respond to new requests from employees for work organization. What may have seemed like a benefit before the pandemic can already be considered a standard by employees. Managers will be required to be willing to take the risk of introducing innovations. However, several new measures and innovations can create reluctance among employees. The psychological pressure that we all experience through a pandemic has a demonstrable effect on people's organizational behavior. The most key challenge, therefore, is to understand the behavior of the members of the organization and to be able to successfully motivate them. Only in this way can the manager effectively manage the organization for a successful reopening.

In the theoretical part of the paper, we gained knowledge about possible changes in the environment. In the practical part, we looked for an answer to the question of what effect these changes will have on behavior. The authors of behavioral science have researched the issue of human crisis behavior in the past. From their empirically obtained knowledge, after considering, we selected 6 deviations from rational, normative behavior. For each of the "biases", we looked in the available literature as an example and a proposal for a solution. This solution was not always from a managerial environment. Analogously, however, we searched for and created a model example, on which we would be able to apply the acquired knowledge to managerial practice by analogy. The solutions are presented in the practical part of the work for each of the "biases" and allow managers to manage, eliminate and often evaluate threats resulting from changes in organizational behavior. Behavioral economics uses descriptive approaches to describe behavior. The combination of economics, decision-making science and psychological disciplines explains the described behavior. This allows the manager to understand this behavior. If the manager gains knowledge about what can cause the occurrence of undesirable patterns of behavior in the workplace, he can rather detect these patterns and respond to them. This contributes to the establishment of the right organizational culture, increases the safety of workers, leads to the elimination of unnecessary conflicts that negatively affect labor productivity. Only a manager who understands employee behavior can become a true leader in a post-pandemic company. And we are confident that the resurgenting economy will definitely need true leaders.

Acknowledgement

This paper is a partial output of the VEGA ME SR project No.1/0017/20 Changes in the application of managerial functions in the context of the fourth industrial revolution and adaptation processes of enterprises in Slovakia.

References

1. WHO, WHO Coronavirus (COVID-19) Dashboard, (2021). https://covid19.who.int (accessed April 19, 2021).
2. M.C. Howard, Understanding face mask use to prevent coronavirus and other illnesses: Development of a multidimensional face mask perceptions scale, Br J Health Psychol. (2020). https://doi.org/10.1111/bjhp.12453.
3. L. Ferretti, C. Wymant, M. Kendall, L. Zhao, A. Nurtay, L. Abeler-Dörner, M. Parker, D. Bonsall, C. Fraser, Quantifying SARS-CoV-2 transmission suggests epidemic control with digital contact tracing, Science, **368** (2020). https://doi.org/10.1126/science.abb6936.
4. A. Baveja, A. Kapoor, B. Melamed, Stopping Covid-19: A pandemic-management service value chain approach, Ann Oper Res., 1–12 (2020). https://doi.org/10.1007/s10479-020-03635-3.
5. M.P. Crayne, The traumatic impact of job loss and job search in the aftermath of COVID-19, Psychol Trauma, **12**, 180–182 (2020). https://doi.org/10.1037/tra0000852.
6. M.B. Omary, J. Eswaraka, S.D. Kimball, P.V. Moghe, R.A. Panettieri, K.W. Scotto, The COVID-19 pandemic and research shutdown: staying safe and productive, J Clin Invest., **130**, 2745–2748 (2020). https://doi.org/10.1172/JCI138646.
7. IMFBlog, The Great Lockdown: Worst Economic Downturn Since the Great Depression, IMF Blog. (2020). https://blogs.imf.org/2020/04/14/the-great-lockdown-worst-economic-downturn-since-the-great-depression/ (accessed April 19, 2021).
8. E. Aguila, Personal Retirement Accounts and Saving, Am Econ J Econ Policy, **3**, 1–24 (2011). https://doi.org/10.1257/pol.3.4.1.
9. M. Griffin, S. Tippins, A case study exploration of blue-collar worker retirement plan investment decisions, International Journal of Applied Management and Technology, **15** (2016) 3.
10. A.E. Sunden, B.J. Surette, Gender differences in the allocation of assets in retirement savings plans, The American Economic Review, **88**, 207–211 (1988).
11. A. Goolsbee, C. Syverson, Fear, lockdown, and diversion: Comparing drivers of pandemic economic decline 2020, Journal of Public Economics, **193** (2021) 104311.
12. C.-C. Topriceanu, A. Wong, J.C. Moon, A.D. Hughes, N. Chaturvedi, G. Conti, D. Bann, P. Patalay, G. Captur, Impact of lockdown on key workers: findings from the COVID-19 survey in four UK national longitudinal studies, J Epidemiol Community Health. (2021). https://doi.org/10.1136/jech-2020-215889.
13. Mesto Nitra, Mesto Nitra - Prípady nerešpektovania nariadení sa opakujú, (2021). https://www.nitra.sk/zobraz/obsah/33325 (accessed April 19, 2021).
14. K. Piovarcsy, Covid-19 spôsobí veľkú vlnu protestov po celom svete, (2020). https://www.trend.sk/nazory-a-komentare/covid-19-sposobi-velku-vlnu-protestov-celom-svete (accessed April 19, 2021).
15. A.D. Meyer, G.R. Brooks, J.B. Goes, Environmental jolts and industry revolutions: Organizational responses to discontinuous change, Strategic Management Journal, 93–110 (1990).
16. M.A. Hitt, J. Arregle, R.M. Holmes, Strategic Management Theory in a Post-Pandemic and Non-Ergodic World, Journal of Management Studies. (2020). https://doi.org/10.1111/joms.12646.
17. R. Adner, C.E. Helfat, Corporate effects and dynamic managerial capabilities, Strategic Management Journal, **24**, 1011–1025 (2003).
18. M.A. Hitt, B.B. Tyler, Strategic decision models: Integrating different perspectives, Strategic Management Journal, **12**, 327–351 (1991).

19. G.P. Hodgkinson, M.P. Healey, Psychological foundations of dynamic capabilities: Reflexion and reflection in strategic management, Strategic Management Journal. 32 (2011) 1500–1516.
20. V. Ratten, Coronavirus (covid-19) and entrepreneurship: changing life and work landscape, Journal of Small Business & Entrepreneurship. 32 (2020) 503–516. https://doi.org/10.1080/08276331.2020.1790167.
21. M. Ahsan, Entrepreneurship and Ethics in the Sharing Economy: A Critical Perspective, J Bus Ethics. 161 (2020) 19–33. https://doi.org/10.1007/s10551-018-3975-2.
22. D. Dimov, From opportunity insight to opportunity intention: The importance of person–situation learning match, Entrepreneurship Theory and Practice. 31 (2007) 561–583.
23. N. Harvey, Behavioral Fatigue: Real Phenomenon, Naïve Construct, or Policy Contrivance?, Front Psychol. 11 (2020). https://doi.org/10.3389/fpsyg.2020.589892.
24. A. vaughanbell, Do we suffer 'behavioural fatigue' for pandemic prevention measures?, Mind Hacks. (2020). https://mindhacks.com/2020/03/20/do-we-suffer-behavioural-fatigue-for-pandemic-prevention-measures/ (accessed April 20, 2021).
25. R.F. Baumeister, E. Bratslavsky, M. Muraven, D.M. Tice, Ego depletion: is the active self a limited resource?, J Pers Soc Psychol. 74 (1998) 1252–1265. https://doi.org/10.1037//0022-3514.74.5.1252.
26. R.F. Baumeister, Ego depletion and self-control failure: An energy model of the self's executive function, Self and Identity. 1 (2002) 129–136.
27. M.S. Hagger, N.L. Chatzisarantis, H. Alberts, C.O. Anggono, C. Batailler, A.R. Birt, R. Brand, M.J. Brandt, G. Brewer, S. Bruyneel, A multilab preregistered replication of the ego-depletion effect, Perspectives on Psychological Science. 11 (2016) 546–573.
28. J. Drury, H. Carter, E. Ntontis, S.T. Guven, Public behaviour in response to the COVID-19 pandemic: understanding the role of group processes, BJPsych Open. 7 (2021) e11. https://doi.org/10.1192/bjo.2020.139.
29. H. Rahmandad, T.Y. Lim, J. Sterman, Behavioral dynamics of COVID-19: estimating under-reporting, multiple waves, and adherence fatigue across 91 nations, MedRxiv. (2020).
30. S. Reicher, J. Drury, Pandemic fatigue? How adherence to covid-19 regulations has been misrepresented and why it matters, BMJ. 372 (2021) n137. https://doi.org/10.1136/bmj.n137.
31. V. Bačová, Rozhodovanie a usudzovanie II, Oblasti a Koncepcie. Bratislava, Ústav Experimentálnej Psychológie SAV. (2011).
32. T. O'Donoghue, M. Rabin, Present bias: Lessons learned and to be learned, American Economic Review. 105 (2015) 273–79.
33. D. Laibson, Golden Eggs and Hyperbolic Discounting, The Quarterly Journal of Economics. 112 (1997) 443–77. https://doi.org/10.1162/003355397555253.
34. G. Loewenstein, D.A. Asch, J.Y. Friedman, L.A. Melichar, K.G. Volpp, Can behavioural economics make us healthier?, Bmj. 344 (2012).
35. M. Van Der Pol, D. Hennessy, B. Manns, The role of time and risk preferences in adherence to physician advice on health behavior change, The European Journal of Health Economics. 18 (2017) 373–386.
36. L.K. John, G. Loewenstein, A.B. Troxel, L. Norton, J.E. Fassbender, K.G. Volpp, Financial incentives for extended weight loss: a randomized, controlled trial, Journal of General Internal Medicine. 26 (2011) 621–626.
37. S.D. Halpern, B. French, D.S. Small, K. Saulsgiver, M.O. Harhay, J. Audrain-McGovern, G. Loewenstein, T.A. Brennan, D.A. Asch, K.G. Volpp, Randomized trial of four financial-incentive programs for smoking cessation, N Engl J Med. 372 (2015) 2108–2117.

38. W. Samuelson, R. Zeckhauser, Status quo bias in decision making, Journal of Risk and Uncertainty. 1 (1988) 7–59.
39. R.H. Thaler, C.R. Sunstein, Nudge: improving decisions about health, wealth, and happiness, Yale University Press, New Haven, 2008.
40. E.J. Johnson, D. Goldstein, Do Defaults Save Lives?, Science. (2003). https://doi.org/10.1126/SCIENCE.1091721.
41. D. Kahneman, A. Tversky, Prospect Theory: An Analysis of Decision under Risk, Econometrica. 47 (1979) 263–291. https://doi.org/10.2307/1914185.
42. J. Luoto, K.G. Carman, Behavioral economics guidelines with applications for health interventions, Washington DC: Inter-American Development Bank. (2014).
43. K.M. Gallagher, J.A. Updegraff, Health message framing effects on attitudes, intentions, and behavior: a meta-analytic review, Annals of Behavioral Medicine. 43 (2012) 101–116.
44. N.D. Weinstein, Unrealistic optimism about future life events., Journal of Personality and Social Psychology. 39 (1980) 806.
45. N.D. Weinstein, Unrealistic optimism about susceptibility to health problems: conclusions from a community-wide sample, J Behav Med. 10 (1987) 481–500. https://doi.org/10.1007/BF00846146.
46. C.A. Roberto, I. Kawachi, eds., Behavioral economics and public health, Oxford University Press, Oxford ; New York, 2016.
47. N. Weinstein, S. Marcus, R. Moser, Smokers' unrealistic optimism about their risk, Tob Control. 14 (2005) 55–59. https://doi.org/10.1136/tc.2004.008375.
48. A.J. Dillard, K.D. McCaul, W.M.P. Klein, Unrealistic Optimism in Smokers: Implications for Smoking Myth Endorsement and Self-Protective Motivation, Journal of Health Communication. 11 (2006) 93–102. https://doi.org/10.1080/10810730600637343.
49. J.L. Matjasko, J.H. Cawley, M.M. Baker-Goering, D.V. Yokum, Applying Behavioral Economics to Public Health Policy: Illustrative Examples and Promising Directions, American Journal of Preventive Medicine. 50 (2016) S13–S19. https://doi.org/10.1016/j.amepre.2016.02.007.
50. M.L. Finucane, A. Alhakami, P. Slovic, S.M. Johnson, The affect heuristic in judgments of risks and benefits, Journal of Behavioral Decision Making. 13 (2000) 1–17. https://doi.org/10.1002/(SICI)1099-0771(200001/03)13:1<1::AID-BDM333>3.0.CO;2-S.
51. P. Slovic, M.L. Finucane, E. Peters, D.G. MacGregor, The affect heuristic, European Journal of Operational Research. 177 (2007) 1333–1352. https://doi.org/10.1016/j.ejor.2005.04.006.
52. J.S. Lerner, Y. Li, P. Valdesolo, K.S. Kassam, Emotion and Decision Making, Annu. Rev. Psychol. 66 (2015) 799–823. https://doi.org/10.1146/annurev-psych-010213-115043.
53. K.T. Fox-Glassman, E.U. Weber, What makes risk acceptable? Revisiting the 1978 psychological dimensions of perceptions of technological risks, Journal of Mathematical Psychology. 75 (2016) 157–169. https://doi.org/10.1016/j.jmp.2016.05.003.
54. V.F. Reyna, W.L. Nelson, P.K. Han, N.F. Dieckmann, How numeracy influences risk comprehension and medical decision making., Psychological Bulletin. 135 (2009) 943–973. https://doi.org/10.1037/a0017327.
55. D. Ariely, Predictably Irrational, Revised and Expanded Edition: The Hidden Forces That Shape Our Decisions, Revised and Expanded ed. edition, Harper Perennial, New York, NY, 2010.
56. R.M. Raafat, N. Chater, C. Frith, Herding in humans, Trends in Cognitive Sciences. 13 (2009) 420–428. https://doi.org/10.1016/j.tics.2009.08.002.
57. S. Coleman, The Minnesota Income Tax Compliance Experiment--State Tax Results, Social Science Research Network, Rochester, NY, 1996. https://papers.ssrn.com/abstract=4419 (accessed April 20, 2021).

Transformation of communication policy of Slovak companies

Robert Vilagi[1,*], *Michal* Konecny[2], and *Frantisek* Pollak[1]

[1]University of Economics in Bratislava, Faculty of Business Management, Dolnozemská cesta 1/b, 852 35 Bratislava, Slovakia
[2]Institute of Technology and Business, Faculty of Corporate Strategy, Okružní 517/10, 37001 České Budějovice, Czech Republic

Abstract. Innovation is a major accelerator of development. This is no different with regard to innovations in business management processes. The communication policy of business entities has been undergoing a transformation for more than two decades, where more and more emphasis is being placed on interactive forms of presentation, with which companies can better target selected markets. The aim of the study is to describe selected characteristics of the Slovak market in the light of the transformation from off-line to on-line. Through the analysis of secondary information sources, it is possible to state that on the Slovak market we record both a continuous increase in the share of social network use by companies, as well as an increase in expenditure on Internet promotion. The presented study maps the period until the outbreak of the COVID-19 pandemic, thus creating a reference framework for subsequent research into the effects of the pandemic on the communication activities of Slovak companies.
Keywords: innovations; communication; Internet; Slovakia; competitiveness

1 Introduction

The transition from off-line to on-line environment can be monitored both on the part of customers and on the part of companies. This shift has a dominantly communicative character, where more and more interactions take place in the virtual world [1-3]. To a large extent, this evolutionary phenomenon can be attributed to the rapid onset of social networks in the first half of the zero years of the new millennium [4-5]. Technology in the form of tablets and smartphones, together with an increasingly available data infrastructure, has offered people a choice where comfort, or rather convenience, has significantly modified common patterns of behavior [6]. Part of the market adapted relatively quickly to changing circumstances, part of the market was skeptical about new trends [7-8]. The process of adopting innovation in any form goes through the usual procedure, which takes the form of a Gaussian curve [9]. At the beginning, there are enthusiasts and innovators who in principle

* Corresponding author : robert.vilagi@euba.sk

adopt any novelty, it is not a rule that innovation will be accepted by the general public, but if it does, it passes to the called early adopters. Due to its number, it is a significantly larger group than the group of early enthusiasts. At this stage, the chance that the mainstream dominant groups will also innovate increases. At this stage, Slovak societies were in the first half of the first decade of the new millennium [10]. The mainstream began to learn the basic procedures of online communication, where both users and content producers approached the possibilities offered by the Internet with a certain degree of caution [11-12]. The second decade of the twenty-first century fully accelerated the deployment of online solutions in almost every aspect of communication [13-15]. From the point of view of accepting innovations, we have moved to the phase where these innovations began to be adopted by late adopters. The end of the decade was held in a relatively lethargic atmosphere, where most mainstream businesses were fully saturated with their digital needs. The beginning of 2020 subsequently examined the genesis of the integration of digital communication tools into the corporate portfolio in full [16-18]. The COVID-19 pandemic and the associated closure of the economy have changed practices almost from day to day in almost every aspect of life. The Internet has become a necessity for business entities, and to be online was for most of them an elementary determinant of market survival. Before the relevant data are available, we decided to examine the extent to which Slovak companies have mastered online approaches in business until the outbreak of the global pandemic. Our goal is to create a reference knowledge base for subsequent in-depth research. Before proceeding to the presentation of selected characteristics, we will briefly describe selected perspectives of the topic as such

1.1 Slovak companies and online communications

Today's world of technology offers companies increasingly accessible opportunities for promotion, while technologies help to create a more competitive environment as such [19-20]. Effective promotion tools are no longer the prerogative of the "rich and big", the focus is on the direct struggle for the customer and the most efficient use of resources. The issue of available resources in the form of funding for promotion is partially losing relevance. Social networks have created a highly competitive environment where the effort made does not necessarily equal the financial resources expended.

Kotler [21] argues that advertising aims to convince the customer of the reason for procuring a particular good or service, while sales promotion aims to motivate the customer to make an immediate purchase.

The simplicity of the promotion of goods or services creates a rapidly growing market flooded with offers and promotional materials from various entities. A mix of goods and services along with a mix of industries.

So how to attract a customer? On this issue, companies turn to marketing agencies and thus support the creative, which has found application in the digital age in almost every industry. There is no validated concept. Quantitative indicators in the online world usually do not play such a significant role, the number of addressed customers does not automatically bring economic benefits. At this point, we asked ourselves the initial question:

To what extent is the Slovak market saturated from the point of view of the Internet and the possibilities that the Internet offers to companies?

2 Objectives and methods

Starting from the main research question, the aim of the study is to describe selected characteristics of the Slovak market in the light of the transformation from off-line to on-line. By decomposing the main goal, we move on to the partial goals of our study, specifically,

we focused our efforts on examining the extent to which Slovak companies had in the second decade of the twenty-first century.:
• internet access,
• own website,
• active use of social networks.

We also considered it important to find out to what extent companies were willing to invest in online promotion during the period under review. As part of the comparison, we proceeded to the analysis of formats outside the promotional mainstream, represented by dailies and television. The mentioned media record a long-term declining trend in terms of readability or viewership, but nevertheless represent the first choice in terms of communicating "traditional" content. For the purposes of our analysis, we have chosen to compare the following media:
• Internet,
• Cinema,
• Outdoor,
• Radio.

From the point of view of information sources, we drew both from the available professional literature, as well as available databases and relevant data of the Statistical Office of the Slovak Republic [22]. The data were supplemented by data on the state of the media market, which were available in the form of infographics from renowned media agencies [23].

From the point of view of the observed period, we used the available data, where the time frame of data represented the years 2017 to 2020 for the data of the statistical office, respectively the years 2018 to 2019 for the data of the selected media agency. The data were collected in the months of March to April 2021. The presented data represent the state of the market in the most up-to-date form from the point of view of available data.

From the point of view of the scientific methods used, we have chosen analysis and synthesis as the main scientific methods. The analyzed connections were decomposed and selected, thoroughly analyzed and subsequently synthesized into graphic and textual form. Illustrations describing the state of the analyzed market are also discussed, in order to interpret the context to a wider professional audience.

Data were processed using the MS Office suite, visualization was performed using an MS Excel spreadsheet.

3 Results and discussion

The Slovak market offers companies a relatively wide portfolio of opportunities for promotion. If we looked at the issue chronologically, the main condition for online promotion of the company is access to the Internet. Of course, we can also use outsourcing, but there is little assumption that something like the overall corporate identity should be outsourced without Internet access. In Slovakia, more than 95% of companies are connected to the Internet. This fact expresses the overall availability of the Internet connection throughout the territory. However, availability does not necessarily reflect quality. Despite this fact, availability is a determinant of quality, when without the availability of a service, any quality is only a hypothetical. The following figure presents the total share of companies with Internet access:

 https://doi.org/10.1051/shsconf/202111503018

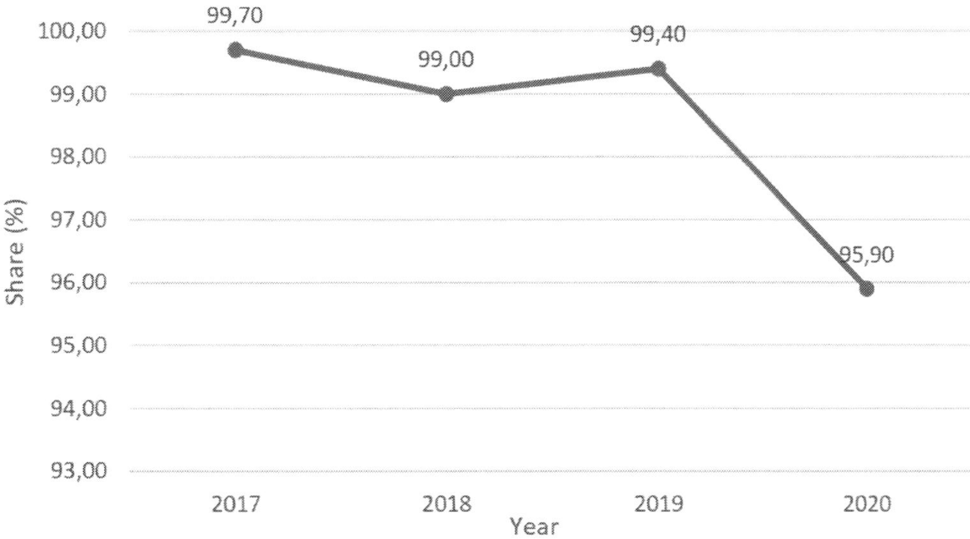

Figure 1. Share of companies with internet access. [22]

One of the initial steps in creating an online communication policy is to create a website. This essential form of communication represents the initial view of society as such. Apart from the industry, it is necessary to have a certain starting point where the customer will find authentic and targeted information about the activity and the offer. Almost 80% of Slovak companies have this form. The following figure provides an overview of trends within the period under review.

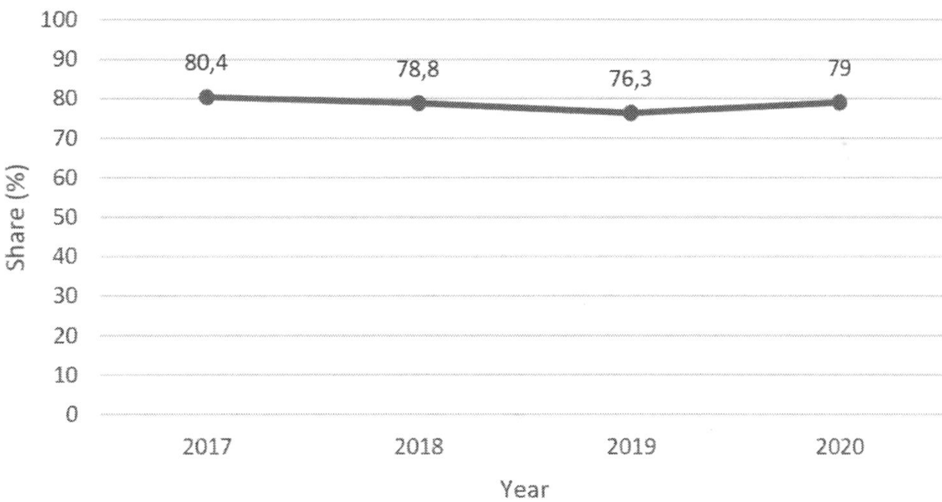

Figure 2. Share of companies with website. [22]

As far as social networks are concerned, from the point of view of the previous parameters, in this case we encounter a relatively sophisticated quantity. Social networks as a medium offer stakeholder a space for direct contact. They also generate data that companies can use to improve their public relations strategy. The motivation of organizations not to use this form of communication can be different. In the case of manufacturing companies or companies in heavy industry, we encounter this form of active non-participation most often. At first glance, networks such as Facebook and Twitter are not at all interesting for companies in the given industries. However, upon a deeper examination of the possibilities, we discover a portfolio of possibilities and motivations for active use. Whether it is regular or crisis communication, internal PR, or strengthening corporate identity in general, research suggests [24-25] that a transparent and open approach is a way to minimize reputational threats. The following figure shows the share of social network use in the monitored market.

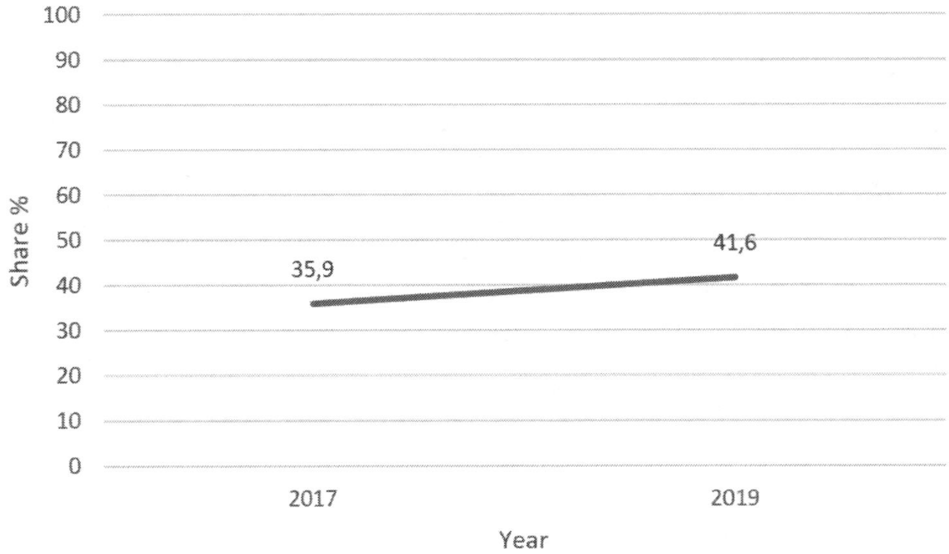

Figure 3. Share of companies with website. [22]

Within the last of the points of the sub-goals defined by us, we will proceed to the presentation of data on the activity of companies within the communication mixes. From the point of view of marketing communication expenditures, companies predominantly use traditional media as a carrier for their target audiences. To a certain extent, this action is rational, especially if we consider the dominantly passive form of such spread messages. Mainstream media such as television and print are a good way for companies to communicate with their market for more than five decades. Therefore, the use of these communication media is a kind of secure game. At this point, it must be stated that mass and unaddressed targeting is in principle ineffective in the hyper information period. From the point of view of the demand side of the market, this fact is confirmed on a daily basis by the continuously declining trend in viewing and reading of traditional media. Media players are well aware of this fact, which is why they are trying to package their offer of media space in various ways. They are increasingly integrating the Internet into their portfolio of services. Ultimately, these facts are also beginning to be understood by companies of traditional thinking, under the pressure of circumstances they are relatively reluctant to change their long-standing and established practices. The following figure presents a shift in spending on selected communication media outside the traditional media stream.

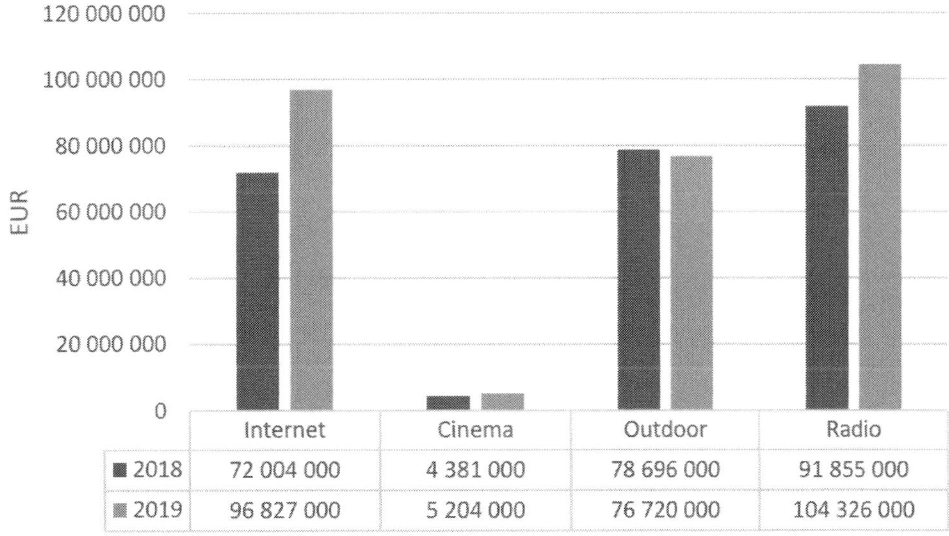

Figure 4. Shift in trends of expenditure on advertising. [23]

As can be seen from the figure, the Internet as a medium record the highest year-on-year growth in terms of companies' advertising spending. This trend only reflects the general state of the market, which we observe in almost all developed economies.

4 Conclusions

The income of each media is largely made up of advertising. Television as such is at the forefront of communicating messages to the market. In this case, customers are often placed in the role of passive consumers of media content. This media space is also the most expensive for advertising. In terms of spending on this promotion type, we come to numbers that are well above all other media. Internet advertising as a form of marketing communication is experiencing a relatively rapid increase in terms of numbers. However, it is necessary to take into account the fact that this type of communication medium places much higher demands on the consumer / user. He is no longer a passive recipient of content, but an active participant and co-creator. From the point of view of the future genesis of the market, it is possible to assume a further continuous increase in the level of online communication via the Internet. It is also possible to assume an increase in the availability of technologies for both the supply and demand side of the market. From the point of view of customers, the Slovak internet market shows the parameters of partial saturation, a jump change and an increase in users cannot be expected. As regards the supply side of the market, market saturation can be expected, but this will be preceded by a continuous growth. From the point of view of advertising expenditures, it is possible to assume that the transition from off-line to on-line will continue. Budgets will be reallocated towards customers, which will lead to increased communication targeting and increased efficiency. An overall increase in the competitiveness of proactive and pro-digitally oriented entities can be expected. The extent to which these facts will be affected by the COVID-19 pandemic is questionable at this time. The market needs to be monitored.

Due to the high dynamics of the market, it is assumed that the research of the topic will continue in the form of a subsequent empirical study performed on a significant sample and data in order to compare the state before and after the pandemic.

From the point of view of the main limitations of the study, it is necessary to mention the dominant regional form of the data, as well as the limited time frames of data within the analysis. The study presents the initial phase of research of the impacts of a pandemic on the national economy.

The article is an output of research project VEGA (1/0240/20) "Financial Aspects of Sustainable Business - Enterprise Succession Solution for Small and Medium-sized Enterprises".

The article is an output of research project VEGA (1/0140/21) "Research on economically important factors of corporate reputation in the context of sustainable industry and low-carbon economy".

References

1. D. Chaffey, *Digital marketing*. London: Pearson (2015)
2. M. Bodiš, Procesy elektronického obchodu. *Trendy v podnikaní* **4**, 4–10 (2014)
3. R. Čiarniené, G. Stankevičiuté, Theoretical Framework of E-Business Competitiveness. Kaunas. *Procedia – Social and Behavioral Sciences* **213**, 734–739 (2015)
4. M. Zeisser, *Unlocking the Exclusive Potential of Social Networks* [online]. Available at: http://www.mckinseyquarterly.com/Unlocking_the_elusive_potential_of_social_networks_2623 (2013)
5. G. L. Robins, Doing Social Network Research: Network-based Research Design for Social Scientists. London: SAGE Publications Ltd. (2015)
6. D. P. Clement, Study Finds 80% of Consumers Shop on Smartphones, M-Commerce Gains [online]. Available at: https://multichannelmerchant.com/ecommerce/study-finds-80-consumers-shop-smartphones/ (2019)
7. O. L. Turcu, A. Feraru, *E-business – A Contemporary Realistic Form Of Successful Business. Economics Edition*, "Vasile Alecsandri" University of Bacau, Faculty of Economic Sciences (2013)
8. R. Rajib, *Learning more ecommerce and online business*. Morrisville: Lulu Press, Inc. (2016)
9. N. Svetozarovova, F. Pollak, Z. Kascakova, The Importance of Key Performance Indicators in the Process of Performance Evaluation of Business Entities Active in the Slovak Republic. *Management 2016 - International Business and Management, Domestic Particularities and Emerging Markets in the Light of Research*. 177-182. (2016)
10. M.A. Stec, P.B. Filip, M. Grzebyk, Pierscieniak, Socio-economic development in EU member states - concept and classifications. *Engineering Economics* **25**, 5 (2014)
11. J. Soviar, Simplification of Marketing Scheme for Business Start-Ups. *Communications - Scientific Letters of the University of Zilina* **13**, 55–57 (2011)
12. R. Delina, Transparency in Electronic Business Negotiations - Evidence Based Analysis. *Quality Innovation Prosperity* **8**, 79–89 (2014)
13. B. Wierzbinski, Communication and information sharing in the process of creating market advantage among small and medium enterprises collaboration. *Hradec Economic Days 2014*. Hradec Králové: Gaudeamus, 437–445 (2014
14. D. Zrakova, M. Kubina, G. Koman, Influence of information-communication system to reputation management of a company. 12th International Scientific Conference of Young Scientists on Sustainable, Modern and Safe Transport. *Procedia Engineering* **192**, 1000–1005 (2017)

15. N. T. Saruc, P. Dorčák, F. Pollák, E-business and its Application in Conditions of Central European Market. *QIP journal*. **17**, 1 (2013)
16. Kalogiannidis, S. Covid Impact on Small Business. *International Journal of Social Science and Economics Invention* **6**, 387-391(2020)
17. Bohn, S., Mejia, M. C., & Lafortune, J. *The Economic Toll of COVID-19 on Small Business*. California: Public Policy Institute of (2020)
18. A. Bhatti, H Akram, H.M Basit, A.U Khan, S.M Raza, M.B Naqvi, Ecommerce trends during COVID-19 pandemic. I*nternational Journal of Future Generation Communication and Networking* **13**, 1449 – 1452 (2020)
19. F. Pollak, P. Dorcak, Analysis of online reputation of elected e-commerce entities operating in the central European market. *IDIMT-2015: Information technology and society - interaction and interdependence: 23nd interdisciplinary information management talks*. 255-262. (2015)
20. F. Sudzina, P. Kmec, The technological paradox and evaluation of the benefits of informatization. *Journal of Economics* **54**, 281-293 (2006)
21. P. Kotler, et al., *Moderní marketing*. Prague: Grada Publishing (2007)
22. Statistical Office of the Slovak Republic. *IKT v podnikoch*. [online]. Available at: http://statdat.statistics.sk/cognosext/cgi-bin/cognos.cgi?b_action=cognosViewer&ui.action=run&ui.object=storeID(%22iF656BB5617D94B22872CB96564161686%22)&ui.name=IKT%20v%20podnikoch%20%5bis1003rs%5d&run.outputFormat=&run.prompt=true&cv.header=false&ui.backURL=%2fcognosext%2fcps4%2fportlets%2fcommon%2fclose.html&run.outputLocale=sk (2020)
23. Kantar. Ako sa vyvíjali výdavky do reklamy v 2019. [2020]. Available at: https://www.kantarmedia.com/sk/blog/infografiky/ako-sa-vyvijali-vydavky-do-reklamy-v-2019 (2020)
24. P. Dorčák, P. Markovič, F. Pollák, Multifactor analysis of online reputation as a tool for enhancing competitiveness of subjects from automotive industry. *Journal of Economics* **65**, 173–186 (2017)
25. M. Fertik, D. Thomson, Wild West 2.0: *How to Protect and Restore Your Reputation on the Untamed Social Frontier*. New York: Amacom (2010)

Theoretical Background to the Role of Trust in Marketing

Annamária Zsigmondová[1], *Tibor* Zsigmond[2*] and *Renáta* Machová[2]

[1] Univerzita J. Selyeho, Fakulta ekonómie a informatiky, Katedra ekonomiky, Bratislavská cesta 3322, Komárno, 94501, Slovenská Republika

[2] Univerzita J. Selyeho, Fakulta ekonómie a informatiky, Katedra manažmentu, Bratislavská cesta 3322, Komárno, 94501, Slovenská Republika

Abstract. The aim of the study is to present the theoretical background of trust and to review its role in marketing among the disciplines mentioned in the article. After the methodological definition, the article provides the reader with a complex overview of the concept of trust in the national and international literature. The need for trust extends to business and market interactions throughout the supply and value chain, as they are interrelated in all areas of production and service. In addition, its existence also plays an important role in online commerce, the lack of which is understood by companies and intermediaries in the reduction of transaction costs and reputation. However, the question arises as to whether trust can be so easily delineated. Is the decision of whether or not to trust the other a simple question to be decided to which there is a clear answer?

Introduction

Consumer behaviour is becoming a vital aspect of customer relationships and marketing strategy. As several studies have pointed out, building consumer confidence is key to long-term success of retailers. However, one of the negative aspects of this is the fear about products and manufacturers.

The role of efficiency is a crucial factor in the private sector. Trust as a prerequisite for successful cooperation is an essential aspect between the actors. We talk about interaction, when the parties pass the uncertainty threshold and balance their intentions. Bilateral trust between the actors is also a kind of "guarantee", as it is supporting achieving the goals. In order to achieve our goals we are put into situations when dependence on others is essential. The consumers nowadays are more consumption-oriented, therefore they also have to rely on others, especially businesses.

The study begins with general definition of trust, which is followed by detailed definition of the term in corporate context and in terms of PR. A chapter of this study is addressing the issue of trust in terms of marketing as well. The chapter emphasizes the importance of trust: developing, maintaining and consequences of trust between the individual market players, i.e. companies and consumers. In this regard, we considered it

* Corresponding author: zsigmondt@ujs.sk

important that the study provide a deeper insight into processing the topic, thus also focusing on the theoretical background of consumer confidence. In addition, the field of e-commerce has also reflected on trust, mistrust and its consequences, as well as its impact on businesses and consumers. At the end of the study a conclusion is provided summarizing the findings.

1 Aims and methodology

The aim of the publication is to examine the role of trust in marketing from the point of view of customer and the business. In order to achieve this goal, the study summarizes the most relevant findings of scientific literature sources. It deals with conceptual definitions of trust and then focuses on business relationships in marketing terms. In consumer society, the development of mistrust is becoming stronger, which is detected on domestic and international level as well. The phenomenon is also reflected in relationship marketing. One of the basic pillars of trust in the field of e-commerce and brick-and-mortar shopping is brand building as a prerequisite of customer action. Both mistrust and reliability are important aspects in terms of the customer. The customer can trust the brand or product, but might have doubts about the intermediary. Satisfaction with online store can also influence the customer trust, which has impact on customer determination to buy the product. The end of the study provides a short conclusion, summarizing the perspective of consumer and the service provider. approached from the provider and consumer side.

2 Theoretical background

2.1 The concept of trust

The concept of trust has to be approached interdisciplinary. It had been studied from the perspective of economics, psychology, sociology and law, but all the mentioned sciences approach the concept of trust from different perspective.

The success of knowledge-based organizations heavily depends on trust. In the case of mistrust, there would not be an effective communication, compatible teamwork, as well as the organizational culture would not function properly. In recent years, the importance of trust has had a major impact on daily life of organizations. Many studies and authors have addressed the issue. Culture is changing as a result of development of modern technologies. Different lifestyles and diversity of value systems require a more sophisticated level of cooperation and interdependence. Czakó [1] found that increased cooperation with the individual influences the level of trust as well. This is reflected in a phenomena how companies adopt to market changes and challenges. Smart devices make the communication effective as well as they introduce some controversial aspects. They enable the efficient flow and exchange of knowledge, but also raise the question of trust and anonymity.

Piotr Sztompka (1999) identified several alternatives for the emergence of trust. Trust may appear only with one of the individuals or in the case when one of the individuals develops an intrinsic motivation to develop trust, as it is necessary to initiate a contact with the other individual. A culture of trust can also be formed, which consists of interrelated values and norms [2]

The concept of trust also plays a significant role in business. Stephen [3] introduces trust as a hidden factor. He is convinced that trust is present in every moment and in all areas of life. The existence of trust strengthens the human relationships, makes communication more efficient, and can improve the quality of organizations. Increasing trust might result in achievement of long-term company goals.

Sajfert et al. [4] describe the importance of trust in the company as a key to employee cooperation and growth. Trust invested into employee-employer cooperation will determine the corporate atmosphere and intergroup cooperation. As a direct result, efficiency to achieve organizational goals will increase. Disagreements, lack of cooperation and adaption, changes in attitude to innovation, strict group stereotypes emerge as a result of mistrust. This type of behaviour creates unfavourable company atmosphere.

2.2 Trust in the market

Trust is the key element of marketing concept. Every relationship based on success depends on the degree of trust between people. Thus, the primary goal of marketing is to satisfy the customer need, maintain and manage the relationship with the customer and deliver promises. Trust is an essential element to achieve these goals.

Trust is a complex behavioural attitude that develops as a multidimensional overlap between the market participants. Trust is a simple spectrum, from established trust to distrust [5].

Trust cannot be built explicitly on direct experience as it would contradict to traditional human values and marketing definitions. However, consumers can prefer the recommendations of a third party in their final decision. Thus, trust can be built on both direct and indirect experiences, provided that the indirect source itself is reliable. People tend to direct their attention to negative experiences conveyed, thus generating contradiction about their own competence. At the same time, most people are unlikely to be aware of this and they set up several trust dilemmas. Expressing opinion can often be seen as an output rather than input to marketing models, so the relative impact of direct or indirect experience on development of trust may be an objective of further research [6].

White argues that conscious positive intentions are the foundations of trust [7]. The hypothesis of Garbarino and Lee are based on a fact that social relations are fundamentally based on benevolence, i.e., the parties do not ignore each other's interests [8]. This kind of behaviour is later reflected in developing organizational structure, as well as it is a determining factor of relationship. Product portfolio and security are part of the corporate structure, while the appropriate incentives and support are related to public relations. Companies should behave if they were the customers themselves expecting a certain behaviour from the company [9].

Trust and mistrust have influence on the behaviour of market participants, including intention and development of future relationships. If one of the participants experiences a negative perception of building trust, the level of cooperation and commitment will decrease [10]. Decreasing level of trust is also reflected in opportunistic behaviour. Initiating a cooperation requires components that will result in development and increase in trust. To Relationship commitment is one of these factors. If one of the parties is generating trust accepting also the vulnerability, while the other party meets the obligation, they have already lay down the cornerstone of trust [11]. It can be assumed that trust is a relevant component of relationship between the parties. [12]. Palmatier, in a paper published more than a decade later, also mentioned that the higher the level of trust, the more committed it can make the consumer [13].

A long-term and fruitful relationship without trust is almost unimaginable. Since customers are in contact with each other, the behaviour of salespeople affects the future confidence of consumers, in which the service provider's further behaviour also plays a role. Increasing level of trust has a direct impact on building long-term relationship which will result in reduction of transaction costs [14]. In the case of online shopping, the brand requires reliability from the online retailer and the information system, through which the transaction takes place. If it is not managed properly, the customer can trust the brand but cannot trust the intermediary (e.g., the Internet). Lewicki et al. [15] and Arnott [16] researched the interaction of trust-related decisions in online context. As a result of e-commerce development, there is a growing need for knowledge and theories about the online consumer behaviour as it is becoming a vital aspect of customer relationships and marketing strategy [17]. Online shopping behaviour needs to be understood [18], and therefore attracts increasing attention of scientists and researchers [19]. As several studies have pointed out, the key to the long-term success of e-retailers is to build consumer confidence [20-22].

2.3 Consumer trust

The aspects of trust had been addressed by several scientific studies in different field of research i.e. economics, management, technology, social studies, psychology and consumer behaviour [23]. Trust is based on the buyer's expectations that the seller will behave in a reliable, ethical and socially appropriate manner, fulfilling the obligations of the buyer, despite the buyer's vulnerability and dependence [24]. Thus, consumer reliability considerations are likely to determine the final purchasing decision between the buyer and seller [25]. Li et al. [26] found that trust is even more important in the case of e-commerce as consumers experience greater risk since they cannot visit brick-and-mortar outlets to have personalized experience and see and touch the products they are intended to buy. It plays a crucial role in determining online shopping intent [27] and purchasing decision [28]. Trust is also a key factor to build customer loyalty and developing strong and lasting relationships between the buyers and sellers [29]. Lack of trust is the biggest barrier of online transactions. s [30]. When negative shopping experience occurs, buyers generate a negative attitude [31], no longer trust the seller, and are likely to turn to further alternatives to meet their needs and desires [32].

Consumer confidence is built on experience, which takes time to develop. It also has a key role in reducing tensions and conflicts, supporting the flow of information, being able to create greater coherence, and also contributing to realization of future transactions [33]. Consumer trust can also be interpreted as a kind of behavioural goal, as a skill to rely on a business partner. Thus, the presence of one party's trust in the reliability and honesty of the other party [11, 12, 34] can lead to positive expectations.

Trust generates a kind of effect among the existing or future partners, a kind of "hope" that the other party does not take advantage of the other party's vulnerability [35]. It is a long and strong relationship that helps to shape the vision of the parties. It contributes to overcome uncertainty and reducing risks [29, 35], initiates balanced relationships, and thus influences the success of firms [36].

In his research, [7] directed the attention to three factors contributing to the development of consumer confidence. These are the following:

- trust,
- problem solving ability,
- loyalty.

Trust means competent expertise in terms of the service provider and a willingness to accept it from the consumer's aspect. Trust based on expertise includes the following three components: knowledge, ability and inclination to skill [7]. In addition to the factors listed, [10] mentions capacity, followed by professional competence and know-how [37]. Trust on the consumer side is a central component that plays a key role in the effectiveness of an organization. It is important to emphasize, that at the same time it is present as a lower risk factor, thus ensuring the feeling of mutual loyalty [29]. Doubting and terminating customer relationship usually happens due to mistrust [38].

Another dimension of consumer confidence is the willingness to solve the problem. The precondition is that all the problems encountered in different phases of the service must show a willingness to be solved. The effect of this will be extended to the management within the service company as well as will increase the consumer confidence. However, it is important to emphasize that the arising problems will offer important opportunities for companies to demonstrate commitment [39]. Overall, it is a constructive relationship in which the service company is detecting the problem and feels motivated to communicate and solve it for the benefit of customer. Reducing risks strengthens consumer cooperation, so the level of cooperation will increase. The service provider thus creates a sustainable relationship based on trust, which is essential to develop customer loyalty [34].

Another important factor in building trust is loyalty. From customer perspective, loyalty is a favourable response for the product portfolio offered by the service provider within a defined period of time. Brand loyalty is closely linked to regular purchase realized by the customer. When defining loyalty, attitude characteristics are essential. It is a strong and intense commitment of the customer, a kind of intention to repeat a purchase [35, 40-42]. In order to maintain the relationship, it generates a viewpoint that involves resistance to various situational factors and also the resistance to competitors' marketing efforts [42, 43].

Because the existence of consumer trust plays a significant role not only in brick-and-mortar shopping but also in e-commerce, online retailers place significant emphasis on consumer trust [44]. Examining the relevance of trust and intent to buy, Komiak and Benbasat [45] concluded that cognitive trust (which focuses on consumer beliefs based on rational expectations of online retailers 'characteristics) affects emotional trust (which affects consumer attitudes and emotions), which further influences the intention to purchase. Moreover, the level of trust of customers exposed to inconsistent product information and reviews significantly influences their purchase intention [46]. As a result, the critical role of trust in determining consumers' purchasing intentions is influenced by satisfaction with products and online stores [47-48]. If retailers want consumers to buy their products, they must go beyond the threshold of reliable behaviour [49-50].

Conclusions

The aim of the study was to present the theoretical background of trust, focusing on its role in marketing, and to emphasize its importance for market players. The authors present the general definition of trust and an overview of research results in the past few decades. Each sub-chapter of the study focuses on a more targeted area of trust, addressing the issue from the perspective of businesses and customers. Trust is a kind of relationship based on experience, but it is also a cornerstone of building customer relationship. It takes time to develop, while a negative experience is enough to ruin trust. It contributes to reducing the costs of manufacturers and service providers. The research presented in this study suggests that trustful business relationships result in greater profits, customer satisfaction, and flexibility. However, it is important to examine other dimension as well, as business relationships are under constant threat due to uncertainty, complexity, specialization, information barriers, etc. It should be emphasized that appropriate advertising can change consumer attitudes towards a particular product and reduce perceptions of product risk. Both direct and indirect marketing can play an important role in consumer decision-making. Direct marketing initiatives have a greater impact on purchasing decisions than media-based methods such as television, radio, and printing. In addition, marketing can significantly influence the consumer awareness about the product performance and ultimately determine the likelihood of the purchase. Product performance and quality are also related to brand building.

E-commerce, as a most affected field nowadays can be characterized by mistrust

 https://doi.org/10.1051/shsconf/202111503019

factors. In e-business environment, businesses have to deal with inappropriate behaviour, identity fraud, and the misuse of personal data. The development of the reliability of online transactions is significantly influenced by branding considerations, as these are key factors in minimizing the risks faced both by products and web manufacturers. As e-commerce is facing higher risks trust is more important for online than offline retailers. In addition, retailers (but certainly not just them) need to focus mainly on direct marketing campaigns, as they have a greater influence on development of risk in online environment. Detailed and accurate product information, acceptable prices for the customer and the consumer's expectation of a favourable purchase can further increase online sales. Companies that sell their products online have to be aware that trustful intermediaries (web sellers) can increase the quality of products they sell. Accordingly, high-quality products are likely to help reduce online uncertainty, increase trust in intermediaries selling these products, and reinforce the feeling that e-channels take care of their users and understand their specific needs.

References

1. Czakó Á. *Szervezetek, szerveződések a társadalomban*. Budapest: BCE Szociológia és Társadalompolitika Intézet (2011)
2. Szabó I. L. Az intézményi, szervezeti bizalom helyzete Magyarországon. *Nemzet és biztonság*, **3**, 3 (2014)
3. Stephen M.R.C. The *Speed of Trust: The One Thing That Changes Everything*. USA: Free Press (2008)
4. Sajfert, D., Nikolić, M., Vukonjanski, J., Terek, E., Vulović, M. The Impact of Ethical Leadership Behavior on Communication Satisfaction and Organizational Earning in Companies in Serbia. *Acta Oeconomica Universitatis Selye*, **8**, 1 (2019)
5. Mari S. *Does Trust Improve Business performance?* Oxford: Oxford University Press (2000)
6. Horecký, J. Operation and action of a trade union (in terms of Czech Republic labour law). *Central European Journal of Labour Law and Personnel Management*, **1**, 1 (2018)
7. White, T. B. Consumer Trust and Advice Acceptance: The Moderating Roles of Benevolence, Expertise, and Negative Emotions. *Journal of Consumer Psychology*, **15**, 2 (2005)
8. Garbarino, E., Lee, O. F. Dynamic Pricing in Internet Retail: Effects on Consumer Trust. *Psychology & Marketing*, **20**, 6 (2003)
9. Mitchell, A. Building consumer trust is the secret of success. *Marketing Week*, **28**, 19 (2005)
10. Cho, J. The mechanism of trust and distrust formation and their relational outcomes. *Journal of Retailing*, **82**, 1 (2006)
11. Morgan, R. M., Hunt, S. D. The commitment-trust theory of relationship marketing. *Journal of Marketing*, **58** (1994)
12. Moorman, C., Zaltman, G., Deshpande, R. Relationships between providers and users of market research: the dynamics of trust within and between organizations. *Journal of Marketing Research*, **29**, 3 (1992)
13. Palmatier, R. W., Jarvis, C. B., Bechkoff, J. R., Kardes, F. R. The Role of Customer Gratitude in Relationship Marketing. *Journal of Marketing*, **73**, 5 (2009)

14. Doney, P. M., Cannon, J. P. An examination of the nature of trust in buyer-seller relationship. *Journal of Marketing*, **61**, 2 (1997)
15. Lewicki, R. J., Mcallister, D. J., Bies, R. J. Trust and distrust: new relationships and realities, *Academy of Management Review*, **23**, 3 (1998)
16. Arnott, D. C. Conceptualizing trust. *Academy of Marketing Conference*, (2005)
17. Close, A. G., Kukar-Kinney, M. Beyond buying: motivations behind consumers online shopping cart use. *Journal of Business Research*, **63**, 9-10, (2010)
18. Herrero, A., San Martin, H. Effects of the risk sources and user involvement on e-commerce adoption: application to tourist services. *Journal of Risk Research*, **15**, 7 (2012)
19. Mosteller, J., Donthu, N., Eroglu, S. The fluent online shopping experience. *Journal of Business Research*, **67**, 11 (2014)
20. Suh, B., Han, I. The impact of customer trust and perception of security control ont he acceptance of electronic commerce. *International Journal of Electronic Commerce*, **7**, 3 (2003)
21. Pavlou, P. A., Fygenson, M. Understanding and predicting electronic commerce adoption: an extension of the theory of planned behavior. *MIS Q*. **30**, 1 (2006)
22. Vos, A., Marinagi, C., Trivellas, P., Eberhagen, N., Skourlas, C., Gianna, G. Risk reduction strategies in online shopping: e-trust perspective. *Procedia Social and Behavioral Sciences*, **147**, 25 (2014)
23. Kim, D. J., Donald, L. F., Rao, H. R. A trust-based consumer decision-making model in electronic commerce: the role of trust, perceived risk and their antecedents. *Decision Support Systems*, **44**, 2 (2008)
24. Gefen, D., Karahanna, E., Straub, D.W. Trust and TAM in online shopping: an integrated model. *MIS Q*. **27**, 1 (2003)
25. Gupta, P., Yaday, M. S., Varadarajan, R. How task-facilitative interactive tools foster buyers' trust in online retailers: a process view of trust development in N. Pappas. *Journal Retailing*, **85**, 2 (2009)
26. Li, H., Jiang, J., Wu, M. The effects of trust assurances on consumers' initial online trust: a two-stage decision-making process perspective. *International Journal of Information Management,* **34**, 3 (2014)
27. Hong, I. B., Cha, H. S. The mediating role of consumer trust in an online merchant in predicting purchase intention. *International Journal of Information Management*, **33**, 6 (2013)
28. Buttner, O. B., Goritz, A. S. Perceived trustworthiness of online shops. *Journal of Consumer Behaviour.*, **7**, 1 (2008)
29. Santos, C. P. dos, Fernandes, D. von der H. Antecedents and Consequences of Consumer Trust in the Context of Service Recovery. *Brazilian Administration Review*, **5**, 3 (2008)
30. Urban, G.L., Amyx, C., Lorenzon, A. Online trust: state of the art, new frontiers, and research potential. *Journal of Interactive Marketing*, **23**, 2 (2009)
31. Gao, L., Bai, X. Online consumer behaviour and its relationship to website atmospheric induced flow: Insights into online travel agencies in China. *Journal of Retailing and Consumer Services*, **21** (2014)
32. Lee, Y. Impacts of decision-making biases on eWOM retrust and risk-reduching strategies. *Computers in Human Behavior*, **40** (2014)

33. Dwyer, F. R., Schurr, P. H., Oh, S. Developing buyer-seller relationships. *Journal of Marketing*, **51**, 2 (1987)
34. Sirdeshmukh, D., Singh, J., Sabol, B. Consumer Trust, Value, and Loyalty in the Relational Exchanges. *Journal of Marketing*, **66**, 1 (2002)
35. Aydin, S., Özer, G., Arasil, Ö. Customer loyalty and the effect of switching costs as moderator variable. *Marketing Intelligence & Planning*, **23**, 1 (2005)
36. Johnson, D., Grayson, K. Cognitive and affective trust in service relationships. *Journal of Business Research*, **58**, 4 (2005)
37. Blomqvist, K. The Many Faces of Trust. *Scandinavian Journal of Management*, **13**, 3 (1997)
38. Harding, R. The Street-level Epistemology of Trust. *Politics & Society*, **21**, 4 (1993)
39. Benda-Prokeinová, R., Dobeš, K., Mura, L., Buleca, J. Engel's Approach as a tool for estimating consumer behaviour. *E & M Ekonomie a Management*, **20**, 2 (2017)
40. Dick, A. S., Basu, K. Customer loyalty: towards an integrated conceptual approach. *Journal of the Academy of Marketing Science*. **22**, 2 (1994)
41. Lichtlé, M-C., Plitchon, V. Understanding better consumer loyalty. *Recherche at Applications en Marketing*, **23**, 4 (2008)
42. Oliver, R. L. Whence Consumer Loyalty? *Journal of Marketing*, **63**, 4 (1999)
43. Bandyopadhyay, S., Martell, M. Does attitudinal loyalty influence behavioural loyalty? A theoretical and empirical study. *Journal of Retailing and Consumer Services*, **14**, 1 (2007)
44. Park, J., Gunn, F., Han, S.L. Multidimensional trust building in e-retailing: cross-cultural differences in trust formation and implications for perceived risk. Journal *of Retailing and Consumer Services*, **19**, 3 (2012)
45. Komiak, S. Y. X., Benbasat, I. Understanding customer trust in agent-mediated electronic commerce, web-mediated electronic commerce, and traditional commerce. *Information Technology and Management*, **5**, 1–2 (2004)
46. Zhang, K. Z. K., Cheung, C. N. K., Lee, M. K. O. Examining the moderating effect of inconsistent reviews and its gender differences on consumers' online shopping decision. *International Journal of Information Management*, **34**, 2 (2014)
47. Wu, I. L. The antecedents of customer satisfaction and its link to complaint intentions in online shopping: an integration of justice, technology, and trust. *International Journal of Information Management,* **33**, 1 (2013)
48. Csiki, O., Horváth, R., Szász, L. A Study of Regional-Level Location Factors of Car Manufacturing Companies in the EU. *Acta Oeconomica*, **69**, S2 (2019)
49. Bente, G., Baptist, O., Leuschner, H. To buy or not to buy: Influence of seller photos and reputation on buyer trust and purchase behaviour. *International Journal of Human – Computer Studies*, **70**, 1 (2012)
50. Bakó, F., Szemerédi, E. Kutatói szeminárium a Széchenyi István Egyetemen: „Családok – generációk - vállalkozások – dilemmák" műhelykonferencia beszámoló. *Tér-Gazdaság-Ember*, **6**, 3 (2018)

AUTHOR INDEX

EDP Sciences – Web of Conferences
17, Avenue du Hoggar
Parc d'Activité de Courtabouf
BP 112
F-91944 Les Ulis Cedex A
France

ISBN 978-1-7138-3473-1

CURRAN ASSOCIATES INC.
proceedings.com

Twelfth International Conference on Advanced Cognitive Technologies and Applications (COGNITIVE 2020)

Held at ComputationWorld 2020

Nice, France
25 - 29 October 2020

IARIA